Forensic Investigation of Blood Evidence

Hirak Ranjan Dash • Noora Rashid Al-Snan
Safia Abdessalem Messaoudi

Forensic Investigation of Blood Evidence

Springer

Hirak Ranjan Dash
Centurion University of Technology and Management
Bhubaneswar, Odisha, India

Safia Abdessalem Messaoudi
Department of Forensic Sciences, College of Forensic and Investigative Sciences
Naif Arab University for Security Sciences (NAUSS)
Riyadh, Saudi Arabia

Noora Rashid Al-Snan
Forensic Science Laboratory, Directorate of Forensic Science
General Directorate of Criminal Investigation and Forensic Science, Ministry of Interior
Manama, Bahrain

ISBN 978-981-95-9182-4 ISBN 978-981-95-9183-1 (eBook)
https://doi.org/10.1007/978-981-95-9183-1

This Springer imprint is published by the registered company Springer Nature Singapore Pte Ltd.
The registered company address is: 152 Beach Road, #21-01/04 Gateway East, Singapore 189721, Singapore

Preface

According to the modern definition of crime, there are four basic elements of a crime: actus reus, mens rea, injury, and humans. The investigating agency explores many conventional techniques to establish the elements of actus reus and mens rea. Forensic biological evidence plays a major role in establishing the last two elements of crime, i.e., injury and human. When a physical injury occurs on a human, the most commonly released body fluid is the blood. Thus, blood is the most commonly found biological evidence found in the crime scene. The forensic investigation of blood follows the order of detection of blood by presumptive tests followed by the confirmatory detection of blood. Once blood is confirmed, its species origin and blood group analysis is performed occasionally. Subsequently, the DNA profiling of blood sample confirms the source attribution and links the perpetrator, victim, and the crime scene. Besides, the blood spatter pattern observed in a crime scene provides crucial investigative leads by detecting the angle of impact and height of origin of the blood stain.

The book *Forensic Investigation of Blood Evidence* is an amalgam of fundamental concepts of blood, its forensic relevance, collection preservation strategies, laboratory analytical techniques, advanced technologies of blood evidence analysis, legal admissibility of blood evidence, and case studies related to blood evidence. The whole book is divided into four modules. Module I describes the Theoretical Foundations of Blood and Forensics and explores the evolution of circulatory system, constituents of blood, basic concept of blood group, species identification, blood transfusion, blood associated diseases, and the role of blood evidence in criminal identification. In Module II, conventional analytical and investigative techniques of blood evidence have been discussed. Analysis of bloodstain pattern, collection and preservation strategies, presumptive and confirmatory detection of blood evidence, estimation of age of blood stain, and the postmortem time interval have been discussed in Module II. Whereas, Module III describes the advanced analytical techniques of blood evidence, where DNA profiling of blood evidence, blood microbiome analysis, and the application of Artificial intelligence in blood evidence examination have been described. Module IV is dedicated to the applications of blood evidence where the crime scene reconstruction methods, case studies related to blood evidence, and the legal admissibility of blood evidence have been discussed.

This book is written in simple English language for the clear understanding of the students, researchers, forensic scientists, investigating officers, advocates, and the judiciary. This book will be a treat for the readers with a basic understanding of forensic biology, zoology, and forensic science. The case studies and legal admissibility section will be highly beneficial for the forensic experts and legal practitioners. The basic concepts of blood and blood tests described in this book will be highly useful for the students and forensic practitioners. Overall, the book will be a treat to the readers and will help them in understanding the fundamental aspects of blood, blood tests, and the advanced methodologies used in forensic investigation of blood.

Wishing a happy reading to the readers!!!!

Bhubaneswar, Odisha, India — Hirak Ranjan Dash
Manama, Bahrain — Noora Rashid Al-Snan
Riyadh, Saudi Arabia — Safia Abdessalem Messaoudi

Acknowledgments

To all the scientists, forensic practitioners, investigators, and legal professionals who dedicate their expertise to the pursuit of truth—this book is for you.

Your commitment to accuracy, integrity, and scientific rigor upholds justice in laboratories, crime scenes, and courtrooms alike. Through your work, evidence speaks louder than assumption, and science remains a pillar of fairness in society.

Special thanks to my colleagues, whose collaboration, insight, and shared dedication enriched every stage of this work. Your professionalism continues to inspire me.

To my family, thank you for your patience, encouragement, and unwavering support. Your belief in this project sustained me through its many challenges.

May this book honor the discipline, diligence, and purpose that define our field—and all those who serve it.

Competing Interests The authors have no competing interests to declare that are relevant to the content of this manuscript.

About the Book

Forensic Investigation of Blood Evidence offers a comprehensive exploration of the scientific, investigative, and legal dimensions of blood as forensic evidence. From the moment blood is released at a crime scene, it becomes a critical biological marker—capable of revealing not only the presence of injury but also the identity of those involved. This book guides the reader through every stage of forensic blood analysis: from presumptive and confirmatory testing to species identification, blood grouping, DNA profiling, and bloodstain pattern interpretation. Emphasizing both foundational principles and cutting-edge techniques, it bridges traditional methodologies with innovations such as microbiome analysis and the application of artificial intelligence in forensic examination.

Structured into four well-defined modules, the book delves into the biological and forensic underpinnings of blood, outlines conventional and advanced analytical strategies, and contextualizes real-world applications through crime scene reconstruction and case studies. Written in clear, accessible language, it serves as a vital resource for students, researchers, forensic practitioners, investigators, and legal professionals. Whether one seeks to understand the science behind blood evidence or its impact in courtroom proceedings, this book provides a vital and engaging reference grounded in both scientific rigor and practical insight.

Contents

About the Authors

Hirak Ranjan Dash is Associate Professor of Forensic Biology and Biotechnology at Centurion University of Technology and Management, Bhubaneswar, affiliated to the National Forensic Sciences University, India. Before joining academia, he worked as a Forensic DNA expert at the Forensic Science Laboratory in Madhya Pradesh, India. He has examined more than 1000 cases by DNA fingerprinting techniques. His research interests include DNA fingerprinting, microbial forensics, microbiome analysis, next-generation sequencing, and mtDNA analysis. He has published more than 60 research papers in various peer-reviewed journals. He has written 14 books in various fields of Biotechnology. He is a life member of International Society of Forensic Genetics, Association of Microbiologists of India, and the Asian Federation of Biotechnology. He served as reviewer for *PLOS One*, *Annals of Human Biology*, *Scientific Reports*, and several other journals. He is the premier scientist from India on NGS-based forensic DNA analysis. He is featured in the list of world's top 2% scientists during 2021–2024.

Noora Rashid Al-Snan is a distinguished forensic DNA analyst with over 18 years of experience in criminal investigations. She joined the Ministry of Interior in the Kingdom of Bahrain in 2008, where she currently serves as the Head of the Biology & DNA Forensic Laboratory at the General Directorate of Criminal Investigation and Forensic Science. She holds a PhD in Molecular Medicine with a specialization in medical genetics, awarded by the Arabian Gulf University in collaboration with Naif Arab University for Security Sciences, Riyadh, Saudi Arabia. Her doctoral research focused on DNA typing of the Bahraini population. She also earned a master's degree in Medical Biotechnology, specializing in genetic engineering and recombinant proteins, from AGU, and holds a bachelor's degree in Molecular Biology and Biochemistry from Kuwait University. Throughout her career, she has played a pivotal role in the resolution of numerous high-profile cases involving terrorism, robbery, and homicide. Her expertise and leadership were internationally recognized when she and her team received the *DNA Hit of the Year* award at the HIDS Conference in 2023 for their work on the Bahrain bomb cases, presented by Gordon Thomas Honeywell. An active contributor to the forensic science community, she frequently represents MOI as a speaker and delegate at international conferences and workshops. She is the corresponding author of multiple peer-reviewed publications and has authored works with leading academic publishers. Her research interests include innovative methods for DNA recovery from complex substrates such as RDX-C4 explosives, teeth, bones, and maggots, as well as advancements in forensic DNA intelligence and the implications of consanguinity in familial DNA searching. She is deeply committed to advancing human resource development, continuous education, and the integration of artificial intelligence and machine learning into forensic workflows, alongside ongoing research and development in forensic science.

Safia Abdessalem Messaoudi is a Tunisian Associate Professor of Molecular Biology and Genetics at Naif Arab University for Security Sciences in Riyadh, Kingdom of Saudi Arabia. She earned her PhD in Biotechnology and Biological sciences from the University of Monastir, Tunisia, in 2013. Her research career began in medical genetics, focusing on molecular diagnostic markers for cardiovascular diseases. In 2015, she transitioned to the field of forensic science at NAUSS, where she now specializes in advancing forensic DNA analysis. She teaches cytogenetics, biostatistics, and bioinformatics, and supervises master's theses on diverse topics including forensic applications of Next-Generation Sequencing, X-STR analysis in Saudi populations, mitochondrial DNA, and the role of circulating DNA in criminal justice and oncology. Her research is dedicated to improving the performance and productivity of DNA testing. Her forward-looking program aims to develop forensic DNA databases, statistical software, and advanced methods for analyzing challenging samples. A key focus is validating NGS technologies to recover reliable genetic information from degraded and ancient DNA. Complementing her scientific work, she is a certified expert in academic quality assurance, with experience developing institutional standards and managing accreditation processes. She has also served as an expert with the UN-NAUSS Regional Expertise Center for Combating Drugs and Crime.

Part I

Theoretical Foundations of Blood and Forensics

Evolution and Structure of the Circulatory System

1

Abstract

As the animal body size and activity levels increased, the circulatory system underwent significant changes that profoundly influenced animal anatomy and physiology. Early life forms lacked complex circulatory structures and relied mainly on simple diffusion and slow bulk flow within rudimentary gastrovascular spaces. Such mechanisms were sufficient only for small organisms with low metabolic demands.

With increasing morphological and physiological complexities, animal lineages independently evolved specialized mechanisms for internal fluid transport. Some groups retained relatively loose, open circulatory systems in which fluid circulates through body cavities, whereas others developed closed networks of vessels that confine blood to defined pathways. Circulatory diversity is extensive, ranging from simple tubular hearts in small crustaceans and accessory pumping structures in earthworms to fully developed, complex, multichambered hearts characteristic of vertebrates such as birds and humans. Among vertebrates, circulatory organization gradually progressed from single circuits to partially divided and fully separated systems, ultimately giving rise to the four-chambered hearts observed in mammals and birds. This evolutionary refinement is essential for warm-blooded animals to sustain high activity levels and maintain elevated metabolic performance.

Modern circulatory systems consist of more than a central pump and connecting vessels; they include the heart, extensive vascular networks, microscopic capillary beds, and lymphatic drainage, all shaped by evolutionary pressures. Examining how these systems have changed across species helps explain the unique traits of different animals, the structural diversity of their hearts, and the relevance of circulatory features in medical and forensic contexts. Rather than focusing on a single aspect, this chapter integrates evolutionary history, structural variation, and functional organization to illustrate the wide range of circulatory designs and the common principles underlying them.

H. R. Dash et al., *Forensic Investigation of Blood Evidence*,
https://doi.org/10.1007/978-981-95-9183-1_1

Keywords

Circulatory System · Cardiovascular Evolution · Open Vs. Closed Circulation · Comparative Anatomy · Microcirculation · Lymphatic System

1 Introduction

1.1 Functions of the Circulatory System

The human circulatory system is an intricate, dynamic, body-wide network that transports oxygen, nutrients, hormones, and immune components, removes carbon dioxide and metabolic waste, and integrates multiple physiological functions essential for cellular survival, systemic homeostasis, and adaptation to environmental changes (López-Manjón and Angón 2009; Attinger et al. 2013; Thiriet 2014; Mahmoud et al. 2021; Guyton and Hall 2021; Vijayaraman et al. 2023; Gómez et al. 2024). Main tasks involve:

- *Carrying respiratory gases*: Blood carries oxygen from the gills or lungs to metabolically active tissues and returns carbon dioxide to organs of excretion. This function is particularly critical in larger, energetically demanding organisms, where simple diffusion alone is insufficient to meet metabolic requirements (Guyton and Hall 2021; Sherwood 2015).
- *Nutrient distribution*: Following digestion, nutrients, including carbohydrates, proteins, lipids, vitamins, and minerals, are absorbed through the gastrointestinal tract and transported via the circulatory system to tissues throughout the body. This distribution supports cellular growth, repair, and energy production (Tortora and Derrickson 2020).
- *Removal of metabolic waste*: Metabolic by-products such as urea and ammonia are conveyed by circulating fluids to specialized excretory organs, including the kidneys or analogous systems, where they are processed and eliminated. This function is essential for preventing toxic accumulation and maintaining cellular integrity (Widmaier et al. 2019).
- *Hormonal transport and regulation*: Hormones are released into the bloodstream and carried to distant target cells, where they regulate growth, metabolism, reproduction, and stress responses. Unlike direct neural signaling, hormonal communication relies on circulation-mediated delivery, enabling coordinated, stepwise physiological adjustments under changing conditions (Hall 2021).
- *Immune surveillance and defense*: White blood cells, immune proteins, and signaling molecules circulate through blood or analogous fluids in invertebrates to detect, respond to, and eliminate pathogens. This system supports both rapid innate immune responses and sustained adaptive immunity (Murphy and Weaver 2016).

- *Body warmth control*: Endothermic organisms regulate temperature by redistributing metabolically generated heat through controlled blood flow. When small blood vessels widen or narrow, they can gain or lose warmth from outside air—thereby contributing to the maintenance of stable internal thermal conditions (Sherwood 2015).
- *Homeostasis plus coordination*: Blood circulation functions like a hub, linking up with breathing, the kidney, hormones, and the defense systems. It keeps body fluids, acidity levels, and pressure in check, thereby maintaining homeostasis despite fluctuations and shifts in external conditions (Guyton and Hall 2021; Tortora and Derrickson 2020).

Illustrative example: In mammals, the heart's rhythmic contractions and arterial elasticity create pulsatile blood flow, ensuring efficient oxygen delivery to tissues and rapid removal of carbon dioxide, a feat impossible with diffusion alone (Hall 2021).

1.2 Overview of Chapter Structure

This chapter aims to provide a comprehensive exploration of circulatory systems from evolutionary, anatomical, and functional perspectives. An integrated overview of the evolutionary progression and structural organization of circulatory systems discussed in this chapter is presented in Fig. 1.1. The structure is as follows:

- *Diffusion-based transport in early metazoans*: Examines how small, thin-bodied animals relied solely on diffusion and how it constrained organismal size and metabolic activity (Schmidt-Nielsen 1997; Brusca et al. 2016)
- *Proto-circulatory mechanisms*: Discusses the transition from diffusion to primitive circulation, including contractile vessels and gastrovascular systems in cnidarians and flatworms (Ruppert et al. 2004; Brusca et al. 2016)
- *Divergence of open and closed circulatory systems*: Compares hemolymph-based open systems in arthropods with blood-based closed systems in annelids and vertebrates, highlighting functional trade-offs (Sherwood 2015; Vogel 1994)
- *Comparative anatomy across phyla*: Surveys structural and functional diversity, including the evolution of hearts, arteries, veins, and capillaries, with emphasis on adaptive specialization (Ruppert et al. 2004; Tortora and Derrickson 2020)
- *Vertebrate circulatory circuits*: Details the evolution of single versus double circulation, pulmonary and systemic circuits, and adaptations in fish, amphibians, reptiles, birds, and mammals (Hall 2021; Widmaier et al. 2019)
- *Specialization of the heart and vessels*: Explores chambered heart evolution, myocardial architecture, valves, and vascular elasticity for efficient blood distribution (Guyton and Hall 2021; Sherwood 2015)

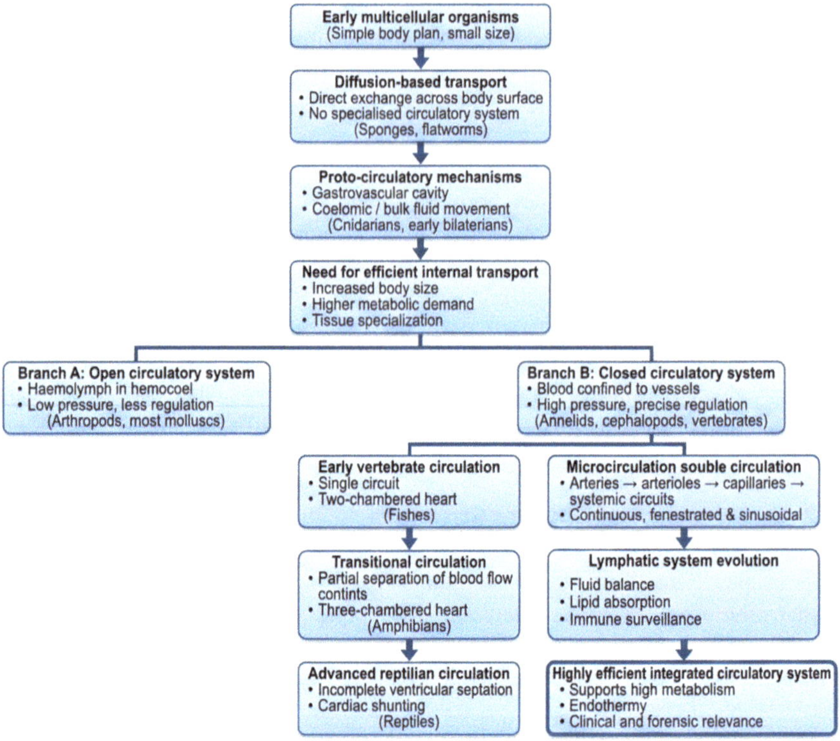

Fig. 1.1 Evolution and organization of the circulatory system

- *Lymphatic system*: Reviews the development, structure, and function of lymphatic vessels in maintaining fluid balance and immunity (Tortora and Derrickson 2020; Murphy and Weaver 2016)
- *Developmental biology of the circulatory system*: Highlights embryological origins, morphogenesis, and regulatory mechanisms of cardiovascular development (Gilbert 2013)
- *Clinical and forensic relevance*: Addresses implications for human health, cardiovascular disorders, and forensic investigations, linking evolutionary principles to applied contexts (Hall 2021; Tortora and Derrickson 2020)
- *Future directions*: Examines emerging research areas, including artificial circulatory devices, organ-on-chip models, and evolutionary physiology studies (Vogel 1994; Schmidt-Nielsen 1997)

By retracing the evolutionary trajectory and functional adaptations of circulatory systems, this chapter provides a foundation for understanding both the biological diversity and clinical significance of this vital physiological network.

2 Pre-Circulatory Strategies and Origins of Circulation

2.1 Organisms Relying on Diffusion and Bulk Flow Without a True Circulatory System

Before the emergence of dedicated circulatory systems, early metazoans depended on simple physical processes, primarily diffusion and bulk fluid movement, to sustain cellular functions.

In sponges (*Phylum Porifera*), the absence of tissues and organs is offset by an elaborate system of water canals. Specialized flagellated cells, choanocytes, generate continuous water currents that drive bulk flow through pores and chambers (Brusca et al. 2016). This movement enables the capture of food particles, gas exchange, and removal of waste, effectively replacing the need for an internal transport network.

Cnidarians including hydra, jellyfish, and sea anemones rely on a gastrovascular cavity that serves dual digestive and circulatory functions. Rhythmic muscular contractions and ciliary activity mix the cavity contents, facilitating nutrient distribution throughout the body (Ruppert et al. 2004). Because cnidarians exhibit radial symmetry and thin tissue layers, diffusion distances remain short enough to meet physiological demands (Schmidt-Nielsen 1997).

Flatworms (*Phylum Platyhelminthes*) demonstrate yet another strategy for compensating for the absence of a circulatory system. Their dorsoventrally flattened morphology increases the surface area–to–volume ratio and minimizes diffusion pathways (Brusca et al. 2016). Gas exchange, nutrient transport, and waste removal occur directly across the body surface or via branching internal canals, enabling them to sustain moderate metabolic activity despite lacking vascular structures (Sherwood 2015).

2.2 Coelomic and Hemocoelic Fluid Movement

As organisms became bigger and more complex, simple diffusion became inadequate for efficient internal transport. Many early bilaterians evolved coelomic cavities, the fluid-filled spaces between body layers that provided structural support and a medium for molecular exchange (Ruppert et al. 2004). Within these cavities, cilia-lined epithelia generated localized currents that enhanced the movement of nutrients, waste products, and signaling molecules (Brusca et al. 2016).

Muscular organisms developed additional mechanisms such as periodic contractions of the body wall, peristaltic movements, and hydrostatic pressure changes that facilitated bulk displacement of coelomic or hemocoel fluids (Vogel 1994). Although these systems lacked directionality, vessel confinement, and regulated pressure seen in true circulatory systems, they functioned as proto-circulatory mechanisms (Schmidt-Nielsen 1997). These innovations represented significant evolutionary steps by providing partial internal distribution and enabling the emergence of more complex tissue architectures.

2.3 Biophysical Constraints, Design Principles, and Evolutionary Drivers of Circulatory Systems

The evolution of circulatory systems has been fundamentally shaped by biophysical and physiological constraints, particularly those associated with diffusion. Diffusive transport is effective only over very short distances, typically on the micrometer scale. As early multicellular organisms increased in body size and internal complexity, diffusion alone became inadequate to meet rising metabolic demands, especially in tissues with high oxygen and nutrient requirements (Schmidt-Nielsen 1997; Sherwood 2015).

One of the most significant selective pressures driving circulatory evolution was increasing body size, which inherently limited diffusion efficiency and delayed internal transport. This limitation was intensified by rising metabolic rates associated with sustained locomotion, predation, and complex ecological behaviors, all of which required rapid and targeted delivery of oxygen, nutrients, and signaling molecules. The emergence of specialized tissues and organs, including nervous and muscular systems, further increased internal heterogeneity, making passive transport insufficient for maintaining physiological homeostasis (Brusca et al. 2016; Sherwood 2015). Additionally, the evolution of protective external coverings such as cuticles, shells, and thickened epithelia reduced direct exchange with the environment, necessitating the development of internal transport mechanisms (Ruppert et al. 2004).

In response to these constraints, several key design principles emerged that shaped the architecture of circulatory systems. Central among these was the evolution of active fluid transport. Specialized pumping mechanisms, such as hearts or contractile tubular structures, enabled the circulation of internal fluids, allowing rapid distribution of oxygen, nutrients, hormones, and signaling molecules while facilitating efficient removal of metabolic waste products from highly active tissues (Tortora and Derrickson 2020; Vogel 1994).

Energy efficiency represented another critical design consideration. Circulatory systems evolved to balance adequate flow rates with minimal energetic cost through optimization of vessel diameter, hierarchical branching patterns, and elastic vessel walls. These features reduce flow resistance and frictional losses, enabling sustained circulation without excessive energy expenditure (Guyton and Hall 2021; Schmidt-Nielsen 1997).

Reliability and fault tolerance emerged as critical evolutionary features of circulatory architecture. The development of redundant pathways, collateral vessels, and unidirectional valves enhanced system resilience by maintaining perfusion despite obstructions, pressure fluctuations, or mechanical stress.

Such features ensured continuity of internal transport and allowed circulatory systems to dynamically adjust blood flow in response to changing tissue demands (Vander et al. 2019).

Evolutionary trade-offs between simplicity and specialization further influenced circulatory design. Open circulatory systems, as seen in many arthropods, are structurally simple and energetically economical but provide limited control over flow

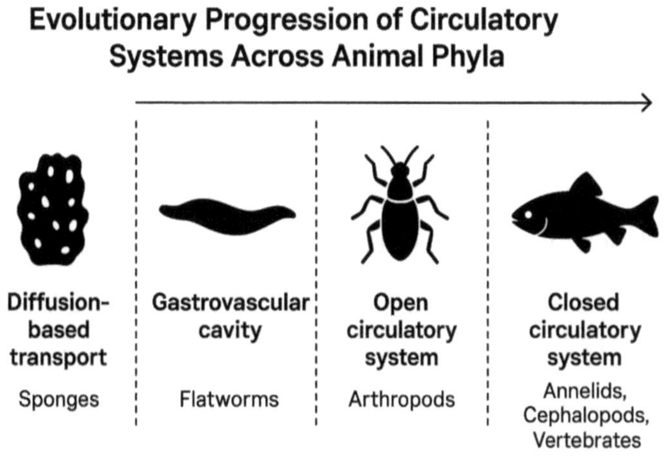

Fig. 1.2 Evolution of the circulatory system in animal phyla

distribution and pressure. In contrast, closed circulatory systems, characteristic of annelids and vertebrates, are more complex but allow precise regulation of blood flow and pressure, supporting higher metabolic rates and sustained activity levels. The distribution of these organizational patterns across animal phyla is illustrated in Fig. 1.2 (Ruppert et al. 2004; Sherwood 2015).

Collectively, these biophysical constraints, selective pressures, and design principles favored the gradual evolution of dedicated circulatory systems, progressing from simple diffusion-based and open arrangements to increasingly regulated and closed networks. This evolutionary trajectory enabled the emergence of high-performance circulatory systems in insects, cephalopods, and vertebrates, capable of supporting larger body sizes, complex organ systems, and elevated metabolic activity (Vogel 1994; Schmidt-Nielsen 1997).

3 Open Versus Closed Circulatory Systems

3.1 Open Circulatory Systems

Open circulatory systems, typical of arthropods, many mollusks, and various other invertebrate taxa, are generally characterized by the lack of a fully enclosed vascular network. In these systems, the circulatory fluid (hemolymph) is pumped from the heart into hemocoelic sinuses, where it directly bathes the tissues and mediates physiological exchange (Brusca et al. 2016; Ruppert et al. 2004).

In arthropods, a dorsal tubular heart equipped with segmental ostia draws hemolymph inward, while pulsatile contractions drive it anteriorly through the hemocoel (Brusca et al. 2016). Some groups possess accessory hearts that aid localized perfusion, such as insect wings or decapod crustacean appendages (Schmidt-Nielsen 1997).

Operating at relatively low pressures, open systems are energetically economical and mechanically simple (Vogel 1994). However, they offer limited capacity for rapid redistribution of flow or fine-tuned regulation, constraining their ability to support sustained high metabolic rates or intense aerobic activity (Sherwood 2015).

3.2 Closed Circulatory Systems

Closed circulatory systems maintain blood within a continuous series of arteries, capillaries, and veins. This arrangement, found in annelids, cephalopods, and vertebrates, allows for high pressures, controlled resistance, and precise modulation of flow to different tissues (Ruppert et al. 2004; Schmidt-Nielsen 1997).

Cephalopods exemplify advanced closed circulation among invertebrates: they possess a systemic heart and paired branchial hearts that support efficient oxygen delivery through gills, an evolutionary convergence with vertebrate cardiovascular specialization (Brusca et al. 2016).

Although closed systems demand greater energetic investment for maintaining vessel integrity and pressure gradients, they provide substantial evolutionary advantages by enabling rapid transport of gases, nutrients, and hormones, thereby supporting elevated metabolic performance (Sherwood 2015; Vogel 1994). A schematic comparison of open and closed circulatory systems is shown in Fig 1.3.

3.3 Convergent and Parallel Evolution

Closed circulatory systems have arisen independently in several lineages, illustrating convergent evolutionary responses to similar physiological pressures such as the

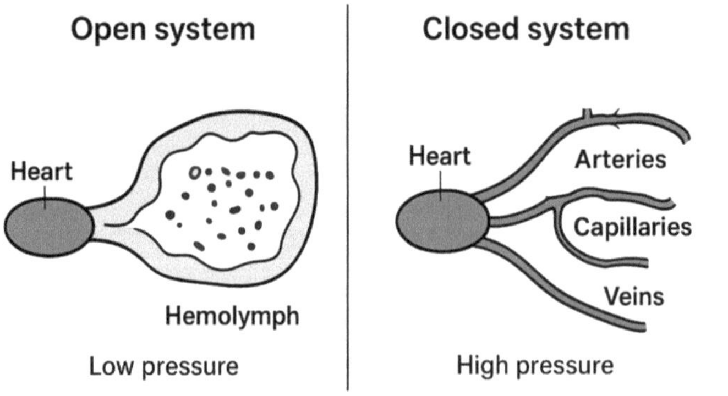

Structural and functional comparison of open anddosed circulatory systems highlighting vessel confinement and pressure regulation.

Fig. 1.3 Open vs closed circulatory system

need for efficient oxygen transport, sustained locomotion, and high metabolic demands (Schmidt-Nielsen 1997; Brusca et al. 2016). Importantly, circulatory architectures do not fall into a strict binary classification. Many organisms exhibit intermediate or transitional patterns, blending features of open and closed designs (Ruppert et al. 2004). These evolutionary trajectories highlight gradual modifications in vessel complexity, pump morphology, contractile specialization, and flow regulation, each shaped by ecological niches and metabolic constraints (Vogel 1994).

4 Comparative Circulatory Anatomy Across Major Animal Groups

4.1 Invertebrates

Invertebrates exhibit extensive and clearly remarkable diversity in circulatory organization, reflecting their broad evolutionary diversification and the wide range of ecological niches they occupy. Arthropods, including insects, crustaceans, and arachnids, possess an open circulatory system in which hemolymph is circulated through a series of interconnected sinuses collectively termed as hemocoels (Chapman 2013; Ruppert et al. 2004). The primary pumping organ is a dorsal tubular heart composed of segmentally arranged chambers with ostia, which allow hemolymph to enter during diastole (Harrison et al. 2012). Contraction propels hemolymph anteriorly through the dorsal vessel, facilitating distribution throughout the body. Many arthropods have additional accessories or auxiliary hearts, such as pulsatile organs associated with insect wings or antennae, which assist in localized perfusion in metabolically demanding regions (Klowden 2013).

Mollusks demonstrate substantial circulatory divergence across classes. Bivalves and most gastropods retain open circulatory systems where hemolymph bathes tissues directly, generating relatively low pressures and slower transport rates (Brusca et al. 2016). In contrast, cephalopods (e.g., squids, octopuses, cuttlefish) have evolved highly efficient closed vascular systems that include a three-part heart complex: a central systemic heart and paired branchial hearts that pump blood through the gills (Budelmann et al. 1997; Wells and Smith 1987). This advanced system supports cephalopods' active predatory lifestyle and elevated metabolic requirements.

Annelids, particularly earthworms and polychaetes, exhibit a closed circulatory system with well-defined vessels (Ruppert et al. 2004). The major longitudinal vessels, dorsal and ventral vessels, are connected by smaller segmental vessels that enhance distribution. Contractile regions or "pseudohearts" in the anterior segments regulate blood pressure and maintain unidirectional flow. The relatively high-pressure annelid system allows efficient nutrient and gas transport compared to many other invertebrate taxa (Curtis and Barnes 1989).

4.2 Basal Chordates

Basal chordates, such as lancelets (cephalochordates) and tunicates (urochordates), provide insights into early circulatory organization preceding vertebrate evolution. Lancelets possess a closed but simple vascular arrangement, lacking a discrete chambered heart. Instead, circulation is driven by peristaltic contractions of a ventral vessel analogous to the vertebrate ventral aorta (Romer and Parsons 1986; Ruppert et al. 2004). Blood flow is continuous but of low pressure, and gas exchange occurs largely across the body surface.

Tunicates, the closest living relatives of vertebrates, possess a small, elongate heart that generates unique bidirectional pumping, periodically reversing the direction of blood flow (Satoh 2014). This feature reflects the simplicity and plasticity of early chordate circulation. Although rudimentary, tunicate and lancelet systems exhibit organizational themes such as a ventral vessel and gill-associated perfusion that foreshadow more complex cardiovascular structures in vertebrates (Nobis et al. 2018).

4.3 Vertebrates—Overview

The vertebrate circulatory system demonstrates a high level of anatomical organization and physiological specialization. Regardless of taxonomic group, vertebrates share a basic structural blueprint that includes:

- A centrally located, muscular heart that drives circulation
- A closed vascular network consisting of arteries, capillaries, and veins
- A hierarchical arterial system delivering oxygenated blood to tissues
- Extensive capillary beds that facilitate gas and metabolite exchange
- Venous pathways returning deoxygenated blood to the heart

This foundational pattern is well documented across vertebrate taxa (Campbell et al. 2020; Randall et al. 2012). However, circulatory architecture varies considerably across lineages. These variations correlate strongly with respiratory strategies, gill-based systems in fish versus lung-based systems in tetrapods, and with metabolic demands tied to ecological niche, chronophysiology (ectothermic vs. endothermic), and activity patterns (Hickman et al. 2017; Schmidt-Nielsen 1997). The evolutionary progression from single-circuit to fully separated dual-circuit systems underpins the diversification of vertebrate life and the emergence of high-performance cardiovascular adaptations in birds and mammals (Else and Hulbert 1987; Olson 2011).

5 Evolution of Vertebrate Circulatory Circuits

5.1 Single-Circuit Circulation in Fishes

The earliest vertebrates exhibit a single, serial circulatory circuit, best exemplified by modern fishes. Their heart comprises two chambers, one atrium and one ventricle (Kardong 2019). Deoxygenated blood flows from the ventricle directly to the gills, where oxygenation occurs (Randall et al. 2014). From the gills, oxygen-rich blood moves to systemic tissues without returning to the heart first. This configuration imposes physiological constraints. Because gill capillaries are delicate, arterial pressure must remain low, reducing systemic blood pressure and limiting oxygen delivery (Hickman et al. 2020). As a result, fish, particularly ectothermic species, have restricted aerobic capacity compared to endothermic vertebrates.

5.2 Transitional Patterns in Amphibians

Amphibians, representing a transition from aquatic to terrestrial life, possess a three-chambered heart consisting of two atria and a single ventricle (Campbell et al. 2020). Although the ventricle is undivided, internal trabeculae reduces mixing of oxygenated and deoxygenated blood (Hill et al. 2022). A defining amphibian feature is reliance on cutaneous respiration, which may supplement or exceed pulmonary gas exchange depending on the species and environment (Burggren and Warburton 2007). Because both the skin and lungs participate in oxygen uptake, amphibians require a flexible circulatory design capable of adjusting blood flow based on oxygen availability and metabolic state.

5.3 Reptiles and Partial Ventricular Separation

Reptiles possess a more refined circulatory structure. Many species have a partially divided ventricle, sometimes described as a “three-and-a-half-chambered” heart, with three functional subdivisions: the cavum arteriosum, cavum venosum, and cavum pulmonale (Hicks and Wang 2012a, b).

This structure enables cardiac shunting, allowing reptiles to regulate blood flow between pulmonary and systemic circuits. Shunting is critical in:

- Thermoregulation, by altering peripheral perfusion (Wang and Hicks 2017)
- Diving or apnea, by reducing pulmonary circulation temporarily (Wang et al. 2019)
- Digestive periods, where systemic circulation increases to support postprandial metabolism (Secor 2008)

These mechanisms place reptiles between amphibians and fully divided systems of birds and mammals.

5.4 Fully Separated Double Circulation in Birds and Mammals

Birds and mammals possess a completely septated, four-chambered heart, allowing full separation of pulmonary and systemic circuits (Campbell et al. 2020). This innovation enables:

- High-pressure systemic circulation, improving oxygen delivery
- Low-pressure pulmonary circulation, protecting delicate lung structures (Hill et al. 2022)
- Sustained high metabolic rates, supporting endothermy and high-energy activities such as flight and endurance locomotion (Farrell and Smith 2017a, b)

Complete separation is foundational to the extraordinary aerobic capacity of endotherms.

5.5 Special Evolutionary Adaptations

Vertebrates entering extreme environments evolved specialized circulatory adaptations:

- Diving mammals (e.g., seals, whales) show bradycardia, selective perfusion, and peripheral vasoconstriction during dives to conserve oxygen (Ponganis 2015).
- High-altitude vertebrates (e.g., bar-headed geese, llamas) possess hemoglobin with increased oxygen affinity, expanded pulmonary surface area, and revised vascular architecture (Scott and Milsom 2016).
- Desert-adapted species employ circulatory thermoregulatory strategies to reduce water loss (Withers and Cooper 2010).
- Burrowing and hypoxia-tolerant animals utilize metabolic depression and enhanced anaerobic capacity to survive low oxygen conditions (Hochachka and Somero 2022).

These examples show the deep integration of circulatory architecture with ecological pressures and life-history strategies.

6 Structural Organization of the Vertebrate Heart

6.1 Gross Anatomy

The vertebrate heart displays remarkable structural diversification, reflecting an evolutionary transition from simple peristaltic pumps to complex, multichambered organs capable of supporting high metabolic rates and specialized physiological demands (Hickman et al. 2017; Randall et al. 2012).

Early vertebrates possess a linear tubular heart composed of sequential compartments: the sinus venosus, atrium, ventricle, and outflow tract, representing the ancestral design from which more complex forms evolved (Romer and Parsons 1986). Over evolutionary time, these components undergo progressive specialization, septation, and morphological refinement.

Most fish retain a two-chambered heart consisting of a single atrium and ventricle, enabling single-circuit circulation in which deoxygenated blood passes from the heart to the gills before entering systemic circulation (Schmidt-Nielsen 1997).

Amphibians exhibit a three-chambered heart with two atria and a partially divided ventricle, producing limited mixing and allowing partial separation of oxygenated and deoxygenated blood (Campbell et al. 2020).

Reptiles display further advancement with an incompletely septate ventricle and specialized outflow tracts such as the cavum pulmonale and cavum arteriosum, enabling differential blood shunting under variable physiological conditions (Hicks and Wang 2012a, b).

In birds and mammals, the heart achieves its most derived state: a fully septated four-chambered structure with two atria and two ventricles, complete division of systemic and pulmonary circulation, and sophisticated valvular architecture (atrioventricular and semilunar valves) (Else and Hulbert 1987; Olson 2011). This design supports high arterial pressures, elevated cardiac output, and intensive metabolic demands associated with endothermy and sustained aerobic activity.

6.2 Myocardial Structure and Coronary Circulation

Myocardial organization among vertebrates demonstrates clear evolutionary gradients corresponding to metabolic intensity and oxygen delivery requirements. In many fishes and amphibians, the myocardium is primarily spongy or trabeculated, enabling oxygen diffusion directly from blood within cardiac chambers and reducing dependence on coronary vessels (Santer 1985; Farrell and Smith 2017a, b).

Reptiles typically possess a mixed myocardium with both spongy and compact layers, representing an intermediate arrangement that balances diffusion-driven and coronary-based oxygenation (Hicks and Wang 2012a, b).

Birds and mammals exhibit predominantly compact myocardium, characterized by densely organized cardiac muscle fibers capable of generating powerful, sustained contractions (Schmidt-Nielsen 1997; Campbell et al. 2020). This compact design necessitates a dedicated coronary circulation comprising coronary arteries, veins, and an extensive microvascular network. The evolution of a robust coronary system is tightly linked to the emergence of endothermy, increased aerobic scope, and high-performance cardiovascular function (Olson 2011; Bruggren et al. 2014). Without coronary perfusion, compact myocardium would be incapable of meeting the oxygen and nutrient demands of continuous, high-intensity activity. The progressive transition from spongy to compact myocardium across vertebrate groups, along with the associated development of coronary circulation, is illustrated in Fig. 1.4.

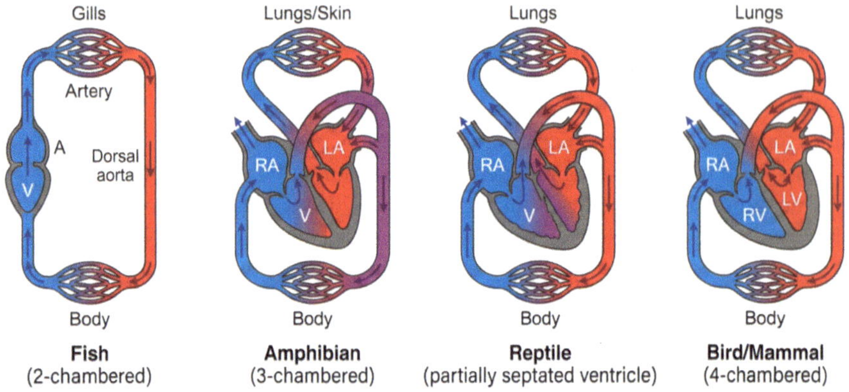

Fig. 1.4 Evolution of vertebrate heart structure

6.3 Cardiac Conduction System

The vertebrate cardiac conduction system shows increasing specialization across evolutionary lineages, paralleling the development of more complex cardiac anatomy. Primitive vertebrates rely on diffuse pacemaker regions and slow, nondiscrete conduction pathways distributed throughout the myocardium (Romer and Parsons 1986).

In advanced vertebrates, especially birds and mammals, the conduction network becomes highly organized and anatomically compartmentalized:

- The sinoatrial (SA) node acts as the primary pacemaker, initiating rhythmic electrical impulses (Irisawa et al. 1993).
- The atrioventricular (AV) node delays impulse transmission, ensuring coordinated atrial contraction before ventricular filling.
- The signal then propagates through the bundle of His, bundle branches, and an extensive network of Purkinje fibers, promoting rapid and synchronous ventricular depolarization (Campbell et al. 2020).

This hierarchical conduction system facilitates precisely timed contractions, improved hemodynamic efficiency, and the high-frequency cardiac cycles necessary for endothermic organisms (Olson 2011). Increased chamber specialization and circulatory demand were major drivers of conduction system evolution.

6.4 Functional Implications

Cardiac structure exerts a profound influence on vertebrate physiological performance. Key functional parameters include:

- Stroke volume, determined by ventricular size, wall thickness, and contractile strength
- Heart rate, regulated by autonomic input and pacemaker activity
- Cardiac output, the product of stroke volume and heart rate

Together, these parameters determine circulatory efficiency and adaptive capacity under varying environmental and metabolic conditions (Randall et al. 2012).

Vertebrates demonstrate varied cardiac adaptations suited to ecological niches and functional demands:

- Athletic mammals and birds develop hypertrophied, high-output hearts optimized for continuous aerobic activity (Else and Hulbert 1987).
- Ectotherms adjust heart rate and cardiac output in response to ambient temperature variation.
- Diving mammals and reptiles exhibit pronounced bradycardia and peripheral vasoconstriction to conserve oxygen stores during submergence (Kooyman 1989).
- High-altitude species modify myocardial oxygen extraction and hemoglobin–oxygen affinity to cope with reduced atmospheric oxygen (Faraci 1991).
- Hypoxia-tolerant organisms such as some amphibians and fish employ biochemical and structural adaptations that preserve cardiac efficiency in low-oxygen environments (Bruggren et al. 2014).

These examples illustrate the tight integration of cardiac design, physiological regulation, and evolutionary adaptation, underscoring the heart's central role in vertebrate survival, specialization, and radiative success.

7 Vascular Architecture and Microcirculation

7.1 Hierarchical Vascular Tree

The vertebrate circulatory system is organized as a branching and highly hierarchical vascular network designed to efficiently distribute oxygen, nutrients, hormones, and immune mediators to all tissues (Guyton and Hall 2021). At the proximal end, arteries transport blood under high pressure and possess thick, elastic walls consisting of a prominent tunica media enriched with smooth muscle and elastic laminae. This elastic recoil helps buffer the pulsatile ejection of blood from the heart, ensuring continuous forward flow despite intermittent ventricular contractions (Boron and Boulpaep 2017).

As arteries branch into smaller arterioles, their diameter progressively decreases, while smooth muscle density increases. Arterioles serve as the primary regulators of peripheral resistance and thus play a central role in maintaining systemic arterial pressure and modulating regional blood flow in response to physiological demand (Hall 2015). The arterioles give rise to capillary networks composed of a single endothelial layer optimized for exchange processes. These networks are strategically distributed according to the metabolic requirements of the underlying tissues (Levick and Michel 2010).

On the venous side, venules collect blood from capillaries and gradually converge into larger veins. Veins possess thinner walls with reduced smooth muscle and frequently contain valves to prevent backflow, particularly in the lower extremities where blood must travel against gravity (Tortora and Derrickson 2018). The hierarchical arrangement of arteries, arterioles, capillaries, venules, and veins ensures optimal hemodynamic regulation, efficient exchange of nutrients and waste products, and effective systemic circulation.

7.2 Capillary Types and Exchange Mechanisms

Capillaries exhibit structural variations that reflect the specific functional demands of the tissues they supply:

- *Continuous capillaries*: These possess uninterrupted endothelial linings joined by tight junctions, restricting permeability and facilitating tightly regulated exchange. They are predominant in the brain, muscle, lung, and skin, where highly selective transport is essential (Boron and Boulpaep 2017).
- *Fenestrated capillaries*: Characterized by circular pores (fenestrae), these capillaries allow rapid exchange of water, solutes, and hormones. They are common in endocrine glands, intestinal villi, and the renal glomeruli, where high-volume exchange is required (Hall 2015).
- *Sinusoidal (discontinuous) capillaries*: These have large inter-endothelial gaps and a discontinuous basement membrane, enabling the movement of large molecules and even entire cells. They are present in the liver, spleen, and bone marrow, where filtration and cellular trafficking occur (Levick and Michel 2010).

Exchange across the capillary wall occurs via several mechanisms:

- Diffusion of gases and lipid-soluble substances along concentration gradients
- Filtration and reabsorption, driven by the balance of hydrostatic and oncotic pressures as described by Starling forces (Levick and Michel 2010)
- Transcytosis, permitting vesicular transport of macromolecules, including proteins and antibodies, across endothelial cells (Boron and Boulpaep 2017)

These complementary mechanisms ensure precise regulation of nutrient delivery and waste removal in accordance with dynamic metabolic needs. The major structural types of capillaries and their roles in microcirculatory exchange are illustrated in Fig. 1.5.

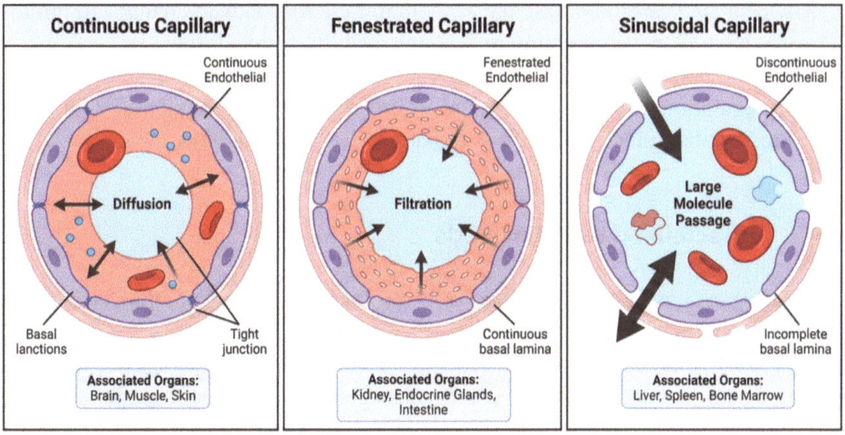

Fig. 1.5 Microcirculation and capillary types

7.3 Regulation of Blood Flow and Pressure

Blood flow and vascular pressure are tightly regulated by neural, hormonal, and local control mechanisms to maintain systemic hemodynamic stability.

7.3.1 Neural Regulation

The autonomic nervous system modulates both cardiac output and vascular tone. Sympathetic activation induces vasoconstriction mediated primarily by α_1-adrenergic receptors, increasing systemic vascular resistance, while parasympathetic influence (largely via the vagus nerve) predominantly affects heart rate with limited direct vascular effects (Guyton and Hall 2021).

7.3.2 Hormonal Regulation

Several circulating hormones exert strong vasoactive effects.

- Adrenaline modulates vasoconstriction or vasodilation depending on receptor distribution.
- Angiotensin II and vasopressin produce potent vasoconstriction and influence fluid balance.
- Nitric oxide, synthesized by endothelial cells, induces vasodilation and plays a key role in maintaining basal vascular tone (Moncada and Higgs 2006).

7.3.3 Local Autoregulation

Tissues adjust their own perfusion in response to metabolic needs:

- *Metabolic control*: Accumulation of metabolites such as CO_2, H^+, adenosine, and lactate induces vasodilation, enhancing local blood flow (Levy et al. 2018).

- *Myogenic mechanisms*: Vascular smooth muscle responds to changes in intraluminal pressure by contracting or relaxing, stabilizing local flow.

Autoregulation is especially critical in organs with high metabolic demand or pressure-sensitive functions, such as the brain, heart, and kidneys.

7.4 Specialized Vascular Beds

Certain organs possess highly specialized vascular architectures optimized for their physiological roles:

- *Cerebral circulation*: Exhibits strong autoregulation and structural specialization through the blood–brain barrier, maintaining stable perfusion critical for neuronal function. It is particularly sensitive to CO_2 levels and intracranial pressure changes (Strandgaard and Paulson 1990).
- *Coronary circulation*: Matches blood supply closely to myocardial metabolic demands, with the majority of perfusion occurring during diastole. Due to limited collateral circulation, even minor obstruction may result in ischemia, highly relevant in clinical and forensic investigations (Klabunde 2017).
- *Renal circulation*: Characterized by high-flow, low-resistance cortical vessels and specialized glomerular capillaries optimized for filtration. Alterations in renal hemodynamics have significant implications in toxicology and systemic disease assessments (Hall 2015).
- *Hepatic portal circulation*: Receives nutrient-rich blood from the gastrointestinal tract via the portal vein, enabling detoxification, metabolism, and nutrient processing before distribution to the systemic circulation. Structural discontinuities in hepatic sinusoids facilitate efficient exchange (Guyton and Hall 2021).

These specialized vascular networks are central to homeostasis and are frequently analyzed in clinical pathology, forensic interpretations of ischemia, trauma analysis, toxicokinetics, and systemic disease investigations.

8 Blood Composition and Biophysical Properties

Blood is a specialized, multifunctional connective tissue comprising approximately 7–8% of total body mass in adult humans. It serves as the primary medium for the transport of respiratory gases, nutrients, metabolic substrates, hormones, and signaling molecules while also contributing to immune defense, thermoregulation, acid–base balance, and waste removal. These integrated functions are essential for maintaining cellular and systemic homeostasis (Silverthorn 2019; Berne et al. 2018).

Structurally, blood consists of cellular elements—erythrocytes, leukocytes, and platelets—suspended in plasma, a protein-rich aqueous matrix. The physical properties of plasma, particularly viscosity and surface tension, together with cellular interactions, govern blood flow under both physiological and pathological conditions (Levick 2018; Popel and Johnson 2005).

From a forensic standpoint, blood assumes additional significance when it escapes the vascular system. Its composition and biophysical behavior influence bloodstain formation, dispersion, clotting, and drying patterns at a scene. Interactions among cellular elements, coagulation factors, plasma proteins, and environmental variables determine stain morphology and temporal changes, enabling reconstruction of injury mechanisms, force direction, and event sequencing (Bevel and Gardner 2008; Brutin and Sobac 2015; James et al. 2014). Thus, understanding blood biology and physics forms a critical foundation for reliable bloodstain pattern analysis (BPA) and medico-legal interpretation.

8.1 Cellular Components and Plasma

Human blood comprises approximately 45% cellular elements and 55% plasma by volume. Erythrocytes are the most abundant cells and are structurally adapted for efficient gas exchange and deformability, allowing passage through narrow microcapillaries. Hemoglobin concentration largely determines oxygen-carrying capacity, while variations in erythrocyte count influence blood viscosity and flow behavior (Levick 2018). Such variations may also affect bloodstain characteristics following hemorrhage.

Leukocytes, though present in lower numbers, are essential for immune surveillance and defense. Their primary forensic relevance lies not in stain morphology but in their role as sources of nuclear DNA, RNA, and immunological markers used in forensic identification and molecular analyses (Anand and Rajagopal 2017).

Platelets play a central role in primary hemostasis by adhering to damaged vascular surfaces and forming an initial plug. Changes in platelet number or function—due to disease, pharmacological agents, or intoxication—can alter clot formation and blood loss, thereby influencing stain appearance and distribution (Pasquereau 2017).

Plasma is composed predominantly of water, along with proteins such as albumin, globulins, and fibrinogen, as well as electrolytes, hormones, and metabolic byproducts. Plasma composition affects blood viscosity, surface tension, and drying behavior, which in turn influence droplet formation, edge characteristics, and satellite spatter patterns observed in BPA (Attinger et al. 2013).

8.2 Coagulation and Clot Formation Dynamics

Hemostasis is achieved through the coordinated interaction of cellular components and plasma-based coagulation factors. Following vascular injury, platelet adhesion and activation trigger the coagulation cascade, ultimately leading to thrombin-mediated conversion of fibrinogen into fibrin, which stabilizes the forming clot (Levi and van der Poll 2010).

In forensic contexts, the dynamics of clot formation are particularly significant. Delayed coagulation—caused by anticoagulant therapy, hypothermia, or underlying coagulation disorders—can result in greater blood spread and thinner stain margins. Conversely, rapid clot formation may limit dispersion and promote localized pooling, thereby affecting interpretations regarding the timing and sequence of injuries (Pasquereau 2017).

8.3 Rheological Properties of Blood

Blood exhibits non-Newtonian, shear-thinning behavior, meaning its viscosity decreases as the shear rate increases. This property is essential for understanding both in vivo blood flow and the formation of stains following vascular injury. Under high shear conditions, such as arterial trauma, blood can atomize into fine droplets, whereas higher viscosity favors the formation of cohesive drops with limited dispersion (Attinger et al. 2013; Laan et al. 2015).

Blood viscosity is influenced by several factors, including hematocrit, plasma protein concentration, temperature, and pathological conditions. Additionally, surface tension and adhesion to substrates affect droplet shape, impact angle determination, and satellite spatter formation. These factors highlight the importance of accounting for biological variability when interpreting bloodstain patterns in forensic analysis.

8.4 Blood Behavior Outside the Body

Once blood leaves the circulation, it undergoes progressive physicochemical changes. Evaporation gradually increases its viscosity, while clot retraction and fibrin crosslinking modify the structure of the stain. The surface characteristics of the receiving substrate—whether porous or nonporous—significantly influence stain appearance, edge definition, and absorption patterns (Bevel and Gardner 2008).

Environmental factors such as temperature, humidity, and airflow also affect drying rates and stain morphology. These variables are routinely considered when estimating the time since deposition and reconstructing the sequence of events at a scene; however, inherent variability requires cautious interpretation to avoid overestimation of precision.

8.5 Molecular and Pathophysiological Considerations

At the molecular level, variations in hemoglobin concentration, oxidative state, and degradation processes such as hemolysis can influence blood color, opacity, and drying behavior. These changes may affect both visual assessment and the performance of presumptive and confirmatory blood tests (James et al. 2014; White and Saferstein 2018).

Pathophysiological conditions, including coagulation disorders, liver disease, and the use of anticoagulant or antiplatelet medications, can significantly modify clotting time and hemorrhage patterns. Failure to account for these systemic factors may lead to misinterpretation of stain size, distribution, or injury severity (Hoffman and Monroe 2018; Saunders et al. 2019). Integrating clinical context with physical evidence therefore enhances the reliability of forensic reconstruction.

9 Circulatory Function and Hemodynamics

The circulatory system operates as a tightly regulated hydraulic network that transforms myocardial contractile activity into effective systemic perfusion. This coordinated process ensures the continuous delivery of oxygen and nutrients to tissues while facilitating the removal of metabolic waste products. Circulatory function is governed by core hemodynamic principles, including pressure gradients, flow velocity, vascular resistance, vessel compliance, and cardiac output. These variables interact dynamically to maintain circulatory homeostasis across resting, exertional, and stress-related states (Levick 2018; Berne et al. 2018; Thiriet 2014).

From a forensic standpoint, these same hemodynamic determinants critically influence the behavior of blood following vascular disruption. Arterial pressure and pulsatile flow shape the distance, directionality, and morphology of projected bloodstains, whereas venous and capillary pressures primarily govern passive flow, gravitational pooling, and substrate saturation. Variations in vessel caliber, wall elasticity, and regional perfusion further modulate blood dispersion and volume loss. Together, these physiological parameters provide a mechanistic foundation for bloodstain pattern analysis (BPA), enabling reconstruction of injury mechanisms, sequencing of events, and estimation of relative timing (Bevel and Gardner 2008; James et al. 2014; Knight and Saukko 2016; Dolinak and Matshes 2015).

9.1 Principles of Blood Flow

Blood flow through the vasculature is primarily described by Poiseuille's law, which relates volumetric flow rate (Q) to the pressure gradient (ΔP), vessel radius (r), blood viscosity (η), and vessel length (L):

$$Q = \frac{\pi r^4 \Delta P}{8\eta L}$$

This relationship demonstrates the disproportionate influence of vessel radius on flow dynamics; even minor reductions in arterial diameter—due to vasoconstriction, atherosclerosis, or postmortem vessel collapse—can markedly reduce flow velocity and blood volume displacement (Levick 2018). When breached, high-pressure arteries with thick muscular walls may eject blood at velocities exceeding 4–6 m/s, producing characteristic pulsatile spurting patterns. In contrast, veins, which operate under lower pressure and possess thin, collapsible walls, typically generate slow, steady bleeding that results in pooling or drip patterns rather than projected spatter (Bevel and Gardner 2008).

9.2 Systemic, Pulmonary, and Coronary Circulation

The cardiovascular system comprises multiple interconnected but functionally distinct circuits, each defined by specific pressure regimes, flow characteristics, and physiological roles. The systemic circulation, driven by the left ventricle, operates under high pressure and high flow, imparting substantial kinetic energy to arterial blood. Consequently, injuries to systemic arteries are commonly associated with forceful, pulsatile blood ejection and well-defined arterial spatter patterns that are central to BPA and trauma reconstruction (Bevel and Gardner 2008; James et al. 2014).

In contrast, the pulmonary circulation is a low-pressure, low-resistance system designed to protect the alveolar–capillary interface while enabling efficient gas exchange. Although pulmonary vessel rupture rarely produces external bloodstains, pulmonary hemorrhage encountered at autopsy may indicate internal trauma, asphyxia mechanisms, or underlying cardiopulmonary disease (Levick 2018; Knight and Saukko 2016). The coronary circulation, supplying the myocardium, operates under intermediate pressures and is critical for sustaining cardiac output. Postmortem ischemic changes or coronary disruption may alter intracardiac blood distribution or produce hemorrhagic findings that can mimic traumatic injury, necessitating careful anatomical and physiological interpretation (Dolinak and Matshes 2015).

The marked pressure differentials among these circulatory circuits directly influence the velocity, volume, and dispersion of blood following vascular breach. Recognition of circuit-specific hemodynamic profiles is therefore essential for accurate interpretation of bloodstain morphology and avoidance of forensic misclassification.

9.3 Pressure Gradients, Cardiac Output, and Pulsatility

Cardiac output, defined as the product of stroke volume and heart rate, is the principal determinant of arterial pulsatility within the systemic circulation. Cyclic fluctuations in arterial pressure during the cardiac cycle generate rhythmic spurting when an artery is severed: systolic peaks produce high-velocity jets, while diastolic phases yield reduced flow or transient cessation. In forensic analysis, the spacing, orientation, and termination of arterial spurts may provide insights into vessel type, anatomical location of injury, and the temporal relationship between cardiac activity and trauma (James et al. 2014).

Elevated systemic blood pressure associated with physical exertion, emotional stress, pain, or fear—conditions frequently present during violent encounters—can increase transmural capillary pressure and predispose fragile microvessels to rupture. The presence of petechiae or micro-hemorrhages may therefore serve as indicators of acute physiological stress or impaired venous return near the time of injury, contributing contextual information regarding the victim's physiological state (Gibbins 2018).

9.4 Vessel Compliance and Hemodynamic Responses

Arteries and veins differ markedly in their mechanical properties, which significantly influence both blood flow and extravasation patterns following injury. Arterial compliance, largely determined by the elastic components of the tunica media, permits accommodation of pulsatile flow while limiting wall stress. Pathological reductions in compliance, as observed in arteriosclerosis or chronic hypertension, increase shear stress and may amplify the velocity and dispersion of arterial blood upon rupture (Levick 2018; Nichols et al. 2011).

Veins, by contrast, exhibit high compliance due to thin walls and low intraluminal pressure, allowing them to collapse readily. Postmortem venous drainage plays a major role in gravity-dependent blood pooling and the development of livor mortis, providing critical information about body position and postmortem interval (Dolinak and Matshes 2015; Knight and Saukko 2016). Vessel-specific biomechanical responses—including elastic recoil, rupture threshold, and transient vasoconstriction—directly affect the volume and pattern of blood loss, underscoring the need for integrated hemodynamic assessment in BPA (Bevel and Gardner 2008; James et al. 2014).

9.5 Shear Stress and Flow Disturbances

Shear stress, generated by friction between flowing blood and the vascular endothelium, is essential for maintaining endothelial integrity under normal laminar conditions. However, regions of disturbed flow—such as bifurcations, curved segments, or stenotic vessels—exhibit localized turbulence that elevates mechanical stress on the vessel wall, increasing susceptibility to rupture during trauma (Anand and Rajagopal 2017; Levick 2018; Nichols et al. 2011). Pathological changes, including atherosclerotic plaque formation, further intensify these disturbances and may alter the kinetics of blood extravasation.

In forensic practice, turbulent flow at the site of injury can manifest as irregular spatter patterns, elongated droplets, satellite stains, and nonlinear dispersion. Such features often reflect intrinsic hemodynamic disturbances rather than post-event contamination. Understanding the relationship between shear forces and flow irregularities enables investigators to distinguish among accidental injury, blunt-force trauma, and intentional assault, thereby refining the reconstruction of injury dynamics (Bevel and Gardner 2008; James et al. 2014; Dolinak and Matshes 2015).

9.6 Hemodynamics and Forensic Bloodstain Patterns

The integration of hemodynamic principles is fundamental to the accurate interpretation of bloodstain patterns. Arterial injuries, characterized by high systolic pressure and pulsatile flow, typically produce rhythmic arcs and projected spurts that correspond to cardiac cycles. In contrast, venous injuries generate continuous, low-velocity bleeding that tends to pool near the wound, while capillary bleeding produces diffuse, low-volume staining and petechial hemorrhages (Bevel and Gardner 2008; James et al. 2014; Dolinak and Matshes 2015).

Impact spatter further illustrates the interaction between hemodynamics, blood rheology, and external forces. Blunt-force trauma or high-velocity projectiles impart kinetic energy that interacts with intravascular pressure and the viscoelastic properties of blood, generating secondary spatter, satellite droplets, and directional patterns crucial for event reconstruction (Attinger et al. 2013; Laan et al. 2015; Brutin and Sobac 2015).

Following cessation of circulation, gravitational redistribution of blood leads to livor mortis, the distribution and fixation of which are influenced by vascular anatomy, residual pressure gradients, and body position (Knight and Saukko 2016; Dolinak and Matshes 2015).

By integrating physiological, biomechanical, and hemodynamic principles, forensic practitioners can more accurately infer vessel type, injury mechanism, temporal relationships, and postmortem changes, thereby strengthening the scientific basis of bloodstain pattern analysis and forensic reconstruction.

10 Lymphatic System and Its Evolution

10.1 Origins and Basic Functions

The lymphatic system arose early in vertebrate evolution as a complementary vascular network responsible for maintaining internal fluid balance and supporting immune defense (Randall et al. 2014). Its emergence was largely driven by the need to counteract continuous plasma leakage from blood capillaries into interstitial spaces, a consequence of capillary hydrostatic pressure and endothelial permeability. The development of lymphatic vessels allowed vertebrates to efficiently recover interstitial fluid and return it to the bloodstream, thereby stabilizing circulatory volume and tissue homeostasis (Hill et al. 2022).

Beyond fluid recovery, the lymphatic system fulfils two key physiological roles:

- *Lipid absorption*: Specialized lymphatic structures in the intestinal villi, known as lacteals, absorb dietary lipids and fat-soluble vitamins. These are transported as chylomicrons via lymphatic vessels to the venous circulation, bypassing the hepatic portal system (Hickman et al. 2020).
- *Immune surveillance*: Lymphatic vessels transport antigens, immune cells, and cytokines from tissues to regional lymph nodes. Within the nodes, immune components such as dendritic cells and lymphocytes initiate adaptive immune responses and filter pathogens or debris (Murphy and Weaver 2016).

Overall, the lymphatic system represents a major evolutionary advancement linking circulatory efficiency with increasing immune complexity in vertebrates (Kardong 2019).

10.2 Structural Organization

The lymphatic system consists of a hierarchical network adapted for the collection, movement, and immunological processing of lymph.

- *Lymph capillaries*: These blind-ended vessels exhibit high permeability due to overlapping endothelial cells that form one-way valve-like junctions, allowing uptake of interstitial fluid, proteins, pathogens, and immune cells (Hill et al. 2022). Their discontinuous basement membrane further enhances transport efficiency.
- *Lymphatic vessels*: Absorbed lymph moves into larger collecting vessels equipped with intraluminal valves that prevent backflow. These vessels contain contractile segments called lymphangions, whose rhythmic smooth muscle contractions generate pulsatile lymph flow (Withers and Cooper 2010).
- *Lymph nodes*: These encapsulated biological filters contain macrophages, dendritic cells, and lymphocytes. As lymph percolates through nodal sinuses, pathogens are trapped, antigens are presented, and adaptive immune responses are activated (Murphy and Weaver 2016).

Unlike the cardiovascular system, the lymphatic system lacks a central pump. Lymph flow depends on:

- Skeletal muscle contraction
- Thoracic pressure changes during breathing
- Intrinsic lymphatic smooth muscle activity
- Body posture and movement

This combination ensures effective transport despite low internal pressure (Hill et al. 2022).

10.3 Comparative Lymphatic Anatomy

Variations in lymphatic structures across vertebrate taxa illustrate the evolutionary integration of circulatory and immune functions.

- *Lymph hearts*: Lower vertebrates, including amphibians, reptiles, and some fish, possess lymph hearts, which actively pump lymph into veins (Kardong 2019). These structures compensate for limited skeletal muscle contributions and lower circulatory pressures typical of ectothermic organisms.
- *Lymph node distribution*: True lymph nodes are a mammalian specialization associated with complex adaptive immunity (Murphy and Weaver 2016). Birds lack discrete nodes but contain diffuse lymphoid aggregates. Reptiles and amphibians rely more heavily on the spleen and mucosal lymphoid tissues for immune filtration (Hickman et al. 2020).
- *Lymphatic duct architecture*: In mammals, major lymphatic ducts such as the thoracic duct and right lymphatic duct reflect evolutionary refinement toward efficient lipid transport, immune coordination, and maintenance of high metabolic rates (Randall et al. 2014). Increasing anatomical specialization across vertebrates parallels rising immunological complexity and metabolic demands. These evolutionary changes are illustrated in Fig. 1.6, which depicts the development of the lymphatic system as a secondary vascular network, showing the progression from simple lymph hearts in lower vertebrates to highly organized lymphatic ducts and nodes in mammals.

These comparative differences reflect a broad evolutionary trend toward enhanced compartmentalization, improved immune surveillance, and tighter integration between lymphatic and cardiovascular systems.

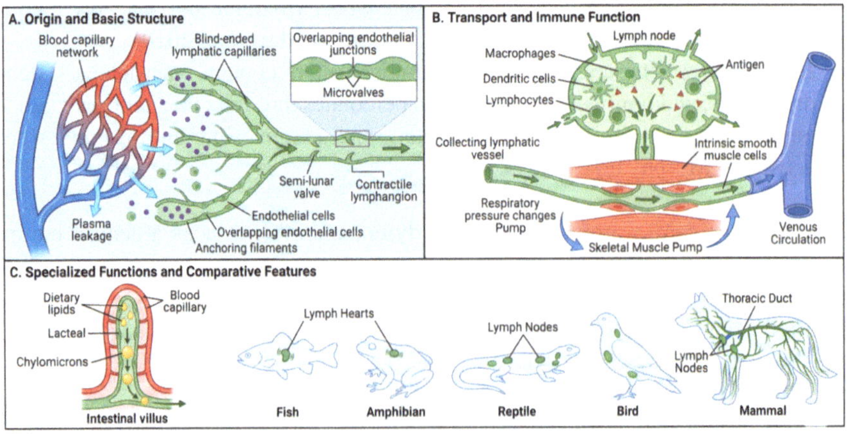

Fig. 1.6 Development of lymphatic system as secondary vascular network

11 Developmental Biology of the Circulatory System

11.1 Embryonic Origin of Heart and Vessels

The circulatory system is among the earliest functional systems to emerge during vertebrate embryogenesis, reflecting its essential role in supporting rapid growth, nutrient distribution, and metabolic demand (Gilbert and Barresi 2020). Its formation involves a highly coordinated sequence of cell migration, tissue patterning, molecular signaling, and morphogenetic remodeling.

11.1.1 Formation of the Primitive Heart Tube

Cardiac development begins with the specification of the cardiogenic mesoderm, derived from the lateral plate mesoderm. Bilateral cardiac progenitor fields migrate toward the embryonic midline and fuse to form the primitive linear heart tube (Sadler 2020). This tube comprises an inner endocardial layer and an outer myocardial layer, separated by cardiac jelly—an extracellular matrix that later contributes to valve and septal formation (Moorman and Christoffels 2003).

11.1.2 Heart Tube Looping

The initially straight heart tube undergoes rightward (dextral) looping, a critical process that establishes the spatial orientation of future chambers (Harvey 2002). Looping transforms the simple tube into a helically shaped structure with anatomically distinct regions destined to form the atria, ventricles, and outflow tract.

11.1.3 Chamber Formation and Septation

As development progresses:

- *Chamber ballooning*: Atrial and ventricular chambers expand through region-specific myocardial growth (Moorman and Christoffels 2003).
- *Septation*: The heart divides into left and right circulatory pathways, progressively separating pulmonary and systemic circuits (Gilbert and Barresi 2020).
- *Valve formation*: Atrioventricular and semilunar valves develop from endocardial cushion remodeling, ensuring unidirectional blood flow (Sadler 2020).

Disruptions in looping, chamber formation, or septation are major contributors to congenital heart defects, underscoring the sensitivity of cardiovascular development.

11.1.4 Vasculogenesis and Angiogenesis

The embryonic vascular system develops through two interrelated processes:

- *Vasculogenesis*: De novo formation of blood vessels from angioblasts, which differentiate into endothelial cells. This process establishes the initial vascular plexus, including major vessels such as the dorsal aorta and cardinal veins (Risau 1997).

- *Angiogenesis*: Expansion and remodeling of existing vessels via endothelial sprouting, guided by growth factors such as VEGF, hemodynamic forces, and oxygen gradients (Potente et al. 2011). Angiogenesis produces the hierarchical vascular network essential for organogenesis and fetal survival.

Together, vasculogenesis and angiogenesis generate the extensive circulatory architecture characteristic of vertebrate embryos.

11.2 Evolutionary Conservation of Developmental Pathways

The fundamental developmental processes governing heart and vessel formation are highly conserved across vertebrates, reflecting their shared ancestry and the essential nature of cardiovascular function (Davidson 2010). Conserved molecular toolkits (NKX2-5, GATA, VEGF, TBX) function as the common evolutionary foundation, modified by regulatory "tinkering" to produce the diverse cardiovascular architectures (from two- to four-chambered hearts) seen across vertebrates, illustrating evo-devo principles.

11.2.1 Core Gene Regulatory Networks

Several deeply conserved gene families orchestrate cardiovascular formation:

- *NKX2-5*: A master transcription factor required for cardiac specification; homologs occur in both vertebrates and invertebrates, underscoring their ancient evolutionary origin (Harvey 2002).
- *GATA factors (GATA4, GATA5, GATA6)*: Regulate early cardiomyocyte differentiation, chamber morphogenesis, and heart tube maturation (Patient and McGhee 2002).
- *VEGF signaling*: Central to endothelial proliferation, migration, and vessel branching during vasculogenesis and angiogenesis (Potente et al. 2011).
- *TBX, HAND, and MEF2 genes*: Key regulators of chamber identity, myocardial development, and conduction system differentiation (Srivastava 2006).

11.2.2 Evolutionary Modifications ("Tinkering")

While the molecular toolkit is conserved, evolution modifies these pathways to generate variation across vertebrates:

- Differences in chamber numbers (two in fish, three in amphibians/reptiles, four in birds and mammals)
- Divergence in septation complexity and valve morphology
- Variation in conduction system organization between taxa
- Specialized vascular adaptations for high endurance, diving, flight, or hypoxic survival (Farrell and Smith 2017a, b)

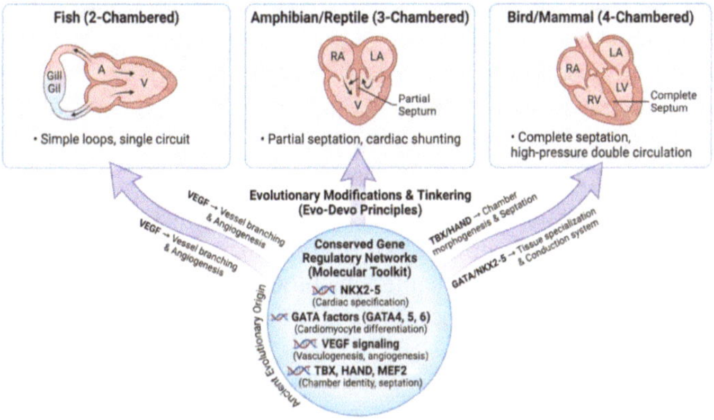

Conserved molecular toolkits (NKX2-5, GATA, VEGF, TBX) function as the common evolutionary foundation, modified by regulatory 'tinkering' to produce the diverse cardiovascular architectures (from 2- to 4-chambered hearts) seen across vertebrates, illustrating evo-devo principles.

Fig. 1.7 Genetic relations and evolutionary modification and tinkering

These examples illustrate evolutionary developmental (evo-devo) principles, where conserved developmental programs are repurposed or elaborated to produce diversity in cardiovascular structure and function. These relationships and modifications are summarized in Fig. 1.7, which depicts the genetic toolkit, conserved pathways, and examples of evolutionary "tinkering" that give rise to structural diversity in vertebrate hearts.

12 Clinical and Forensic Perspectives

12.1 Circulatory System and Patterns of Injury

A detailed understanding of cardiac and vascular anatomy is essential for interpreting injury patterns, reconstructing events, and distinguishing antemortem from postmortem findings in forensic examinations (Spitz and Spitz 2020). Because the circulatory system distributes blood under pressure, disruptions to vessels and cardiac structures produce characteristic and often diagnostic patterns of hemorrhage and tissue damage.

12.1.1 Hemorrhage Patterns

Different hemorrhage types provide valuable insights into the mechanism, force, and chronology of trauma:

- External hemorrhage results from breaches in the skin or mucosal surfaces and may reflect the characteristics of the wounding instrument, differentiating between sharp-force and blunt-force trauma (DiMaio and DiMaio 2001).

- Internal hemorrhage, including hemothorax, hemoperitoneum, or mediastinal bleeding, is frequently associated with injury to major vessels or vital organs such as the liver, spleen, or heart (Saukko and Knight 2015).
- Petechiae and ecchymoses reflect rupture of small vessels in tissues. Petechiae are particularly significant in cases of asphyxia, strangulation, and compressive neck trauma (Dolinak and Matshes 2015).
- Arterial spurting patterns are produced when high-pressure arterial vessels are breached, causing pulsatile blood ejection that leaves distinctive projected and wave-form stains valuable in bloodstain pattern analysis (James et al. 2014).

12.1.2 Vessel Rupture

Vessel rupture may occur due to:

- Blunt-force trauma, which may produce intimal tears, lacerations, pseudoaneurysms, or complete vascular transection. The aortic isthmus is particularly vulnerable due to fixation points and deceleration forces (Spitz and Spitz 2020).
- Penetrating injuries, such as stab or gunshot wounds, which frequently damage deep vessels, leading to rapid exsanguination and collapse (DiMaio and DiMaio 2001).
- Spontaneous or pathological rupture, including aneurysms, dissections, or rupture of atherosclerotic vessels, which are the major causes of sudden natural death (Saukko and Knight 2015).

Identification of vessel damage allows forensic experts to determine the direction and magnitude of force, survival intervals, and potential mechanisms of death.

12.1.3 Blunt and Penetrating Trauma

The circulatory system responds differently to varying mechanisms of injury:

- Blunt trauma may result in contusions, lacerations, and shearing injuries. Cardiac contusions, chamber ruptures, coronary artery tears, and pericardial tamponade are critical findings in motor-vehicle collisions and crush injuries (Dolinak and Matshes 2015).
- Penetrating trauma produces linear or perforating wound tracts, with predictable vascular injuries based on the object's velocity and trajectory. Gunshot wounds may disrupt major vessels, produce cavitation effects, and lead to fatal exsanguination or embolization of projectiles (DiMaio 2016).

These trauma patterns assist in determining intent (homicidal, suicidal, accidental) and reconstructing event sequences.

12.2 Postmortem Circulatory Changes

Following cessation of cardiac activity, circulation stops abruptly, triggering passive blood redistribution and structural alterations that provide valuable information during death investigations.

12.2.1 Livor Mortis (Postmortem Hypostasis)

When the heart ceases pumping, blood settles in the dependent regions of the body due to gravity, producing characteristic discoloration patterns:

- *Fixed lividity*: Indicates the body's position after death
- *Areas of blanching*: Denote external compression from ligatures, bedding, or objects (Saukko and Knight 2015)
- *Shift or absence of lividity*: Suggests postmortem repositioning of the body

These features are essential for reconstructing postmortem movement and estimating the postmortem interval (PMI).

12.2.2 Postmortem Clot Formation

After death, blood coagulates and separates into distinct forms:

- *Chicken fat clots*: Yellow and gelatinous
- *Currantjelly clots*: Dark red

Unlike antemortem thrombi, postmortem clots are unattached, smooth, and lack Lines of Zahn, reducing the risk of misinterpretation during autopsy (Spitz and Spitz 2020).

12.2.3 Implications for PMI Estimation

Postmortem circulatory changes contribute to PMI estimation through multiple indicators:

- The degree and fixation of livor mortis provide approximate time ranges.
- Vascular collapse, clot morphology, and hemolysis serve as supplementary markers (Saukko and Knight 2015).
- Blood distribution in organs and tissues aids in distinguishing antemortem hemorrhage from postmortem pooling.

In forensic toxicology, circulatory cessation may also cause postmortem redistribution of substances, particularly in highly perfused organs such as the liver, lungs, and myocardium (Pounder and Jones 2021).

12.3 Integration with Other Forensic Domains

Understanding the circulatory system enhances multiple subfields of forensic science, enabling comprehensive interpretation of evidence.

12.3.1 Bloodstain Pattern Analysis (BPA)

The biomechanics of blood influenced by vessel pressure, breach characteristics, and tissue structure determines:

- Impact spatter
- Arterial spurting
- Cast-off stains
- Expirated patterns (James et al. 2014)

Accurate differentiation of these patterns requires knowledge of vascular anatomy and hemodynamics.

12.3.2 Forensic Toxicology

The circulatory system governs xenobiotic absorption, distribution, and elimination. Postmortem interpretation relies on:

- Identifying tissue-specific drug partitioning
- Recognizing organs prone to redistribution (e.g., heart, lungs, liver)
- Comparing central vs. peripheral blood concentrations (Pounder and Jones 2021)

Understanding circulatory physiology is essential for distinguishing true intoxication from postmortem artifacts.

12.3.3 Postmortem Microbiome and Thanatomicrobiomics

Postmortem circulatory collapse affects microbial migration and decomposition dynamics:

- Bacterial translocation from the gut into systemic circulation
- Gas formation and vascular distension
- Tissue colonization patterns useful for PMI estimation (Metcalf et al. 2016)

These emerging microbial markers are increasingly integrated into advanced forensic models of postmortem interval estimation.

These diverse forensic applications are illustrated in Fig. 1.8, which highlights the clinical and forensic relevance of the circulatory system across BPA, toxicology, and postmortem microbiome analyses.

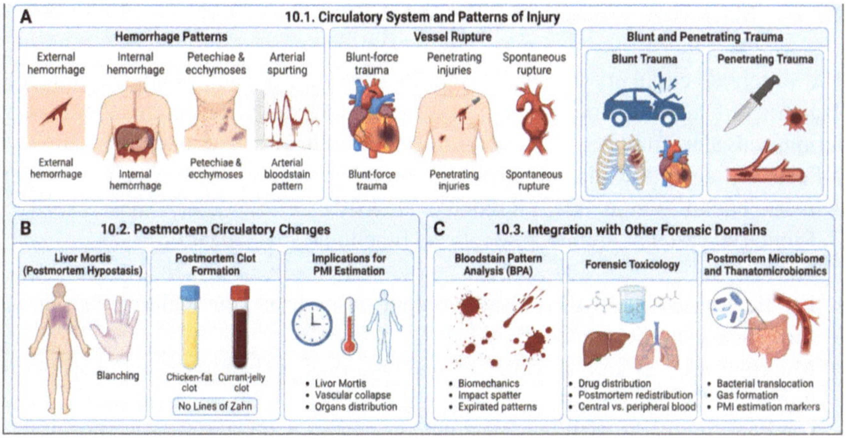

Fig. 1.8 Clinical and forensic perspectives of the circulatory system

13 Summary and Future Directions

13.1 Key Evolutionary Transitions

The evolution of the vertebrate circulatory system represents one of the most transformative physiological developments in animal history. Early multicellular organisms depended primarily on diffusion for gas and nutrient exchange, a constraint that limited body size, metabolic rate, and ecological complexity (Schmidt-Nielsen 1997). The appearance of open circulatory systems in many invertebrates marked a significant step toward internal transport efficiency, though these systems remained limited by low pressure, non-directional flow, and restricted capacity to support high metabolic demands (Brusca et al. 2016).

The origin of a closed circulatory system in vertebrates was a decisive evolutionary event. The formation of enclosed vessels, directional blood flow, and increasingly specialized cardiac chambers allowed for:

- Higher and more regulated systemic pressure
- Substantially improved oxygen transport efficiency
- Progressive separation of oxygenated and deoxygenated blood in advanced taxa
- Greater aerobic capacity and sustained locomotor activity

These innovations reached their zenith in the fully septated, four-chambered hearts of birds and mammals, enabling high cardiac output and supporting the energetic demands of endothermy (Else and Hulbert 1987; Olson 2011). Such cardiovascular sophistication played a central role in facilitating flight, endurance running, long-distance migration, and the broad ecological diversification that characterizes modern vertebrate clades (Campbell et al. 2020).

13.2 Structure–Function Relationships

The vertebrate circulatory system exemplifies the fundamental biological principle that structure and function evolve in concert. At the cardiac level, chamber organization, valvular architecture, myocardial morphology, and conduction pathways collectively determine the efficiency of blood propulsion and systemic oxygen delivery (Hickman et al. 2017; Randall et al. 2012).

Similarly, vascular architecture reflects functional specialization:

- Elastic arteries buffer the pulsatile output of the heart, maintaining continuous flow (Schmidt-Nielsen 1997).
- Muscular arteries and arterioles regulate regional perfusion through vasomotor control.
- Capillary networks facilitate metabolic exchange, with structural variants (continuous, fenestrated, sinusoidal) suited to organ-specific demands.

Microcirculation integrates these features, ensuring that perfusion dynamically matches tissue metabolic requirements (Olson 2011). Environmental pressures including temperature extremes, hypoxia, diving depths, altitudes, and locomotor intensities further shape cardiovascular adaptations, unfolding the dynamic interplay between anatomical design and physiological performance (Bruggren et al. 2014).

13.3 Open Questions and Research Frontiers

Although significant progress has been made, numerous aspects related to circulatory evolution and vertebrate cardiovascular function remain poorly understood. Ongoing advances in comparative physiology, genomics, and developmental biology continue to reveal and highlight critical areas for future investigation and research.

13.3.1 Extreme Cardiovascular Adaptations

Several vertebrates demonstrate extraordinary modifications to cardiovascular structure and function:

- Diving mammals, such as seals and whales, withstand prolonged hypoxia and redistribute blood selectively during deep dives (Kooyman 1989).
- Birds in sustained flight maintain exceptionally high cardiac output, efficient coronary perfusion, and rapid conduction capacity (Campbell et al. 2020).
- Gigantic vertebrates, including whales and extinct sauropods, exhibit unique cardiovascular scaling strategies to maintain perfusion across large body dimensions (Faraci 1991).

Studying these systems provides models for resilience, hypoxia tolerance, and stress physiology with implications for human cardiac medicine.

13.3.2 Comparative Genomics of Circulatory Evolution

Genomic and transcriptomic studies now enable researchers to trace:

- Conserved developmental pathways (e.g., NKX2-5, TBX, GATA factors)
- Gene duplications and regulatory innovations
- Origins of cardiac chambers, valves, conduction pathways, and vascular networks

These data strengthen connections between genotype, phenotype, and the evolutionary history of the cardiovascular system (Irisawa et al. 1993; Nobis et al. 2018).

13.3.3 Evo-Devo and Its Biomedical and Forensic Applications

The integration of evolutionary biology with developmental genetics (evo-devo) offers new insights into mechanisms underlying congenital cardiac defects. Regenerative strategies, including myocardial repair, play a crucial role in advancing cardiovascular treatment. Additionally, the interpretation of pathological changes in forensic investigations, such as trauma-related vascular injuries or age-related cardiac remodeling, offers important insights for accurate analysis and cause-of-death determination. Understanding how circulatory structures form, change, and fail deepens both biomedical and forensic interpretability (Bruggren et al. 2014).

13.3.4 Implications for Medicine, Conservation, and Forensic Science

Future research continues to explore and to expand the applied significance of complex cardiovascular biology. In the realm of clinical medicine, evolutionary perspectives offer valuable and solid frameworks for understanding susceptibility to cardiovascular disease. Moreover, and in conservation physiology, circulatory insights are instrumental as they aid in evaluating species' resilience to several environmental stressors such as climate change, pollutants, and natural habitat degradation. Within forensic sciences, the principles of circulatory physiology enhance the interpretation of traumatic injury, postmortem interval estimation, hemodynamic patterns associated with shock, and hemorrhage, in addition to the distribution of toxins throughout the vascular system (Randall et al. 2012).

As multidisciplinary approaches expand, the vertebrate circulatory system remains a vital framework for understanding adaptation, resilience, and biological diversity.

References

Anand M, Rajagopal KR (2017) A short review of advances in the modelling of blood rheology and clot formation. Fluids 2(3):35

Attinger D, Moore C, Donaldson A, Jafari A, Stone HA (2013) Fluid dynamics topics in bloodstain pattern analysis: comparative review and research opportunities. Forensic Sci Int 231(1–3):375–396
Berne RM, Levy MN, Koeppen BM, Stanton BA (2018) Physiology, 7th edn. Elsevier
Bevel T, Gardner RM (2008) Bloodstain pattern analysis: with an introduction to crime scene reconstruction, 3rd edn. CRC Press
Boron WF, Boulpaep EL (2017) Medical physiology, 3rd edn
Bruggren W, Farrell AP, Lillywhite HB (2014) Cardiovascular physiology in vertebrates: adaptations in form and function. CRC Press
Brusca RC, Moore W, Shuster SM (2016) Invertebrates, 3rd edn. Oxford University Press
Brutin D, Sobac B (2015) Pattern formation in drying drops of blood. J Fluid Mech 776:1–32
Budelmann BU, Schipp R, Boletzky SV (1997) Cephalopoda. In: Harrison FW, Kohn AJ (eds) Microscopic anatomy of invertebrates, vol 6. Wiley-Liss, pp 119–414
Burggren WW, Warburton SJ (2007) Amphibian cardiovascular development and function. Comp Biochem Physiol Part A 148(4):665–676
Campbell NA, Reece JB, Urry LA, Cain ML, Wasserman SA, Minorsky PV, Orr RB (2020) Biology, 12th edn. Pearson
Chapman RF (2013) The insects: structure and function, 5th edn. Cambridge University Press
Curtis H, Barnes NS (1989) Biology, 5th edn. Worth Publishers
Davidson EH (2010) The regulatory genome: gene regulatory networks in development and evolution, 2nd edn. Academic Press
DiMaio VJ, DiMaio D (2001) Forensic pathology. 2nd edn. CRC Press, p. 452
Dolinak D, Matshes E (2015) Forensic pathology: principles and practice. Elsevier
Else PL, Hulbert AJ (1987) Evolution of mammalian endothermic metabolism: "leaky" membranes as a source of heat. Am J Physiol 253(1):R1–R7
Faraci FM (1991) Adaptations to hypoxia in birds: how to thrive in a low-oxygen world. Annu Rev Physiol 53:59–70
Farrell AP, Smith F (2017a) Cardiac physiology in fish. In: Farrell AP (ed) Encyclopedia of fish physiology. Academic Press, pp 546–559
Farrell AP, Smith FM (2017b) Cardiovascular systems in vertebrates. In: Randall D, Burggren W, French K (eds) Eckert animal physiology, 8th edn. W. H. Freeman, pp 345–389
Gibbins I (2018) Autonomic nervous system and stress. Handb Clin Neurol 157:109–125
Gilbert SF (2013) Developmental biology, 10th edn. Sinauer Associates
Gilbert SF, Barresi MJF (2020) Developmental biology, 12th edn. Sinauer Associates
Gómez JT, González Rios JL, Dressler F (2024) Electric circuit representation of the human circulatory system to estimate the position of nanosensors in vessels. Nano Commun Netw 40:100499
Guyton AC, Hall JE (2021) Textbook of medical physiology, 14th edn. Elsevier
Hall JE (2015) Guyton and Hall: textbook of medical physiology, 13th edn. Elsevier
Hall JE (2021) Pocket companion to Guyton and Hall textbook of medical physiology, 14th edn. Elsevier
Harrison JF, Woods HA, Roberts SP (2012) Ecological and environmental physiology of insects. Oxford University Press
Harvey RP (2002) Patterning the vertebrate heart. Nat Rev Genet 3(7):544–556
Hickman CP, Roberts LS, Keen SL, Larson A, Eisenhour DJ, Ober WC (2017) Integrated principles of zoology, 17th edn. McGraw-Hill
Hickman CP, Roberts LS, Keen SL, Eisenhour DJ, Larson A, I'anson H (2020) Integrated principles of zoology, 18th edn. McGraw-Hill
Hicks JW, Wang T (2012a) Cardiovascular regulation in reptiles. Physiology 27(1):14–27
Hicks JW, Wang T (2012b) The functional significance of the reptilian heart. J Exp Biol 215:663–671
Hill RW, Wyse GA, Anderson M (2022) Animal physiology, 5th edn. Sinauer Associates
Hochachka PW, Somero GN (2022) Biochemical adaptation, 3rd edn. Oxford University Press

Hoffman M, Monroe DM (2018) Coagulation 2018: a modern view of hemostasis. Hematology 1:1–11
Irisawa H, Brown HF, Giles W (1993) Cardiac pacemaking in the sinoatrial node. Physiol Rev 73(1):197–227
James SH, Kish PE, Sutton TP (2014) Principles of bloodstain pattern analysis: theory and practice. CRC Press
Kardong KV (2019) Vertebrates: comparative anatomy, function, evolution, 8th edn. McGraw-Hill
Klabunde RE (2017) Cardiovascular physiology concepts, 2nd edn. Wolters Kluwer
Klowden MJ (2013) Physiological systems in insects, 3rd edn. Academic Press
Knight B, Saukko P (2016) Knight's forensic pathology, 4th edn. CRC Press
Kooyman GL (1989) Diverse divers: physiology and behavior. Springer
Laan N, Smith F, Nicloux C, Brutin D (2015) Morphology of bloodstains: from drop impact physics to forensic analysis. Phys Rev Appl 4(4):044017
Levi M, van der Poll T (2010) Coagulation and sepsis. Thromb Res 125(Suppl 1):S38–S40
Levick JR (2018) Cardiovascular physiology, 6th edn. CRC Press
Levick JR, Michel CC (2010) Microvascular fluid exchange and the revised Starling principle. Cardiovasc Res 87(2):198–210
Levy MN, Koeppen BM, Stanton BA (2018) Berne & Levy physiology, 7th edn. Elsevier
López-Manjón A, Angón P (2009) Circulatory system physiology and pathology. J Physiol Biochem
Mahmoud AM, Al-Maghrabi J, Allemailem KS, El-Shehawi AM (2021) Cardiovascular homeostasis and metabolic regulation. Front Physiol 12:640146
Moncada S, Higgs A (2006) Nitric oxide and vascular control. Pharmacol Rev 43(2):109–142
Moorman AFM, Christoffels VM (2003) Cardiac chamber formation: developmental mechanisms and evolutionary insights. Dev Biol 258(1):1–18
Murphy K, Weaver C (2016) Janeway's immunobiology, 9th edn. Garland Science
Nichols WW, O'Rourke MF, Vlachopoulos C (2011) McDonald's blood flow in arteries, 6th edn. Hodder Arnold
Nobis W, Maier VH, Christiaen L (2018) Chordate evolution and the origin of vertebrates. Annu Rev Genomics Hum Genet 19:69–93
Olson KR (2011) Circulatory system. In: Heath AG (ed) Comparative physiology of animals. Cambridge University Press, pp 55–91
Pasquereau B (2017) Clotting dynamics in forensic reconstruction. Int J Legal Med 131:1207–1216
Patient RK, McGhee JD (2002) The GATA family of transcription factors. Development 129(14):3215–3226
Ponganis PJ (2015) Diving physiology of marine mammals and seabirds. Cambridge University Press
Popel AS, Johnson PC (2005) Microcirculation and hemorheology. Annu Rev Fluid Mech 37:43–69
Potente M, Gerhardt H, Carmeliet P (2011) Basic and therapeutic aspects of angiogenesis. Cell 146(6):873–887
Randall D, Burggren W, French K (2012) Eckert's animal physiology: mechanisms and adaptations, 6th edn. W.H. Freeman
Randall D, Burggren W, French K, Eckert R (2014) Eckert animal physiology, 7th edn. W. H. Freeman
Risau W (1997) Mechanisms of angiogenesis. Nature 386(6626):671–674
Romer AS, Parsons TS (1986) The vertebrate body, 6th edn. Saunders College Publishing
Ruppert EE, Fox RS, Barnes RD (2004) Invertebrate zoology, 7th edn. Brooks/Cole
Sadler TW (2020) Langman's medical embryology, 14th edn. Wolters Kluwer
Santer R (1985) Morphology and innervation of the fish heart. Springer
Satoh N (2014) Chordate origins and evolution: the molecular evolutionary road to vertebrates. Academic Press
Saukko P, Knight B (2015) Knight's forensic pathology, 4th edn. CRC Press. https://doi.org/10.1201/b13266

Saunders S et al (2019) Age-related vascular changes and forensic implications. Forensic Sci Int 301:107–114
Schmidt-Nielsen K (1997) Animal physiology: adaptation and environment, 5th edn. Cambridge University Press
Scott GR, Milsom WK (2016) Control of breathing and adaptation to high altitude in vertebrates. Annu Rev Physiol 78:473–495
Secor SM (2008) Digestive physiology of the Burmese python. Physiol Biochem Zool 81(4):373–385
Sherwood L (2015) Human physiology: from cells to systems, 9th edn. Cengage Learning
Silverthorn DU (2019) Human physiology: an integrated approach, 8th edn. Pearson
Srivastava D (2006) Making or breaking the heart: from lineage determination to morphogenesis. Cell 126(6):1037–1048
Strandgaard S, Paulson OB (1990) Cerebral autoregulation. Stroke 21(12):1647–1649
Thiriet M (2014) Anatomy and physiology of the circulatory and ventilatory systems. Springer. https://doi.org/10.1007/978-1-4614-9469-0
Tortora GJ, Derrickson B (2018) Principles of anatomy and physiology, 15th edn. Wiley
Tortora GJ, Derrickson BH (2020) Principles of anatomy and physiology, 16th edn. Wiley
Vander AJ, Widmaier EP, Raff H, Strang KT (2019) Vander's human physiology: the mechanisms of body function, 15th edn. McGraw-Hill Education
Vijayaraman P, Ahmed A, Baranchuk A, Ahsan M (2023) Systems-level regulation of cardiovascular dynamics. Nat Rev Cardiol
Vogel S (1994) Life in moving fluids: the physical biology of flow, 2nd edn. Princeton University Press
Wang T, Hicks JW (2017) The role of cardiac shunts in reptilian circulation. J Exp Biol 220(2):115–124
Wang T, Burggren W, Hicks JW (2019) The cardiovascular physiology of ectothermic vertebrates. Cambridge University Press
Wells MJ, Smith PJS (1987) The performance of the octopus circulatory system: a comparison of octopods and decapods. J Zool 213:283–299
White PC, Saferstein R (2018) Forensic science handbook, 3rd edn. Pearson
Widmaier EP, Raff H, Strang KT, Vander AJ (2019) Vander's human physiology: the mechanisms of body function, 15th edn. McGraw-Hill Education
Withers PC, Cooper CE (2010) Metabolic, thermoregulatory and ventilatory physiology of desert mammals. J Arid Environ 74(9):1083–1090

Constituents of Blood: Functions and Disorders

2

Abstract

Blood, as a specialized connective tissue, embodies a complex interplay between its fluid matrix and cellular components. Plasma provides the biochemical environment that sustains and regulates physiological processes, while red blood cells ensure efficient gas transport, white blood cells orchestrate immune defense, and platelets safeguard vascular integrity through hemostasis. Together, these constituents maintain homeostasis and enable the survival of multicellular organisms. Disruptions in any of these components, whether in concentration, structure, or function, are central to numerous pathologies including anemia, coagulopathies, leukemias, and thrombotic disorders. The clinical importance of blood is thus reflected not only in diagnostics but also in therapeutic strategies, ranging from transfusion medicine and immunotherapy to anticoagulant and antiplatelet therapies. High-resolution multi-omics approaches are uncovering novel biomarkers and therapeutic targets for systemic diseases such as cancer, cardiovascular disorders, and autoimmune conditions. Furthermore, precision medicine and stem cell-based therapies hold promise for tailoring treatments to individual patients, potentially transforming hematology from a primarily supportive discipline to one at the forefront of personalized healthcare.

Keywords

Blood components · RBC · WBC · Platelets · Hemoglobin · Blood disorders

1 Introduction

Blood is a specialized connective tissue essential for maintaining homeostasis and ensuring the survival of multicellular organisms. In adults, it constitutes approximately 7–8% of total body weight, a proportion consistent with clinical and physiological observations (Guyton and Hall 2020). Structurally, blood consists of a fluid matrix, the plasma, and three major cellular elements: red blood cells (RBCs), white blood cells (WBCs), and platelets.

H. R. Dash et al., *Forensic Investigation of Blood Evidence*,
https://doi.org/10.1007/978-981-95-9183-1_2

Each component performs distinct yet highly coordinated functions. Red blood cells are primarily responsible for oxygen and carbon dioxide transport, thereby supporting tissue metabolism and contributing to acid–base balance. White blood cells serve as the primary mediators of immune defense, providing protection against pathogens and maintaining immune surveillance. Platelets are central to hemostasis, enabling vascular repair and clot formation following injury. Plasma, in turn, acts as the transport medium for these cells while carrying electrolytes, proteins, nutrients, and signaling molecules that regulate systemic physiology.

The balance and quality of these components are critical for maintaining health. Disruptions in their concentration, structure, or function underlie a wide spectrum of diseases, ranging from anemia and immunodeficiencies to coagulopathies and thrombotic disorders (Hoffman et al. 2018). Recent advances in single-cell genomics have refined our understanding of hematopoiesis and blood cell regulation, offering new frameworks for studying lineage hierarchies and molecular drivers of dysfunction (Safina et al. 2024). Understanding the constituents of blood, their physiological roles, and their involvement in disease remains a cornerstone of modern hematology and clinical medicine.

The study of blood constituents is fundamental not only for understanding normal physiology but also for the diagnosis, management, and treatment of systemic diseases. Disorders such as anemia, leukemia, thrombocytopenia, and plasma protein deficiencies are prevalent worldwide and represent major public health burdens (GBD 2021 Hemoglobinopathies and Hemolytic Anemias Collaborators et al. 2025). Updated criteria for myelodysplastic syndromes and new guidelines for anemia highlight how diagnostic frameworks continue to evolve with molecular and clinical advances (Nachtkamp et al. 2024). Large-scale registries of bleeding disorders further demonstrate the importance of global monitoring in improving access to care (Coffin et al. 2023; Mohamed et al. 2025). In parallel, technological innovations such as artificial intelligence are being applied to optimize the diagnosis and monitoring of conditions such as sickle cell disease (Kuan et al. 2025).

Advances in hematology and molecular biology have revealed the genetic and biochemical underpinnings of many disorders, offering novel therapeutic strategies including targeted monoclonal antibodies, gene therapy, and stem cell transplantation (Koury 2014; Orkin and Zon 2008).

This chapter synthesizes current knowledge on the major constituents of blood plasma, red blood cells, white blood cells, and platelets, focusing on their physiological roles, regulatory mechanisms, and the clinical consequences of their dysfunction. Emphasis is placed on linking molecular and cellular insights with clinical manifestations to highlight the translational importance of hematological research.

2 Plasma

2.1 Composition of Plasma

Plasma constitutes about 55% of total blood volume and serves as the aqueous medium in which cells are suspended. It is composed primarily of water (~90%) but also contains electrolytes, proteins, lipids, glucose, amino acids, hormones, dissolved gases, and metabolic waste products (Guyton and Hall 2020). The protein fraction, which accounts for ~7% of plasma, is dominated by albumin, globulins, and fibrinogen. Albumin, synthesized in the liver, is the most abundant plasma protein; it plays a critical role in maintaining colloid osmotic pressure and in transporting various lipophilic and small molecules (Peters 1996). Globulins include immunoglobulins and transport proteins (e.g., transferrin, haptoglobin), while fibrinogen and other clotting factors are essential for the coagulation cascade. Electrolytes such as sodium, potassium, chloride, calcium, magnesium, and bicarbonate regulate pH, osmotic balance, and neuromuscular excitability (Palmer 2015).

Recent large-scale proteomic surveys have extended our understanding of plasma composition. An "atlas" of the plasma proteome in over 53,000 individuals has enabled the correlation of plasma protein abundances with phenotypes and disease traits, enhancing the identification of biomarker candidates and functional networks (Deng et al. 2025). Moreover, methodological advances in mass spectrometry (MS) and sample preparation have improved the depth of coverage for low-abundance proteins in plasma, allowing the detection and quantification of thousands of proteins in a single run (Menon et al. 2025).

2.2 Functions of Plasma

Plasma serves as the principal transport medium for nutrients, metabolites, hormones, gases, and waste products, thereby enabling systemic integration of organ systems. It helps maintain acid–base homeostasis via the bicarbonate buffer system and transports regulatory proteins and enzymes that mediate vascular tone, coagulation, and immune responses. Plasma proteins also function as reservoirs of amino acids, carry lipophilic molecules, and exert antioxidant, antimicrobial, and immunomodulatory effects (Baynes and Dominiczak 2019). Because the plasma compartment reflects the biochemical and physiological milieu, changes in plasma glucose, lipids, hormones, or protein patterns are widely used as biomarkers in diagnostics and prognosis.

Advances in proteomics have underscored the dynamic and informative nature of the plasma proteome. Clinical proteomics is transitioning toward multi-omic integration and high-throughput platforms, which allow for longitudinal monitoring and discovery of disease signatures (Deng et al. 2025). MS-based techniques, especially data-independent acquisition (DIA) methods, have gained prominence for their reproducibility and expanded protein coverage in complex matrices such as plasma (Menon et al. 2025).

2.3 Plasma-Related Disorders

Abnormalities in plasma composition are implicated in numerous clinical conditions. Hypoproteinemia, often resulting from malnutrition, liver dysfunction, or nephrotic syndrome, leads to a reduction in oncotic pressure and is frequently associated with edema formation (Guyton and Hall 2020; Baynes and Dominiczak 2019). Hypoalbuminemia is a negative prognostic marker in chronic liver disease and systemic illness. Reduced fibrinogen or deficiency of coagulation factors predisposes to bleeding diatheses (Deng et al. 2025).

Inherited coagulation factor deficiencies, such as hemophilia A (factor VIII) or hemophilia B (factor IX), produce severe bleeding phenotypes. Acquired deficiencies may develop in disseminated intravascular coagulation (DIC), vitamin K deficiency, or advanced hepatic failure. Conversely, hypercoagulability may arise from elevated fibrinogen or dysregulated clotting factor expression (Triplett 2000).

Electrolyte imbalances in plasma can have systemic consequences. Disorders such as hyponatremia, hyperkalemia, hypocalcemia, and disturbances in magnesium or phosphate homeostasis affect neuromuscular function, cardiac conduction, and metabolic stability (Palmer 2015). In intensive care settings, such electrolyte disorders are common and are strongly associated with adverse outcomes (Hew-Butler et al. 2018). In hypertensive populations, combined electrolyte abnormalities and renal biomarker changes correlate with disease severity and cardiovascular risk (Upadhyay 2015).

2.4 Clinical and Research Implications

Because plasma mirrors internal physiology and pathology, it remains central to diagnostics and translational research. Proteomics and metabolomics of plasma are being leveraged to discover novel biomarkers for cancers, cardiovascular disease, neurological disorders, and aging (Anderson and Anderson 2002; Coenen et al. 2023). The development of ultrasensitive MS workflows capable of detecting low-abundance proteins directly in undepleted plasma promises further expansion of diagnostic potential (Kirsher et al. 2025).

In therapeutics, therapeutic plasma exchange is employed in autoimmune, neurologic, and hematologic diseases to modulate pathological plasma components. The dual role of plasma—as both a diagnostic reservoir and a therapeutic target—underscores its centrality in modern precision medicine.

3 Red Blood Cells (RBCs)

3.1 Structure and Lifespan

Red blood cells (RBCs), or erythrocytes, are the most abundant cellular constituents of blood, averaging about 4.5–6.0 × 10^6/μL in healthy adults (Kaushansky et al. 2021). Mature RBCs are anucleate, biconcave discs (~7–8 μm in diameter),

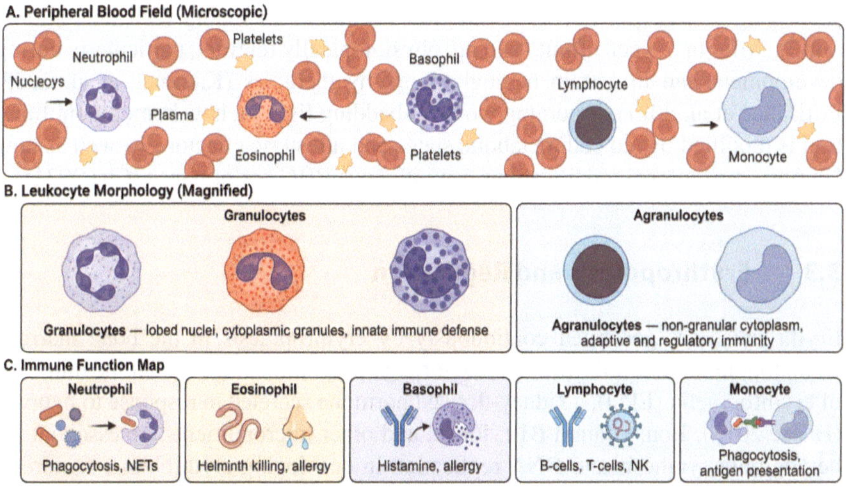

Fig. 2.1 Peripheral blood and leukocyte classification. (**a**) *Peripheral blood field (microscopic)* illustrates erythrocytes, platelets, and major leukocyte types, including neutrophils, eosinophils, basophils, lymphocytes, and monocytes. (**b**) *Leukocyte morphology (magnified)* distinguishing granulocytes—characterized by lobulated nuclei and cytoplasmic granules and primarily involved in innate immune defense—from agranulocytes, which exhibit nongranular cytoplasm and play key roles in adaptive and regulatory immunity. (**c**) *Immune function map* summarizing principal functions: neutrophils (phagocytosis, NET formation), eosinophils (helminth defense and allergic responses), basophils (histamine release and allergy), lymphocytes (B-cell-, T-cell-, and NK-cell-mediated immunity), and monocytes (phagocytosis and antigen presentation)

optimized for maximal gas exchange and deformability to traverse narrow capillaries. Their unique cytoskeletal network, comprising spectrin, ankyrin, band 3, and other membrane proteins, confers resilience and flexibility (Mohandas and Gallagher 2008). The average lifespan of an RBC is ~120 days, after which senescent cells are removed primarily in the spleen by macrophages via erythrophagocytosis (Hoffman et al. 2017). Various constituents of blood are illustrated in Fig. 2.1.

Recent insights from *Advances in Red Blood Cells Research* highlight that novel ion channels (e.g., Piezo1) and mechanosensitive pathways are involved in RBC homeostasis, affecting volume regulation, aging processes, and responses to shear stress (Lyu et al. 2024). Also, omics and metabolomic studies have begun to elucidate how metabolic shifts within RBCs contribute to aging and function (Cha 2024).

3.2 Hemoglobin and Gas Transport

The hallmark feature of RBCs is hemoglobin (Hb), a tetrameric protein consisting of globin chains and heme prosthetic groups capable of reversible O_2 binding. Hemoglobin facilitates oxygen delivery from lungs to peripheral tissues and supports CO_2 return as carbaminohemoglobin as well as via the bicarbonate buffering system (Ahmed et al. 2020). It also contributes to the buffering capacity of blood, stabilizing arterial pH (≈7.4). Hemoglobin variants, including fetal hemoglobin (HbF), differ in oxygen affinity and are physiologically relevant, particularly during development and in certain hemoglobinopathy therapies (Kaufman et al. 2025; Pritišanac et al. 2021). Emerging work is shedding light on how hemoglobin function is modified by red cell metabolic states and allosteric effectors, as well as how Hb post-translational modifications can influence RBC performance (Cha 2024).

3.3 Erythropoiesis and Regulation

Erythrocytes are generated continuously by erythropoiesis in the bone marrow. Hematopoietic stem cells differentiate through progenitor stages under the influence of erythropoietin (EPO), a kidney-derived hormone secreted in response to hypoxia (Haase 2013). Iron, vitamin B12, folate, and other micronutrients are essential for hemoglobin synthesis and DNA replication in developing erythroblasts (Andrews 2008). Recent reviews outline that transcription factors (e.g., GATA1, KLF, BCL11A) and epigenetic regulators play critical roles in orchestrating erythroid differentiation and lineage commitment (Tang and Wang 2023; Hewitt 2023; Cha 2024). Moreover, the interplay between iron metabolism and erythropoiesis mediated by factors such as erythroferrone (ERFE) and hepcidin is now better appreciated (Coffey and Ganz 2018; Pirotte et al. 2021).

Work in in vitro and ex vivo systems is also advancing: stem cell-derived RBC generation (e.g., from iPSCs or HSCs) is being optimized, though enucleation efficiency, yield, and scalability remain challenges (Noordin et al. 2025). Cultured reticulocytes derived from modified progenitors have potential as transfusion alternatives or drug carriers (Olivier et al. 2024).

3.4 Disorders of Red Blood Cells

3.4.1 Anemias

Anemia, defined as a deficiency in hemoglobin concentration or RBC count, is among the most prevalent hematologic conditions globally (WHO 2021). Its causes are classically grouped into decreased RBC production, increased RBC destruction, or blood loss.

Iron deficiency anemia (IDA) remains the most common form of anemia globally, primarily caused by chronic blood loss (Yu et al. 2025), insufficient dietary iron intake, or increased iron demands such as during pregnancy. The global burden of iron deficiency anemia continues to be substantial, particularly affecting women of reproductive age, pregnant women, and young children. Key causes include menstrual bleeding, gastrointestinal bleeding, low iron availability in the diet, malabsorption, and increased physiological needs in pregnancy and growth. The prevalence of IDA is higher in low- and middle-income regions due to nutritional deficiencies and limited access to healthcare. Public health strategies such as iron supplementation programs for women and children, fortification of staple foods with iron, and addressing underlying causes like chronic bleeding or parasitic infections have proven effective in reducing IDA prevalence. The condition leads to symptoms like fatigue, weakness, pale skin, cognitive impairment, and increased morbidity, underscoring the need for continued targeted interventions (Wang et al. 2024).

Megaloblastic anemia arises primarily from deficiencies of vitamin B12 (cobalamin) or folate (vitamin B9), which are essential for DNA synthesis. The deficiency in either vitamin leads to impaired DNA synthesis, causing asynchronous maturation of the nucleus and cytoplasm in red blood cell precursors (Hariz and Bhattacharya 2025). This results in the production of large, macrocytic red blood cells (macrocytes) with characteristic features such as macro-ovalocytes and hypersegmented neutrophils on blood smears. Vitamin B12 deficiency is often secondary to pernicious anemia, an autoimmune condition causing impaired intrinsic factor secretion and poor B12 absorption, but can also arise from malabsorption, dietary insufficiency (e.g., vegan diet), or gastrectomy. Folate deficiency may result from inadequate intake, increased requirements (pregnancy, hemolysis), or malabsorption. Megaloblastic anemia presents clinically with fatigue, pallor, and sometimes neurological symptoms (in B12 deficiency). Effective treatment requires correcting the underlying deficiency through supplementation and addressing any causative factors (Torrez et al. 2022).

Aplastic anemia is characterized by failure of bone marrow hematopoiesis, resulting in reduced production of red blood cells, white blood cells, and platelets. The most common cause is immune-mediated destruction, where autoreactive T lymphocytes attack hematopoietic stem and progenitor cells, leading to apoptosis and bone marrow failure. This autoimmune process involves increased levels of pro-inflammatory cytokines like interferon-γ and tumor necrosis factor-α, which suppress hematopoiesis (Young 2018; Kulasekararaj et al. 2025). Other causes include exposure to toxins such as pesticides, benzene, and certain drugs; radiation; viral infections (e.g., hepatitis, Epstein–Barr virus, cytomegalovirus, parvovirus B19, HIV); and inherited bone marrow failure syndromes (Lundgren et al. 2025). The pathogenesis also involves defects in regulatory T-cell populations, increased apoptotic signaling via Fas receptor overexpression, and macrophage-mediated stem cell loss (Groarke et al. 2021).

Anemia of inflammation/chronic disease involves suppressed erythropoiesis, iron sequestration, and blunted EPO responses; newly characterized interactions include elevated ERFE and altered hepcidin dynamics (Chen et al. 2024).

Hemoglobinopathies are inherited disorders characterized by abnormalities in hemoglobin structure or production. Sickle cell disease arises from a point mutation in the β-globin gene that produces hemoglobin S (HbS), which polymerizes under low oxygen conditions and leads to vaso-occlusion and chronic hemolysis (Rees et al. 2010). Thalassemias, in contrast, result from reduced synthesis of α- or β-globin chains, causing globin-chain imbalance, ineffective erythropoiesis, and microcytic anemia (Cao and Galanello 2010). Recent advances in gene therapy and genome editing are transforming the management of these disorders; for instance, exagamglogene autotemcel (exa-cel/Casgevy), a CRISPR-based therapy, received FDA approval in 2023/2024 for transfusion-dependent β-thalassemia and severe sickle cell disease, enabling sustained transfusion independence in many patients. Current reviews emphasize key considerations for these novel approaches, including optimal timing of intervention, safety profiles, and patient eligibility criteria.

Polycythemia

An excessive RBC mass (polycythemia) may be primary (e.g., polycythemia vera, often associated with JAK2 mutations) or secondary (due to chronic hypoxia or inappropriate EPO production). Increased viscosity and thrombosis risk are central clinical concerns (Tefferi and Barbui 2017).

Hemolytic Disorders

Hemolysis is the premature destruction of RBCs and may be intrinsic (e.g., membrane defects such as hereditary spherocytosis; enzymatic defects such as G6PD deficiency) or extrinsic (e.g., autoimmune hemolysis, infections, toxins). Biomarkers include elevated lactate dehydrogenase (LDH), indirect bilirubin, reduced haptoglobin, and reticulocytosis (Berentsen and Tjønnfjord 2012). Advances in molecular diagnostics are enabling more precise classification of hemolytic anemias, including next-generation sequencing of membrane and enzyme genes (Agarwal and Rets 2023).

3.5 Clinical and Research Implications

RBC disorders contribute significantly to global morbidity, particularly in low- and middle-income settings. Advances in molecular diagnostics have enabled finer subclassification of anemias and hemoglobinopathies, guiding targeted therapeutics such as hydroxyurea (for SCD) and gene therapy trials (for β-thalassemia and SCD) (Cavazzana et al. 2017). The success of CRISPR-based therapies, such as Casgevy and beticel, is pushing the frontier toward curative approaches (Almasoudi 2025). Beyond therapy, RBCs are being explored as drug delivery vehicles and as biomarkers of systemic physiology (e.g., oxidative stress, metabolic status) because of their widespread distribution and exposure to circulation (Vincy et al. 2022). As in vitro erythropoiesis matures as a technology, cultured RBCs may one day supplement the transfusion supply or serve as "designer" RBCs for therapeutic delivery (Ling et al. 2024).

4 White Blood Cells (Leukocytes)

4.1 Overview and Classification

White blood cells (WBCs), or leukocytes, are central to host defense and immune surveillance. They constitute less than 1% of total blood volume but exert disproportionate influence on health and disease (Parham 2015). Leukocytes are broadly classified into granulocytes—neutrophils, eosinophils, and basophils—and agranulocytes—lymphocytes and monocytes. Each subtype arises from distinct progenitors in the bone marrow under the control of specific transcription factors and cytokines, reflecting the complexity of hematopoiesis (Akashi et al. 2000).

4.2 Granulocytes

4.2.1 Neutrophils

Neutrophils are the most abundant leukocyte subtype, representing 50–70% of circulating WBCs. They are first responders in acute inflammation, rapidly migrating to infection sites through chemotaxis mediated by chemokines such as IL-8 and complement fragments (Kolaczkowska and Kubes 2013). Their antimicrobial arsenal includes phagocytosis, degranulation, and neutrophil extracellular trap (NET) formation (Brinkmann et al. 2004).

Disorders: Neutropenia, defined as <1500 cells/μL, predisposes to severe bacterial and fungal infections. It may arise from chemotherapy, congenital bone marrow failure syndromes, or autoimmune destruction (Newburger and Dale 2013). Conversely, neutrophilia is observed in bacterial infections, inflammation, and myeloproliferative disorders.

4.2.2 Eosinophils

Eosinophils account for 1–4% of circulating WBCs and are specialized in defense against multicellular parasites. They contain granules rich in major basic protein, eosinophil cationic protein, and eosinophil peroxidase, which mediate cytotoxicity against helminths (Rothenberg and Hogan 2006). Eosinophils are also key effector cells in allergic diseases such as asthma.

Disorders: Eosinophilia is commonly associated with parasitic infections, allergic conditions, and certain neoplasms (e.g., Hodgkin lymphoma). Hypereosinophilic syndromes represent a group of disorders with persistent eosinophilia leading to organ damage (Klion 2009).

4.2.3 Basophils

Basophils are the least abundant leukocytes (<1%) but are crucial mediators of allergic responses. They release histamine, leukotrienes, and heparin upon activation, contributing to vasodilation, bronchoconstriction, and anticoagulation (Karasuyama et al. 2018).

Disorders: Basophilia is rare and usually indicates myeloproliferative neoplasms, particularly chronic myeloid leukemia.

4.3 Agranulocytes

Lymphocytes: Lymphocytes constitute 20–40% of WBCs and orchestrate adaptive immunity. They are subdivided into B cells, T cells, and natural killer (NK) cells: B cells differentiate into plasma cells and secrete antibodies, providing humoral immunity. T cells include $CD4^+$ helper T cells, which regulate immune responses, and $CD8^+$ cytotoxic T cells, which eliminate virus-infected or malignant cells (Murphy and Weaver 2016). NK cells provide rapid responses to infected and transformed cells, bridging innate and adaptive immunity.

Disorders of lymphocytes include

Immunodeficiencies: Congenital (e.g., severe combined immunodeficiency, SCID) or acquired (e.g., HIV infection, which depletes $CD4^+$ T cells).

Lymphoproliferative disorders: Leukemias and lymphomas represent malignant transformations of lymphocytes. Chronic lymphocytic leukemia (CLL) is the most common leukemia in adults, while the non-Hodgkin and Hodgkin lymphomas are major global cancers (Siegel et al. 2022).

Autoimmune diseases: Aberrant T- or B-cell activity underlies systemic lupus erythematosus, rheumatoid arthritis, and type 1 diabetes.

4.3.1 Monocytes and Macrophages

Monocytes constitute ~5–10% of circulating WBCs and differentiate into macrophages or dendritic cells upon tissue migration. These cells are pivotal in phagocytosis, antigen presentation, and orchestration of inflammatory responses (Geissmann et al. 2010). Macrophages display remarkable plasticity, polarizing into pro-inflammatory (M1) or anti-inflammatory (M2) phenotypes depending on microenvironmental cues.

Disorders: Monocytosis is observed in chronic infections (e.g., tuberculosis, endocarditis), autoimmune conditions, and myelodysplastic syndromes. Rare genetic disorders such as chronic granulomatous disease impair macrophage killing of pathogens.

4.4 Clinical Implications and Research Advances

Leukocyte disorders reflect perturbations of immune regulation and hematopoiesis. Clinically, leukocyte counts and differentials are essential diagnostic tools. For example, neutrophilia suggests bacterial infection, lymphocytosis often indicates viral infection, and eosinophilia prompts evaluation for allergy or parasitic disease. Advances in flow cytometry, immunophenotyping, and next-generation sequencing have revolutionized the classification of leukemias and lymphomas, enabling precision therapy (Arber et al. 2016).

Immunotherapies targeting leukocytes have transformed oncology and autoimmunity. Checkpoint inhibitors (e.g., anti-PD-1, anti-CTLA-4) unleash T cells against cancer, while CAR-T cell therapy reprograms lymphocytes to attack malignant B cells (June et al. 2018). At the same time, biologics targeting eosinophils (e.g., anti-IL-5 monoclonal antibodies) have provided breakthroughs in asthma management (FitzGerald et al. 2019).

These advances highlight leukocytes as not only key effectors of immunity but also therapeutic targets in modern medicine.

5 Platelets (Thrombocytes)

5.1 Structure and Function

Platelets, or thrombocytes, are small anucleate cell fragments derived from the cytoplasm of bone marrow megakaryocytes. They circulate at concentrations of 150,000–450,000/μL and have an average lifespan of 7–10 days (Ruggeri 2002). Despite lacking nuclei, platelets contain mitochondria, secretory granules (α-granules, dense granules, lysosomes), and a robust cytoskeletal system that enables rapid morphological changes upon activation.

Platelets are essential for hemostasis. Upon vascular injury, they adhere to exposed subendothelial collagen via von Willebrand factor (vWF), become activated, and release granule contents such as ADP, serotonin, and thromboxane A_2 to recruit additional platelets. Activated platelets form aggregates stabilized by fibrin, creating the primary hemostatic plug (Jackson 2011).

Beyond hemostasis, platelets play roles in inflammation, wound healing, and immunity by interacting with leukocytes and endothelial cells and releasing growth factors (Gawaz et al. 2005).

5.2 Platelet Disorders

Platelet abnormalities can be quantitative (abnormal counts) or qualitative (functional defects).

Thrombocytopenia (low platelet count): Defined as platelet counts <150,000/μL, thrombocytopenia predisposes to bleeding and can be caused by decreased production (e.g., bone marrow failure, aplastic anemia), increased destruction (immune thrombocytopenia, disseminated intravascular coagulation), or sequestration in the spleen (Cines and Blanchette 2002). Severe cases (<20,000/μL) are associated with spontaneous bleeding, including intracranial hemorrhage.

Thrombocytosis (high platelet count): Platelet counts >450,000/μL may be reactive (secondary to inflammation, infection, or iron deficiency) or clonal (associated with myeloproliferative neoplasms such as essential thrombocythemia) (Tefferi and Barbui 2019). Thrombocytosis increases the risk of thrombosis and, paradoxically, bleeding due to dysfunctional platelets.

Functional platelet disorders: Even with normal counts, platelets may fail to function properly. Examples include inherited disorders such as Glanzmann thrombasthenia (deficiency of GPIIb/IIIa integrin) and the Bernard–Soulier syndrome (deficiency of GPIb) (Nurden 2011). Acquired dysfunction occurs with antiplatelet drugs (e.g., aspirin, P2Y12 inhibitors) or uremia.

5.3 Clinical and Therapeutic Implications

Platelet counts are routinely assessed in complete blood counts, while platelet function is evaluated with bleeding time, platelet aggregation tests, or flow cytometry. Platelet transfusion remains the mainstay of therapy for severe thrombocytopenia or platelet dysfunction in bleeding patients (Estcourt et al. 2017).

Pharmacologic modulation of platelets has revolutionized medicine. Antiplatelet agents such as aspirin, clopidogrel, and glycoprotein IIb/IIIa inhibitors are essential in the prevention and treatment of cardiovascular diseases, particularly myocardial infarction and stroke (Patrono et al. 2017). Emerging therapies target platelet signaling pathways or immune-mediated destruction, offering novel avenues for managing thrombocytopenia and thrombotic diseases.

Thus, platelets are not only guardians of vascular integrity but also central players in cardiovascular pathology and therapeutic innovation.

6 Integration of Blood Constituents in Physiology

6.1 Coordinated Functions of Plasma, Cells, and Platelets

Although each blood constituent has distinct roles, their physiological importance arises from their integration as a system. Plasma provides the transport medium for red and white blood cells and platelets while simultaneously maintaining electrolyte and protein balance. Red blood cells (RBCs) deliver oxygen and remove carbon dioxide, functions that directly influence leukocyte activity and platelet metabolism. White blood cells (WBCs) circulate within the plasma compartment, ready to extravasate into tissues upon inflammatory signaling, while platelets act as sentinels, detecting vascular injury and interacting with both leukocytes and endothelium to coordinate immune and repair responses (Semple et al. 2011).

6.2 Blood in Systemic Diseases

Disruption of the delicate balance between these elements underlies numerous systemic diseases. In infection, plasma proteins such as complement, together with leukocytes and platelets, mount coordinated responses to neutralize pathogens, though excessive activation can lead to sepsis (Hotchkiss et al. 2016). In cancer, platelets and coagulation factors facilitate tumor angiogenesis and metastasis, while leukocytes shape the tumor microenvironment (Labelle and Hynes 2012). Cardiovascular diseases exemplify the integration of blood components, where dysregulated lipoproteins in plasma, hyperactive platelets, and inflamed leukocytes converge to promote atherosclerosis and thrombosis (Libby et al. 2019).

6.3 Emerging Perspectives

Modern hematology increasingly emphasizes this systems view of blood. Advances in multi-omics, including proteomics, transcriptomics, and metabolomics, allow simultaneous profiling of plasma proteins, RBC-derived metabolites, leukocyte gene expression, and platelet signaling. This integrated perspective provides insights into complex conditions such as autoimmune diseases, thrombo-inflammation, and metabolic syndromes (Eicher et al. 2020). Understanding the interplay among blood constituents is therefore not only fundamental for physiology but also crucial for developing precision medicine approaches.

7 Conclusions and Perspectives

Blood, as a specialized connective tissue, embodies a complex interplay between its fluid matrix and cellular components. Plasma provides the biochemical environment that sustains and regulates physiological processes, while red blood cells ensure efficient gas transport, white blood cells orchestrate immune defense, and platelets safeguard vascular integrity through hemostasis. Together, these constituents maintain homeostasis and enable the survival of multicellular organisms.

Disruptions in any of these components, whether in concentration, structure, or function, are central to numerous pathologies including anemia, coagulopathies, leukemias, and thrombotic disorders. The clinical importance of blood is thus reflected not only in diagnostics but also in therapeutic strategies, ranging from transfusion medicine and immunotherapy to anticoagulant and antiplatelet therapies.

Looking ahead, advances *in* genomics, proteomics, and metabolomics are reshaping our understanding of hematology. High-resolution multi-omics approaches are uncovering novel biomarkers and therapeutic targets for systemic diseases such as cancer, cardiovascular disorders, and autoimmune conditions. Furthermore, precision medicine and stem cell-based therapies hold promise for tailoring treatments to individual patients, potentially transforming hematology from a primarily supportive discipline to one at the forefront of personalized healthcare.

In sum, blood is more than a transport medium: it is a dynamic and integrative system whose study continues to provide insights into physiology, pathology, and future medical innovations.

References

Agarwal AM, Rets AV (2023) Molecular diagnosis of hereditary hemolytic anemias: recent updates. Int J Lab Hematol 45(Suppl 2):79–86. https://doi.org/10.1111/ijlh.14106. PMID: 37290893

Ahmed MH, Ghatge MS, Safo MK (2020) Hemoglobin: structure, function, and allostery. Subcell Biochem 94:345–382. https://doi.org/10.1007/978-3-030-41769-7_14. PMID: 32189307; PMCID: PMC7370311

Akashi K, Traver D, Miyamoto T, Weissman IL (2000) A clonogenic common myeloid progenitor that gives rise to all myeloid lineages. Nature 404(6774):193–197

Almasoudi HH (2025) Therapeutic promise of CRISPR-Cas9 gene editing in sickle cell disease and β-thalassemia: a current review. Curr Res Transl Med 73(3):103513. https://doi.org/10.1016/j.retram.2025.103513. Epub 2025 Apr 12. PMID: 40252393

Anderson NL, Anderson NG (2002) The human plasma proteome: history, character, and diagnostic prospects. Mol Cell Proteomics 1(11):845–867

Andrews NC (2008) Forging a field: the golden age of iron biology. Blood 112(2):219–230

Arber DA, Orazi A, Hasserjian R et al (2016) The 2016 revision to the World Health Organization classification of myeloid neoplasms and acute leukemia. Blood 127(20):2391–2405

Baynes J, Dominiczak MH (2019) Medical biochemistry, 5th edn. Elsevier

Berentsen S, Tjønnfjord GE (2012) Diagnosis and treatment of cold agglutinin mediated autoimmune hemolytic anemia. Blood Rev 26(3):107–115

Brinkmann V, Reichard U, Goosmann C et al (2004) Neutrophil extracellular traps kill bacteria. Science 303(5663):1532–1535

Cao A, Galanello R (2010) Beta-thalassemia. Genet Med 12(2):61–76

Cavazzana M, Antoniani C, Miccio A (2017) Gene therapy for hemoglobinopathies: a long and successful story. EMBO Mol Med 9(11):1347–1351

Cha HJ (2024) Erythropoiesis: insights from a genomic perspective. Exp Mol Med 56:2099–2104. https://doi.org/10.1038/s12276-024-01311-1

Chen S, Xiao J, Cai W, Lu X, Liu C, Dong Y, Zheng Y, Song G, Sun Q, Wang H, Xiao Z (2024) Association of the systemic immune-inflammation index with anemia: a population-based study. Front Immunol 15:1391573. https://doi.org/10.3389/fimmu.2024.1391573

Cines DB, Blanchette VS (2002) Immune thrombocytopenic purpura. N Engl J Med 346(13):995–1008

Coenen L, Lehallier B, de Vries HE, Middeldorp J (2023) Markers of aging: unsupervised integrated analyses of the human plasma proteome. Front Aging 4:1112109. https://doi.org/10.3389/fragi.2023.1112109

Coffey R, Ganz T (2018) Erythroferrone: an erythroid regulator of hepcidin and iron metabolism. Hemasphere. 2(2):e35. https://doi.org/10.1097/HS9.0000000000000035. PMID: 31723763; PMCID: PMC6745900

Coffin D et al (2023) The World Federation of Hemophilia World Bleeding Disorders Registry: insights from the first 10,000 patients. Res Pract Thrombosis Haemostasis 7(2):e100145. https://doi.org/10.1002/rth2.10145

Deng YT, You J, He Y, Zhang Y, Li HY, Wu XR, Cheng JY, Guo Y, Long ZW, Chen YL, Li ZY, Yang L, Zhang YR, Chen SD, Ge YJ, Huang YY, Shi LM, Dong Q, Mao Y, Feng JF, Cheng W, Yu JT (2025) Atlas of the plasma proteome in health and disease in 53,026 adults. Cell 188(1):253–271.e7. https://doi.org/10.1016/j.cell.2024.10.045. Epub 2024 Nov 22. PMID: 39579765

Eicher T, Dehghani M, Assenov Y et al (2020) Blood multi-omics in population-based cohorts and their potential to decipher complex diseases. Nat Rev Genet 21(9):586–602

Estcourt LJ, Malouf R, Doree C, Trivella M, Hopewell S (2017) Prophylactic platelet transfusions prior to surgery for people with a low platelet count. Cochrane Database Syst Rev 9:CD012779

FitzGerald JM, Bleecker ER, Nair P et al (2019) Benralizumab, an anti–interleukin-5 receptor α monoclonal antibody, as add-on treatment for severe asthma. Lancet Respir Med 7(1):20–34

Gawaz M, Langer H, May AE (2005) Platelets in inflammation and atherogenesis. J Clin Invest 115(12):3378–3384

GBD 2021 Hemoglobinopathies and Hemolytic Anemias Collaborators, Ojo TT, Amegbor PM, Islam F, Gyamfi J, Mai A et al (2025) Burden of hemoglobinopathies and hemolytic anemias in the World Health Organization African region, 2000–2021: findings from the Global Burden of Disease 2021 study. PLOS Glob Public Health 5(9):e0005197. https://doi.org/10.1371/journal.pgph.0005197

Geissmann F, Manz MG, Jung S, Sieweke MH, Merad M, Ley K (2010) Development of monocytes, macrophages, and dendritic cells. Science 327(5966):656–661

Groarke EM, Young NS, Calvo KR (2021) Distinguishing constitutional from acquired bone marrow failure in the hematology clinic. Best Pract Res Clin Haematol 34(2):101275. https://doi.org/10.1016/j.beha.2021.101275. Epub 2021 Jun 2. PMID: 34404527; PMCID: PMC8374085

Guyton AC, Hall JE (2020) Textbook of medical physiology, 14th edn. Elsevier

Haase VH (2013) Regulation of erythropoiesis by hypoxia-inducible factors. Blood Rev 27(1):41–53

Hariz A, Bhattacharya PT (2025) Megaloblastic anemia. In: StatPearls [Internet]. StatPearls Publishing, Treasure Island (FL). PMID: 30725939

Hew-Butler TD, Eskin C, Bickham J, Rusnak M, VanderMeulen M (2018) Dehydration is how you define it: comparison of 318 blood and urine athlete spot checks. BMJ Open Sport Exerc Med 4:e000297. https://doi.org/10.1136/bmjsem-2017-000297

Hewitt KJ (2023) Core transcription balancing erythropoiesis. Blood 142(25):2130–2131. https://doi.org/10.1182/blood.2023022304. PMID: 38127413

Hoffman R, Benz EJ, Silberstein LE, Heslop HE, Anastasi J, Weitz JI, Abutalib SA (2018) Hematology: basic principles and practice, 7th edn. Elsevier

Hoffman R, Benz EJ, Silberstein LE, Heslop HE, Weitz JI, Anastasi J (2017) Hematology: basic principles and practice, 7th edn. Elsevier

Hotchkiss RS, Moldawer LL, Opal SM, Reinhart K, Turnbull IR, Vincent JL (2016) Sepsis and septic shock. Nat Rev Dis Prim 2:16045

Jackson SP (2011) Arterial thrombosis—insidious, unpredictable and deadly. Nat Med 17(11):1423–1436

June CH, O'Connor RS, Kawalekar OU, Ghassemi S, Milone MC (2018) CAR T cell immunotherapy for human cancer. Science 359(6382):1361–1365

Karasuyama H, Miyake K, Yoshikawa S, Yamanishi Y (2018) How do basophils contribute to Th2 cell differentiation and allergic responses? Int Immunol 30(10):437–444

Kaufman DP, Khattar J, Lappin SL (2025) Physiology, fetal hemoglobin [Updated 2023 Mar 20]. In: StatPearls [Internet]. StatPearls Publishing, Treasure Island (FL). Available from: https://www.ncbi.nlm.nih.gov/books/NBK500011/

Kaushansky K, Lichtman MA, Prchal JT, Levi M, Press OW, Burns LJ, Caligiuri M (2021) Williams hematology, 10th edn. McGraw-Hill

Kirsher DY, Chand S, Phong A et al (2025) Current landscape of plasma proteomics from technical innovations to biological insights and biomarker discovery. Commun Chem 8:279. https://doi.org/10.1038/s42004-025-01665-1

Klion AD (2009) How I treat hypereosinophilic syndromes. Blood 114(18):3736–3741

Kolaczkowska E, Kubes P (2013) Neutrophil recruitment and function in health and inflammation. Nat Rev Immunol 13(3):159–175

Koury MJ (2014) Blood cell production and its regulation: hematopoiesis. In: Kaushansky K et al (eds) Williams hematology, 9th edn. McGraw Hill

Kuan YKI, Kok YJ, Liu NSH, Ong BJA, Chee YJ, Xu C, Chow M, Ramanathan K, Dalan R, Ho P, Fan BE (2025) Artificial intelligence in clinical thrombosis and hemostasis: a review. Res Pract Thromb Haemost 9(5):102984. https://doi.org/10.1016/j.rpth.2025.102984. PMID: 40837028; PMCID: PMC12362677

Kulasekararaj AG, Austin G et al (2025) Bone marrow failure: causes and complications. Medicine 53(5):273–277

Labelle M, Hynes RO (2012) The initial hours of metastasis: the importance of cooperative host–tumor cell interactions during hematogenous dissemination. Cancer Discov 2(12):1091–1099
Libby P, Buring JE, Badimon L et al (2019) Atherosclerosis. Nat Rev Dis Primers 5:56
Lundgren S, Huuhtanen J, Keränen M, Feng X, Patel BA, Ryland GL, Fox LC, Bravo-Perez C, Clemente M, Kerr C, Walldin G, Dufva O, Zaimoku Y, Tuononen T, Myllymäki M, Ebeling F, Jokinen E, Heinonen M, Kasanen T et al (2025) Single-cell analysis of aplastic anemia reveals a convergence of NK and NK-like CD8+ T cells with a disease-associated TCR signature. Sci Transl Med 17(787):eadl6758. https://doi.org/10.1126/scitranslmed.adl6758
Lyu J, Ni M, Weiss MJ, Xu J (2024) Metabolic regulation of erythrocyte development and disorders. Exp Hematol 131:104153. https://doi.org/10.1016/j.exphem.2024.104153. Epub 2024 Jan 17. PMID: 38237718; PMCID: PMC10939827
Menon AA, Gansen B, Mulder H, Neely ML, Papavasileiou P, Salisbury ML, Southern BD, Hesslinger C, Leonard TB, Meissner F, Todd JL (2025) Mass spectrometry-based peripheral blood proteomics for biomarker discovery in idiopathic pulmonary fibrosis. Respir Res 26(1):294. https://doi.org/10.1186/s12931-025-03377-5. PMID: 41126189; PMCID: PMC12548219
Mohamed FRR, Aldabbagh Z, Kalou W, Hamsho K, Aldabbagh A, Kalou A, Sajid MR (2025) The use of artificial intelligence in the prevention and management of bleeding disorders: a systematic review. Front Med (Lausanne) 12:1606788. https://doi.org/10.3389/fmed.2025.1606788. PMID: 41140650; PMCID: PMC12546342
Mohandas N, Gallagher PG (2008) Red cell membrane: past, present, and future. Blood 112(10):3939–3948
Murphy K, Weaver C (2016) Janeway's immunobiology, 9th edn. Garland Science
Nachtkamp K, Strupp C, Vukelja M et al (2024) The new WHO 2022 and ICC proposals for the classification of myelodysplastic neoplasms. Validation based on the Düsseldorf MDS registry and proposals for a merged classification. Leukemia 38:442–445. https://doi.org/10.1038/s41375-024-02157-2
Newburger PE, Dale DC (2013) Evaluation and management of patients with isolated neutropenia. Semin Hematol 50(3):198–206
Noordin SS, Al-Amoudi BHO, Mohamed R, Aziz MY, Noor SNFM, Zabidi MA (2025) Unraveling common stem cell sources and key reporting parameters in studies related to stem cell-derived red blood cells: a review. Biomed Res Ther 12(5):7372–7385. https://doi.org/10.15419/bmrat.v12i5.976
Nurden AT (2011) Inherited disorders of platelets: molecular insights. Blood Rev 25(4):191–199
Olivier E, Zhang S, Yan Z, Bouhassira EE (2024) Stem cell factor and erythropoietin-independent production of cultured reticulocytes. Haematologica 109(11):3705–3720. https://doi.org/10.3324/haematol.2023.284427
Orkin SH, Zon LI (2008) Hematopoiesis: an evolving paradigm for stem cell biology. Cell 132(4):631–644. https://doi.org/10.1016/j.cell.2008.01.025
Palmer BF (2015) Regulation of potassium homeostasis. Clin J Am Soc Nephrol 10(6):1050–1060
Parham P (2015) The immune system, 4th edn. Garland Science
Patrono C, Morais J, Baigent C et al (2017) Antiplatelet agents for the treatment and prevention of atherothrombosis. Eur Heart J 38(45):3202–3211
Peters T (1996) All about albumin: biochemistry, genetics, and medical applications. Academic Press
Pirotte M, Fillet M, Seidel L, Jaspers A, Baron F, Beguin Y (2021) Erythroferrone and hepcidin as mediators between erythropoiesis and iron metabolism during allogeneic hematopoietic stem cell transplant. Am J Hematol 96(10):1275–1286. https://doi.org/10.1002/ajh.26300. Epub 2021 Aug 24. PMID: 34310730; PMCID: PMC9291814
Pritišanac E, Urlesberger B, Schwaberger B, Pichler G (2021) Fetal hemoglobin and tissue oxygenation measured with near-infrared spectroscopy—a systematic qualitative review. Front Pediatr 9:710465. https://doi.org/10.3389/fped.2021.710465
Rees DC, Williams TN, Gladwin MT (2010) Sickle-cell disease. Lancet 376(9757):2018–2031
Rothenberg ME, Hogan SP (2006) The eosinophil. Annu Rev Immunol 24:147–174

Ruggeri ZM (2002) Platelets in atherothrombosis. Nat Med 8(11):1227–1234
Safina K et al (2024) New frameworks for hematopoiesis derived from single-cell genomics. Blood 144(10):1039–1052. https://doi.org/10.1182/blood.2023027434
Semple JW, Italiano JE, Freedman J (2011) Platelets and the immune continuum. Nat Rev Immunol 11(4):264–274
Siegel RL, Miller KD, Jemal A (2022) Cancer statistics, 2022. CA: A Cancer J Clin 72(1):7–33
Tang P, Wang H (2023) Regulation of erythropoiesis: emerging concepts and therapeutic implications. Hematology 28(1):2250645. https://doi.org/10.1080/16078454.2023.2250645. PMID: 37639548
Tefferi A, Barbui T (2017) Polycythemia vera and essential thrombocythemia: 2017 update on diagnosis, risk-stratification, and management. Am J Hematol 92(1):94–108
Tefferi A, Barbui T (2019) Polycythemia vera and essential thrombocythemia: 2019 update on diagnosis, risk-stratification and management. Am J Hematol 94(1):133–143
Torrez M, Chabot-Richards D, Babu D, Lockhart E, Foucar K (2022) How I investigate acquired megaloblastic anemia. Int J Lab Hematol 44(2):236–247. https://doi.org/10.1111/ijlh.13789. Epub 2022 Jan 3. PMID: 34981651
Triplett DA (2000) Coagulation and bleeding disorders: review and update. Clin Chem 46(8 Pt 2):1260–1269. PMID: 10926920
Upadhyay RK (2015) Emerging risk biomarkers in cardiovascular diseases and disorders. J Lipids 2015:971453. https://doi.org/10.1155/2015/971453. Epub 2015 Apr 8. PMID: 25949827; PMCID: PMC4407625
Vincy A, Mazumder S, Amrita, Banerjee I, Hwang KC, Vankayala R (2022) Recent progress in red blood cells-derived particles as novel bioinspired drug delivery systems: challenges and strategies for clinical translation. Front Chem 10:905256. https://doi.org/10.3389/fchem.2022.905256. PMID: 35572105; PMCID: PMC9092017
Wang L, Liang D, Huangfu H, Shi X, Liu S, Zhong P, Luo Z, Ke C, Lai Y (2024) Iron deficiency: global trends and projections from 1990 to 2050. Nutrients 16(20):3434. https://doi.org/10.3390/nu16203434. PMID: 39458430; PMCID: PMC11510637
WHO (2021) Global anaemia estimates, 2021 edition. World Health Organization
Young NS (2018) Aplastic anemia. N Engl J Med 379(17):1643–1656
Yu D, Ni Y, Chen K, Xu H, Huang X, He Y (2025) Global burden of anemia attributable to non-communicable diseases: GBD 2021 analysis and projections. Front Nutr 12:1557986. https://doi.org/10.3389/fnut.2025.1557986. PMID: 40843193; PMCID: PMC1236464

Blood Groups: Genetics and Physiology Basis

3

Abstract

Human blood groups are critical for transfusion medicine, organ transplantation, and understanding susceptibility to infectious and autoimmune diseases. The ABO and Rhesus (Rh) systems are the most clinically relevant, defined by surface antigens on erythrocytes. Here, we review the genetic mechanisms governing blood group expression and their physiological and clinical consequences. The ABO locus on chromosome 9 encodes glycosyltransferases responsible for the A and B antigens, with co-dominant inheritance patterns, while the O allele produces a nonfunctional enzyme. The Rh system, primarily determined by the RHD gene on chromosome 1, dictates Rh positivity and negativity, with implications for the hemolytic disease of the newborn. We discuss antigen–antibody interactions, natural and alloimmune responses, and the molecular basis of rare blood group systems such as Kell, Duffy, and Kidd. Integration of genomic, biochemical, and population studies provides insight into the evolution of blood group diversity, their functional roles in pathogen resistance, and the molecular basis of transfusion reactions. Understanding the genetics and physiology of blood groups not only informs clinical practice but also enhances the knowledge of human evolution and disease susceptibility.

Keywords

Blood groups · ABO group · Rh system · Kell · Duffy · Kidd

1 Introduction

Blood groups represent inherited antigenic profiles expressed primarily on erythrocytes and various tissues, including endothelial and epithelial cells (Watkins 2011; Hosoi 2008; Yamamoto 2021; Clausen and van der Schoot 2025; Ramsey 2025). To date, more than 360 red cell antigens organized into 45 blood group systems have been recognized by the International Society of Blood Transfusion (ISBT), reflecting remarkable genetic and molecular diversity (Daniels et al. 2022; ISBT Working

H. R. Dash et al., *Forensic Investigation of Blood Evidence*,
https://doi.org/10.1007/978-981-95-9183-1_3

Party 2023). Among these, the ABO and Rh systems account for over 80% of clinically significant hemolytic transfusion reactions and hemolytic disease of the fetus and newborn (HDFN), underscoring their dominant role in transfusion safety and perinatal medicine (Hall et al. 2023; Westhoff 2022; van't Oever et al. 2025).

The discovery of the ABO system by Landsteiner and the subsequent characterization of the Rh system revolutionized modern medicine, forming the foundation of safe blood transfusion practices, solid organ and hematopoietic stem cell transplantation, and immune compatibility testing (Zhou et al. 2020; Anstee 2021; Arthur et al. 2024; Ramsey 2025). Advances in molecular immunohematology have further refined blood group typing, shifting from serological methods to genotype-based approaches, particularly in transfusion-dependent patients and complex alloimmunization settings (Westhoff et al. 2019; Flegel 2023).

Beyond their clinical relevance, ABO and Rh polymorphisms provide valuable insights into human evolutionary history, population genetics, and pathogen-driven selection. Genome-wide association studies (GWAS) and large-scale sequencing efforts have demonstrated strong associations between ABO blood groups and susceptibility to infectious diseases, including malaria, cholera, Helicobacter pylori infection, and SARS-CoV-2, as well as noncommunicable diseases such as thromboembolism, cardiovascular disease, and cancer (Cooling 2015; Ellinghaus et al. 2020; Paré et al. 2021; Dentali et al. 2012). These associations highlight the pleiotropic effects of blood group antigens on host–pathogen interactions, inflammation, and coagulation pathways.

Recent comparative genomic and phylogenetic analyses suggest that balancing selection and convergent evolution have shaped the global distribution of ABO and Rh alleles, maintaining diversity across populations despite strong selective pressures (Saitou and Yamamoto 1997; Villanea et al. 2022). Moreover, emerging evidence indicates that blood group antigens may modulate immune signaling, endothelial function, and glycosylation patterns, extending their biological significance far beyond erythrocytes (Storry et al. 2023; Patnaik and Spitalnik 2024).

2 ABO Blood Group System

2.1 Genetic Architecture of the ABO Locus

2.1.1 Gene Location and Structure

The ABO gene is located on chromosome 9q34.2 and spans approximately 18 kb, comprising seven exons, with exons 6 and 7 encoding the majority of the catalytic domain of the glycosyltransferase (Yamamoto et al. 1990; Chester and Olsson 2001). The gene encodes a Golgi-resident glycosyltransferase responsible for modifying the H antigen, a precursor oligosaccharide widely expressed on erythrocytes, epithelial cells, platelets, and vascular endothelium (Hakomori 1999; Storry et al. 2023). The ABO locus is highly polymorphic, with extensive allelic heterogeneity arising from point mutations, deletions, recombination events, and regulatory variants, making it one of the most genetically diverse loci in the human genome (Daniels 2020; Flegel 2023).

Three major alleles—A, B, and O—account for the classical ABO phenotypes. These alleles differ primarily in their coding sequences, which directly determine enzymatic specificity and activity, thereby defining the terminal carbohydrate structure expressed on the cell surface (Yamamoto et al. 1990; Anstee 2010).

2.1.2 Molecular Basis of A, B, and O Alleles

The A allele encodes an α-1,3-N-acetylgalactosaminyltransferase (GTA) that catalyzes the transfer of N-acetylgalactosamine (GalNAc) to the H antigen, generating the A antigen. In contrast, the B allele encodes an α-1,3-galactosyltransferase (GTB) that transfers D-galactose, resulting in the B antigen (Yamamoto et al. 1990). Remarkably, these two enzymes differ by only four critical amino acid substitutions—Arg176Gly, Gly235Ser, Leu266Met, and Gly268Ala—which alter the active site conformation and dictate donor sugar specificity (Yamamoto et al. 1992; Patenaude et al. 2002). Structural and crystallographic studies have confirmed that these substitutions modulate substrate binding and catalytic efficiency without affecting overall protein folding (Patenaude et al. 2002; Seto et al. 1999).

The O allele most commonly results from a single nucleotide deletion (261delG) in exon 6, causing a frameshift and premature termination of translation, yielding a nonfunctional enzyme. Consequently, individuals with the O phenotype express only the unmodified H antigen on their cell surfaces (Yamamoto et al. 1990; Daniels 2020). Additional rare O variants arise from nonsense mutations or splice-site alterations, further contributing to ABO diversity across populations (Flegel 2023).

2.1.3 ABO Subgroups

Beyond the principal alleles, the ABO system encompasses numerous subgroups that influence antigen density and serologic behavior. The A1 subgroup, present in approximately 80% of A individuals, exhibits high antigen density and complex branched glycan structures, whereas the A2 subgroup (~20%) expresses fewer A antigenic sites due to reduced transferase activity and altered substrate utilization (Olsson and Chester 2001; Daniels 2020). Weak A and B subgroups (e.g., A3, Ax, B3) arise from missense mutations or regulatory variants that partially impair enzyme function or expression, leading to diminished antigen presentation (Reid and Lomas-Francis 2019).

These subgroups are clinically significant, as they may cause discrepant serologic typing, unexpected alloantibody formation, or transfusion incompatibilities, particularly in immunocompromised or transfusion-dependent patients (Westhoff 2022; Flegel 2023).

2.1.4 Mutations and Functional Consequences

Mutations within the ABO gene—including single-nucleotide substitutions, small insertions or deletions, and splice-site defects—can profoundly influence enzyme activity, antigen density, and glycan structure. For example, A2-associated variants reduce catalytic efficiency, whereas rare O alleles result in a complete loss of function (Daniels 2020; Reid and Lomas-Francis 2019). Beyond transfusion medicine, these molecular alterations affect hemostasis, endothelial function, inflammation, and pathogen adhesion, linking ABO polymorphism to the risks of thrombosis, cardiovascular disease, and infection (Anstee 2010; Franchini and Lippi 2015; Patnaik and Spitalnik 2024). Thus, the ABO locus represents a paradigmatic example of how subtle genetic variation can exert wide-ranging biological and clinical effects.

2.2 Biosynthesis of ABO Antigens

The H antigen constitutes the essential precursor for all ABO blood group determinants. Its synthesis is mediated by FUT1, an α-1,2-fucosyltransferase encoded by the FUT1 gene, which adds L-fucose to type 2 precursor chains on erythrocytes and various epithelial tissues (Kelly et al. 1995; Koda et al. 1997). This fucosylation step is indispensable, as the H antigen serves as the acceptor substrate for ABO glycosyltransferases.

In individuals lacking functional FUT1—most notably those with the Bombay phenotype (hh genotype)—H antigen synthesis is completely absent, resulting in the failure to express A or B antigens regardless of the underlying ABO genotype (Bhende et al. 1952; Oriol et al. 1981). These individuals produce potent anti-H antibodies, often of both IgM and IgG classes, making them incompatible with all conventional ABO blood groups and necessitating transfusion exclusively from Bombay phenotype donors. This rare phenotype underscores the central role of glycan biosynthesis in transfusion compatibility and immunogenicity (Daniels 2020; Storry et al. 2023).

Following H antigen formation, the individual's ABO genotype determines which terminal sugar is added within the Golgi apparatus. The A-transferase catalyzes the addition of GalNAc, producing the A antigen, while the B-transferase transfers D-galactose, yielding the B antigen (Yamamoto et al. 1990). These enzymatic reactions rely on precise donor–acceptor recognition governed by the four critical amino acid differences between GTA and GTB (Yamamoto et al. 1992; Patenaude et al. 2002).

In contrast, the nonfunctional O-transferase fails to modify the H antigen, resulting in exclusive expression of the H structure on erythrocytes. This molecular mechanism explains both the universal donor status of group O red blood cells and the presence of naturally occurring anti-A and anti-B antibodies in group O individuals (Anstee 2010; Daniels 2020). Collectively, the biosynthetic pathway of ABO antigens exemplifies how tightly regulated glycosylation processes translate genetic variation into clinically critical phenotypes.

2.3 Possible Blood Types and Their Determinants

The ABO blood group system is determined by three principal alleles—A, B, and O—located at the ABO locus on chromosome 9q34.2. These alleles encode distinct glycosyltransferases that catalyze the modification of the H antigen, a fucosylated precursor oligosaccharide expressed on erythrocytes and multiple tissue types (Yamamoto et al. 1990; Daniels 2013; Anstee 2010).

The A allele encodes an α-1,3-N-acetylgalactosaminyltransferase that adds N-acetylgalactosamine (GalNAc), whereas the B allele encodes an α-1,3-galactosyltransferase that adds D-galactose to the terminal fucose of the H antigen. In contrast, the O allele most commonly harbors a frameshift mutation (261delG), resulting in a truncated, nonfunctional enzyme and leaving the H antigen unmodified (Yamamoto et al. 1990; Patenaude et al. 2002).

As a consequence of codominant inheritance, an individual's genotype directly determines the ABO phenotype:

AA or AO → phenotype A
BB or BO → phenotype B
AB → phenotype AB (coexpression of A and B antigens)
OO → phenotype O (expression of H antigen only)

These genotype–phenotype correlations are robust, evolutionarily conserved, and form the molecular foundation of modern transfusion medicine, transplantation matching (Zeng et al. 2025), forensic genetics, and population studies (Daniels 2013; Flegel 2023).

2.4 Global Population Distribution of ABO Blood Types

2.4.1 Phenotypes and Global Distribution

The ABO blood group system exhibits pronounced geographical and population-specific variation, reflecting the combined effects of evolutionary selection, genetic drift, founder effects, and human migration. Globally, blood group O is the most prevalent phenotype (Garratty et al. 2004), reaching 90–100% in certain Indigenous populations of Central and South America—an extreme distribution attributed to founder effects and prolonged genetic isolation (Mourant et al. 1976; Rodrigues et al. 2021).

Blood group A occurs at high frequencies in Northern and Central Europe, as well as in populations of European ancestry in North America and Australia (Dean 2005). Blood group B is most prevalent in South Asia, Central Asia, and parts of Africa, consistent with ancient migration routes and region-specific selective pressures (Sigamani and Gajulapalli 2022; Vuhahula et al. 2023). The AB phenotype remains universally rare, typically accounting for 2–5% of individuals worldwide, reflecting the lower combined frequencies of A and B alleles (Roychoudhuri and Nei 1988).

2.4.2 Worldwide Distribution Patterns and Population Genetics

The global distribution of ABO blood types is shaped by pathogen-mediated selection in addition to neutral evolutionary processes. In malaria-endemic regions, blood group O is overrepresented, likely due to reduced rosetting and cytoadherence of Plasmodium falciparum-infected erythrocytes, conferring partial protection against severe malaria (Cserti and Dzik 2007; Rowe et al. 2007). Conversely, susceptibility to pathogens such as *Vibrio cholerae*, norovirus, *Helicobacter pylori*, and smallpox has been associated with non-O blood groups, supporting the hypothesis of balancing selection maintaining ABO polymorphism across populations (Cooling 2015; Saitou and Yamamoto 1997).

Founder effects further explain extreme allele frequencies, while admixture events have produced complex ABO patterns in populations with mixed European, African, and Indigenous ancestry (Garratty et al. 2004). Collectively, these observations establish the ABO system as a classical model of human population genetics shaped by both biology and environment.

2.5 Blood Type Compatibility in Transfusion Medicine

2.5.1 Rationale for ABO Compatibility

ABO compatibility is the single most critical determinant of transfusion safety, owing to the presence of naturally occurring anti-A and anti-B antibodies that develop early in life. These antibodies are predominantly IgM, arise independently of transfusion or pregnancy, and are thought to be induced by exposure to environmental glycan antigens expressed by gut microbiota and common pathogens (Springer and Horton 1969; Cooling 2015).

Because ABO antibodies efficiently activate the complement cascade, transfusion of ABO-incompatible red blood cells can result in acute intravascular hemolysis, disseminated intravascular coagulation, acute renal failure, and fatal shock (Reid and Lomas-Francis 2004; Harmening 2012). This immunological mechanism underpins the absolute requirement for ABO matching prior to transfusion.

2.5.2 ABO Red Blood Cell Compatibility

Group O individuals, possessing both anti-A and anti-B antibodies, are restricted to O RBCs, whereas group AB individuals, lacking these antibodies, are universal RBC recipients (Daniels 2013). The compatibility of ABO red blood cells is summarized in Table 3.1.

Table 3.1 ABO red blood cell compatibility

Recipient blood type	Compatible donor RBCs
O	O only
A	A or O
B	B or O
AB	AB, B, or O

2.5.3 Plasma Compatibility: The Reverse Principle

Plasma compatibility follows the inverse logic of RBC transfusion, as it depends on donor antibodies rather than recipient antigens. Group AB plasma, lacking both anti-A and anti-B antibodies, is the true universal plasma donor, while group O plasma, containing both antibodies, can only be safely transfused to O recipients (Harmening 2012). This distinction is particularly critical in massive transfusion protocols and emergency medicine.

2.5.4 Bombay Phenotype (hh): A Special Case

The Bombay phenotype (Oh) represents a rare but clinically critical exception. Caused by mutations in FUT1, individuals lack the H antigen, precluding the expression of A, B, or O antigens regardless of the ABO genotype (Bhende et al. 1952; Soejima and Koda 2023). These individuals produce potent anti-H antibodies, rendering them incompatible with all standard ABO blood units. Consequently, they can only receive blood from other Bombay phenotype donors. Although globally rare (~1:250,000), the phenotype occurs more frequently in parts of India (~1:7000–10,000), reflecting founder effects (Shahshahani et al. 2013). This highlights the operational challenges posed by rare blood groups in transfusion services.

2.6 Physiological Roles of ABO Antigens

ABO antigens are now recognized as biologically active glycan determinants, with effects extending far beyond transfusion compatibility. They are expressed on vascular endothelium, platelets, epithelial cells, and plasma proteins, integrating ABO biology into hemostatic, vascular, and immune pathways (Murray et al. 2019; Ward et al. 2020).

2.6.1 Hemostasis and Coagulation

One of the most robust physiological associations involves von Willebrand factor (vWF) and factor VIII. ABO glycans are covalently attached to vWF, influencing its plasma concentration and clearance. Individuals with blood group O exhibit approximately 25–30% lower vWF and factor VIII levels, attributed to increased proteolysis by ADAMTS13 and accelerated clearance of vWF lacking terminal A/B glycans (Murray et al. 2019; Ward et al. 2020; Franchini and Lippi 2015).

These differences translate into clinically meaningful variation in bleeding risk and thrombotic susceptibility, with non-O blood groups displaying increased risk of venous thromboembolism and cardiovascular disease (Broni et al. 2025).

2.6.2 Endothelial Biology and Vascular Function

ABO antigens expressed on endothelial cells influence cell–cell adhesion, inflammatory signaling, and vascular permeability. Non-O blood groups, particularly A, are associated with higher levels of vWF, soluble thrombomodulin, ICAM-1, and

P-selectin, markers of endothelial activation implicated in sepsis, ARDS, and microvascular dysfunction (Ward et al. 2020; Paré et al. 2021).

2.6.3 Host–Pathogen Interactions

ABO glycans also function as pathogen attachment factors. Norovirus susceptibility is strongly influenced by ABO and secretor status, while Helicobacter pylori preferentially binds A antigens on gastric epithelium (Cooling 2015; Liao et al. 2020). In malaria, non-O erythrocytes form rosettes more readily, contributing to severe disease, whereas group O confers partial protection (Rowe et al. 2007).

Emerging evidence also links ABO type to SARS-CoV-2 susceptibility and severity, potentially via effects on viral binding, coagulation, and endothelial inflammation, although mechanistic pathways remain under investigation (Ellinghaus et al. 2020; Paré et al. 2021).

2.7 ABO Blood Groups and Associated Health Risks

Clinically, ABO blood groups are associated with a spectrum of health risks and disease outcomes. Non-O individuals demonstrate elevated vWF levels and increased risk of thrombosis, while group A and B antigens are linked to higher susceptibility to gastric and pancreatic cancers through mechanisms involving H. pylori colonization, glycan-mediated signaling, and inflammation (O'Donnell and Laffan 2001; Franchini and Mannucci 2014; Borén et al. 1993; Wolpin et al. 2010).

ABO type also modulates infectious disease susceptibility, including malaria, COVID-19, norovirus, and smallpox (Rowe et al. 2007; Guillon et al. 2008; Zhao et al. 2021; Tan and Jiang 2005; Duncan et al. 1987), as well as metabolic and gastrointestinal conditions such as peptic ulcer disease in group O and dyslipidemia or type 2 diabetes associations in A and B groups (Borén et al. 1993; Ewald and Sumner 2016; Barbalic et al. 2010). Furthermore, ABO compatibility is a critical determinant in organ transplantation (Zeng et al. 2025), as preformed anti-A or anti-B antibodies can trigger hyperacute rejection via complement activation and endothelial injury, emphasizing the necessity of accurate ABO typing for transplant success (Montgomery et al, 2009; Zeng et al. 2025). Collectively, these findings underscore that ABO antigens are not merely serologic markers but active modulators of coagulation, vascular biology, host–pathogen interactions, disease susceptibility, metabolism, and transplantation outcomes (Zeng et al. 2025).

3 Rhesus (Rh) Blood Group System

The Rhesus (Rh) blood group system is the second most clinically significant system after ABO (Avent and Reid 2000). It comprises more than 50 defined antigens, with the D antigen (RhD) being the most immunologically important due to its strong immunogenicity (Knowles and Regan 2006). Rh antigens are expressed on the erythrocyte membrane as part of a multiprotein complex, including the Rh-associated glycoprotein (RHAG), and play a central role in transfusion safety, hemolytic disease of the fetus and newborn (HDFN), and organ transplantation (Zeng et al. 2025). Beyond immunological relevance, emerging evidence suggests that Rh proteins may contribute to erythrocyte physiology, gas transport, and nitrogen metabolism (Daniels 2013; Westhoff 2004).

3.1 Molecular Genetics of RhD and RhCE

3.1.1 Genomic Organization and Gene Structure

The expression of Rhesus (Rh) blood group antigens is governed primarily by two closely linked and highly homologous genes, RHD and RHCE, located on chromosome 1p36.11. These genes originated from an ancestral duplication event and share more than 90% sequence identity, yet they encode distinct antigenic determinants that constitute the molecular foundation of the Rh blood group system (Avent and Reid 2000; Westhoff 2004; Daniels 2020).

The RHD gene encodes the highly immunogenic D antigen, which is second only to ABO in clinical importance. In contrast, RHCE encodes the C/c and E/e antigens, which arise from specific single-nucleotide polymorphisms resulting in amino acid substitutions within the extracellular domains of the RhCE protein. Both genes consist of 10 exons and encode proteins containing 12 transmembrane helices, which integrate into the erythrocyte membrane as part of a multisubunit Rh complex essential for red blood cell membrane integrity and function (Agre et al. 2002; Huang et al. 2009).

3.1.2 Molecular Structure and Mechanisms of Allelic Diversity

The RHD–RHCE locus is embedded within a genomic region enriched in low-copy repeats (LCRs), rendering it particularly susceptible to gene conversion, homologous recombination, and unequal crossover events. These mechanisms drive extensive allelic diversification and account for the remarkable phenotypic variability of the Rh blood group system observed worldwide (Wagner and Flegel 2000; Moulds et al. 2011).

Within RHCE, polymorphisms—particularly in exons 2, 5, and 7—define the C/c and E/e antigenic specificities. In contrast, sequence variation throughout the RHD gene gives rise to a broad spectrum of weak D, partial D, and DEL phenotypes, characterized by quantitatively reduced or qualitatively altered D antigen expression. These variants are of major clinical relevance due to their impact on alloimmunization risk and transfusion compatibility (Westhoff 2004; Flegel 2007; Daniels 2020).

3.1.3 Population-Specific Variation in RHD

A defining characteristic of the Rh blood group system is its pronounced population-specific genetic diversity, particularly regarding the molecular determinants of the D-negative phenotype. Unlike many blood group antigens that show relatively uniform genetic mechanisms across populations, RHD exhibits marked allelic heterogeneity shaped by genomic architecture, recombination dynamics, and human population history (Avent and Reid 2000; Westhoff 2004; Flegel 2011).

Complete RHD Gene Deletion (European Populations)
In individuals of European ancestry, D-negativity is predominantly caused by a complete deletion of the RHD gene, mediated by homologous recombination between flanking Rhesus boxes. This ancestral deletion event accounts for approximately 15–17% prevalence of RhD negativity in European populations and results in a true absence of D antigen expression on erythrocytes (Wagner and Flegel 2000; Westhoff 2004; Flegel 2011).

Nonfunctional or Inactive RHD Alleles (African Populations)
In African populations, D-negativity more frequently arises from nonfunctional or partially functional RHD alleles rather than complete gene deletion. These alleles may contain missense mutations, splice-site defects, frameshifts, or small insertions and deletions, leading to truncated, unstable, or intracellularly retained proteins (Singleton et al. 2000; Reid and Lomas-Francis 2019). This molecular complexity contributes to a higher prevalence of partial D phenotypes and explains the increased risk of anti-D alloimmunization when RhD typing relies solely on serologic methods (Flegel 2007; Westhoff 2022).

RHD–RHCE Hybrid Alleles (East Asian Populations)
In East Asian populations, D-negativity and weak D expression are most commonly associated with RHD–RHCE hybrid alleles generated by gene conversion events. A prominent example is the DEL phenotype, characterized by extremely low levels of D antigen that are undetectable by routine serology but identifiable by adsorption–elution techniques or molecular genotyping. DEL alleles are particularly frequent in East Asian populations and represent a significant source of transfusion-related alloimmunization risk when misclassified as RhD-negative (Okuda et al. 1997; Flegel 2007; Wagner et al. 2018).

Importantly, individuals carrying partial D alleles, regardless of ethnic background, lack one or more D epitopes and remain capable of producing allo-anti-D antibodies upon exposure to conventional RhD-positive red blood cells. This has major implications for transfusion medicine and obstetric care, particularly in increasingly multiethnic societies (Avent and Reid 2000; Westhoff 2022; Flegel 2023).

Collectively, these population-specific molecular patterns underscore the limitations of serologic RhD typing and highlight the growing necessity of molecular RHD genotyping for accurate RhD classification and alloimmunization risk assessment (Daniels 2020; Flegel and Denomme 2021).

3.1.4 Role of RHAG and the Rh Null Phenotype

Surface expression and membrane integration of RhD and RhCE proteins depend on RHAG (Rh-associated glycoprotein), encoded on chromosome 6p21. RHAG acts as both a structural chaperone and a gas transporter, ensuring correct assembly and stabilization of the Rh complex within the erythrocyte membrane (Marini et al. 2000; Huang and Peng 2005).

Mutations in RHAG give rise to the Rh null phenotype, characterized by the complete absence of all Rh antigens. Clinically, this rare condition is associated with membrane fragility, stomatocytosis- or spherocytosis-like morphology, chronic hemolytic anemia, abnormal cation permeability, and increased osmotic fragility (Chérif-Zahar et al. 1996; McGann et al. 2012; Daniels 2020). These observations demonstrate that the Rh complex plays a fundamental structural and physiological role in erythrocyte biology beyond immunohematologic antigenicity.

3.1.5 Clinical Implications

The extraordinary genetic diversity of RHD and RHCE, together with the RHAG-dependent assembly of the Rh complex, has profound implications for transfusion safety, alloimmunization risk, and obstetric management. This is particularly relevant in populations with a high prevalence of partial D or DEL alleles, where conventional serologic typing may be insufficient. Consequently, molecular characterization of Rh variants has become an essential component of modern transfusion medicine, supporting the transition toward precision-based transfusion and prenatal care.

3.2 Protein Structure and Physiological Function

RhD and RhCE proteins are integral membrane proteins composed of 12 transmembrane helices, forming part of a larger Rh–RhAG complex with channel-like properties. Once considered functionally inert aside from antigenicity, Rh proteins are now recognized as gas and ammonium transporters, facilitating NH_3/NH_4^+ and CO_2 movement across the erythrocyte membrane (Kustu and Inwood 2006; Endeward et al. 2008).

Experimental and physiological studies indicate that Rh proteins contribute to erythrocyte deformability, membrane stability, and systemic acid–base homeostasis. By participating in nitrogen metabolism and CO_2 transport, the Rh complex influences pH regulation and gas exchange, underscoring its broader physiological importance beyond transfusion medicine (Bruce et al. 2009; Huang et al. 2009).

3.3 Rhesus Factor Compatibility

The D antigen is one of the most immunogenic antigens in human blood group systems. Rh-negative individuals exposed to Rh-positive red blood cells can develop IgG anti-D antibodies, which may cause delayed hemolytic transfusion reactions or severe hemolytic disease of the fetus and newborn (HDFN) (Daniels 2013; Westhoff 2022). The compatibility of Rhesus factor is summarized in Table 3.2.

Table 3.2 Rhesus factor compatibility

		Donor blood type							
		O–	O+	B–	B+	A–	A+	AB–	AB+
Recipient blood type	AB+	✓	✓	✓	✓	✓	✓	✓	✓
	AB–	✓		✓		✓		✓	
	A+	✓	✓			✓	✓		
	A–	✓				✓			
	B+	✓	✓	✓	✓				
	B–	✓		✓					
	O+	✓	✓						
	O–	✓							

Clinical guidelines therefore recommend that

- Rh-negative recipients receive Rh-negative RBCs.
- Rh-positive recipients may receive either Rh-positive or Rh-negative RBCs.
- Females of childbearing potential should always receive Rh-negative RBCs, regardless of current sensitization status, to prevent future anti-D formation.
- These practices are critical for minimizing alloimmunization and safeguarding reproductive outcomes (Bowman et al. 2013; Flegel 2023).

3.4 Hemolytic Disease of the Fetus and Newborn (HDFN)

HDFN occurs when an Rh-negative mother carries an Rh-positive fetus. Transplacental passage of fetal erythrocytes into the maternal circulation can induce maternal anti-D IgG production. In subsequent pregnancies, these antibodies cross the placenta, opsonize fetal red cells, and cause hemolysis, leading to fetal anemia, hyperbilirubinemia, hydrops fetalis, or intrauterine death in severe cases (Zipursky et al. 2018; Daniels 2020).

The introduction of anti-D immunoglobulin prophylaxis represents one of the most successful preventive strategies in obstetric medicine. Prophylaxis is routinely administered:

At 28 weeks' gestation
Within 72 h postpartum
After miscarriage, trauma, or invasive obstetric procedures

This intervention has reduced the incidence of RhD-associated HDFN by more than 90% in countries with universal prophylaxis programs (Bowman et al. 2013; Zipursky et al. 2018).

3.5 Rh Variants

Several RhD variants have major clinical importance, particularly in transfusion medicine and pregnancy management. Weak D phenotypes are characterized by reduced D antigen expression and are generally considered non-immunogenic; however, individualized molecular assessment is recommended to avoid inappropriate anti-D prophylaxis or transfusion errors. In contrast, partial D phenotypes lack specific D epitopes and can produce allo-anti-D antibodies upon exposure to RhD-positive red blood cells, posing significant risks in transfusion and obstetric settings.

The DEL phenotype, defined by extremely low surface expression of the D antigen, is especially prevalent in East Asian populations and may be mistyped as RhD-negative by routine serology, leading to inadvertent alloimmunization (Westhoff 2004; Flegel 2015; Wagner et al. 2018). The growing application of molecular Rh genotyping has greatly improved the detection and classification of these variants, enabling more accurate donor–recipient matching and supporting the transition toward precision transfusion medicine (Daniels 2013; Flegel and Denomme 2021).

3.6 Clinical Significance Beyond Transfusion

Beyond transfusion and obstetric implications, the Rh system appears to influence additional physiological and pathological traits. Epidemiological and experimental studies have linked RhD status to blood pressure regulation, red cell gas transport, and cardiovascular risk. Moreover, RhD polymorphisms have been associated with behavioral, neurocognitive, and reaction–time differences, particularly in the context of latent *Toxoplasma gondii* infection (Flegr et al. 2008, 2010; Kaňková et al. 2012).

At the molecular level, Rh proteins function as ammonium or gas transporters and play a role in erythrocyte membrane integrity, ion transport, and acid–base homeostasis (Endeward et al. 2008; Bruce et al. 2003). These findings support the concept that the Rh system has broader biological relevance beyond its classical immunohematologic role, encompassing systemic physiological functions and evolutionary adaptations.

4 Combined ABO–Rh Phenotypes (Clinical Relevance)

Combined ABO and Rhesus (Rh) blood group phenotyping is essential in clinical practice, as incompatibilities in either system can result in severe immunological reactions. ABO mismatches are the most dangerous, triggering rapid complement-mediated intravascular hemolysis that may lead to shock, acute renal failure, and disseminated intravascular coagulation (Daniels 2013; Cooling 2015).

Rh incompatibility, particularly in Rh-negative individuals exposed to RhD-positive erythrocytes, typically results in delayed hemolytic reactions due to IgG-mediated alloimmunization. Although less immediate than ABO reactions, anti-D antibodies can have devastating consequences in subsequent transfusions and pregnancies, especially through the hemolytic disease of the fetus and newborn (HDFN) (Moise 2008; Izetbegovic 2013).

Consequently, accurate ABO and Rh typing is mandatory across transfusion medicine, prenatal care, organ transplantation, and emergency medicine. In obstetrics, combined typing allows timely anti-D immunoprophylaxis and prevents HDFN. In transplantation, ABO compatibility is a major determinant of graft survival. In trauma and emergency settings, rapid blood group identification is lifesaving. These considerations highlight the necessity of integrating both ABO and Rh information wherever blood exposure may occur.

5 Evolutionary and Population Genetics Perspective

The ABO blood group system provides a paradigmatic example of pathogen-driven balancing selection in humans. ABO polymorphisms influence susceptibility to numerous infectious diseases, including *Plasmodium falciparum*, *Vibrio cholerae*, norovirus, *Helicobacter pylori*, and historical smallpox outbreaks. Blood group O confers partial protection against severe malaria by reducing rosetting of infected erythrocytes, whereas group A has been associated with increased susceptibility to smallpox and gastric colonization by *H. pylori* (Rowe et al. 2007; Tan and Jiang 2005; Cooling 2015).

Balancing selection, genetic drift, and founder effects have shaped the global distribution of ABO alleles, resulting in high frequencies of group O in Central and South America, group A in Europe and North America, group B in South and Central Asia, and low worldwide prevalence of AB (Yamamoto et al. 2012; Anstee 2010).

In contrast, the Rh system—particularly the RhD antigen—shows a distinct evolutionary trajectory. The RhD deletion underlying the Rh-negative phenotype likely arose once in Europe approximately 30,000–40,000 years ago and spread through founder effects and population bottlenecks (Westhoff 2004; Flegel 2011). Unlike ABO, strong evidence for pathogen-mediated selection on Rh polymorphisms remains limited, although a proposed heterozygote advantage has been hypothesized but not definitively demonstrated.

6 ABO and Rh in Advanced Genomics and Precision Medicine

Genomic technologies have profoundly transformed the clinical management of ABO and Rh blood groups, enabling precision medicine approaches in transfusion, obstetrics, and transplantation. Next-generation sequencing (NGS) allows comprehensive identification of weak, rare, hybrid, or silent alleles that often escape conventional serology. A major clinical application is noninvasive prenatal RHD genotyping using maternal cell-free fetal DNA, which enables early detection of Rh incompatibility and optimizes anti-D prophylaxis strategies (Claeys et al. 2012; van der Schoot et al. 2018).

In chronically transfused patients, such as those with sickle cell disease or thalassemia, molecular blood group genotyping significantly reduces alloimmunization risk by improving donor–recipient matching at the allelic level (Flegel 2015; Chou et al. 2018). These advances address key limitations of serologic testing and support the integration of blood group genomics into personalized transfusion medicine. As sequencing costs decline and clinical implementation expands, routine molecular profiling of ABO and Rh systems is expected to become standard practice in high-risk patient populations.

7 Conclusion

The ABO and Rh blood group systems represent complex genetic, biochemical, and evolutionary frameworks with profound clinical and biological significance. Beyond their indispensable role in transfusion safety and obstetric care, these systems influence host–pathogen interactions, population genetics, and fundamental aspects of erythrocyte physiology. Continued integration of genomics, immunohematology, and evolutionary biology will further refine personalized transfusion strategies and deepen our understanding of how blood group polymorphisms shape human health and disease.

References

Agre P et al (2002) Proc Natl Acad Sci 99(4):1399–1402. https://doi.org/10.1073/pnas.042694999

Anstee DJ (2010) The relationship between blood groups and disease. Blood 115(23):4635–4643. https://doi.org/10.1182/blood-2010-01-261859

Anstee DJ (2021) Transfus Med Hemother 48(2):75–83. https://doi.org/10.1159/000514198

Arthur CM et al (2024) N Engl J Med 390(2):123–134. https://doi.org/10.1056/NEJMra2303877

Avent ND, Reid ME (2000) The Rh blood group system: a review. Blood 95(2):375–387. https://doi.org/10.1182/blood.V95.2.375

Barbalic M, Dupuis J, Dehghan A, Bis JC, Hoogeveen RC, Schnabel RB, Nambi V, Bretler M, Smith NL, Peters A, Lu C, Tracy RP, Aleksic N, Heeriga J, Keaney JF Jr, Rice K, Lip GY, Vasan RS, Glazer NL, Larson MG, Uitterlinden AG, Yamamoto J, Durda P, Haritunians T, Psaty BM, Boerwinkle E, Hofman A, Koenig W, Jenny NS, Witteman JC, Ballantyne C, Benjamin EJ (2010) Large-scale genomic studies reveal central role of ABO in sP-selectin and sICAM-1 levels. Hum Mol Genet 19(9):1863–1872

Bhende YM, Deshpande CK, Bhatia HM, Sanger R, Race RR, Morgan WTJ, Watkins WM (1952) A "new" blood group character related to the ABO system. Lancet 259(6714):903–904

Borén T, Falk P, Roth KA, Larson G, Normark S (1993) Attachment of helicobacter pylori to human gastric epithelium mediated by blood group antigens. Science 262(5141):1892–1895

Bowman JM et al (2013) Innate immune activation after transfusion of stored red blood cells. Transfus Med Rev 27(2):84–92. https://doi.org/10.1016/j.tmrv.2013.01.001

Broni F et al (2025) PLoS One 20(10):e0333547. https://doi.org/10.1371/journal.pone.0333547

Bruce LJ et al (2009) Blood 113(1):109–114. https://doi.org/10.1182/blood-2008-06-165654

Bruce LJ, Beckmann R, Ribeiro ML, Peters LL, Chasis JA, Delaunay J, Mohandas N, Anstee DJ, Tanner MJ (2003) A band 3-based macrocomplex of integral and peripheral proteins in the RBC membrane. Blood 101(10):4180–4188. https://doi.org/10.1182/blood-2002-09-2824

Chérif-Zahar B et al (1996) Nat Genet 12(2):168–173. https://doi.org/10.1038/ng0296-168

Chester MA, Olsson ML (2001) Transfus Med Rev 15(3):177–200. https://doi.org/10.1053/tmrv.2001.24641

Chou ST et al (2018) Blood 132(2):119–130. https://doi.org/10.1182/blood-2018-01-828392

Claeys S et al (2012) Blood 119(2):526–528. https://doi.org/10.1182/blood-2011-08-372037

Clausen FB, van der Schoot CE (2025) Noninvasive fetal blood group antigen genotyping. Blood Transfus 23(2):101–108. https://doi.org/10.2450/BloodTransfus.712

Cooling L (2015) Clin Microbiol Rev 28(3):801–870. https://doi.org/10.1128/CMR.00109-14

Cserti CM, Dzik WH (2007) The ABO blood group system and plasmodium falciparum malaria. Blood 110(7):2250–2258

Daniels G (2013) Human blood groups, 3rd edn. Wiley-Blackwell

Daniels G (2020) Human blood groups, 4th edn. Wiley-Blackwell

Daniels G et al (2022) Transfus Med 32(1):4–15. https://doi.org/10.1111/tme.12815

Dean L (2005) Blood groups and red cell antigens. National Center for Biotechnology Information

Dentali F et al (2012) J Thromb Haemost 10(9):1723–1732. https://doi.org/10.1111/j.1538-7836.2012.04801.x

Ellinghaus D et al (2020) N Engl J Med 383(16):1522–1534. https://doi.org/10.1056/NEJMoa2020283

Endeward V et al (2008) Proc Natl Acad Sci 105(9):3799–3804. https://doi.org/10.1073/pnas.0711401105

Ewald DR, Sumner SCJ (2016) Blood type biochemistry and human disease. WIREs Syst Biol Med 8:517–535

Flegel WA (2007) Transfus Apher Sci 36(1):5–17. https://doi.org/10.1016/j.transci.2006.09.002

Flegr J, Novotná M, Lindová J, Havlicek J (2008) Neurophysiological effect of the Rh factor. Protective role of the RhD molecule against toxoplasma-induced impairment of reaction times in women. Neuroendocrinol Lett 29(4):4

Flegr J, Novotná M, Fialová A, Kolbeková P, Gasová Z (2010) The influence of RhD phenotype on toxoplasmosis-and age-associated changes in personality profile of blood donors. Folia Parasitol 57(2):143

Flegr J, Prandota J, Sovičková M, Israili ZH (2014) Toxoplasmosis—a global threat. Correlation of latent toxoplasmosis with specific disease burden in a set of 88 countries. PLoS One 9(3):e90203

Flegel WA (2011 Feb) Molecular genetics and clinical applications for RH. Transfus Apher Sci 44(1):81–91. https://doi.org/10.1016/j.transci.2010.12.013

Flegel WA (2015) Blood Transfus 13(3):319–329. https://doi.org/10.2450/2015.0001-15

Flegel WA (2023) Transfus Med Rev 37(1):25–36. https://doi.org/10.1016/j.tmrv.2022.11.002

Flegel WA, Denomme GA (2021) Transfusion 61(S1):S54–S62. https://doi.org/10.1111/trf.16299

Franchini M, Lippi G (2015) Semin Thromb Hemost 41(5):545–549. https://doi.org/10.1055/s-0035-1550439

Franchini M, Mannucci PM (2014) ABO blood group and thrombotic vascular disease. Thromb Haemost 112(6):1103–1109

Garratty G, Glynn SA, McEntire R (2004) ABO and Rh(D) phenotype frequencies of different racial/ethnic groups in the United States. Transfusion 44(5):703–706

Guillon P, Clément M, Sébille V, Rivain JG, Chou CF, Ruvoën-Clouet N, Le Pendu J (2008) Inhibition of the interaction between the SARS-CoV spike protein and its cellular receptor by anti-histo-blood group antibodies. Glycobiology 18(12):1085–1093. https://doi.org/10.1093/glycob/cwn093

Hakomori S (1999) Biochim Biophys Acta 1473(1):247–266. https://doi.org/10.1016/S0304-4165(99)00182-4

Hall AM et al (2023) Blood 141(4):357–368. https://doi.org/10.1182/blood.2022017432

Hosoi E (2008) J Med Investig 55(3–4):174–182. https://doi.org/10.2152/jmi.55.174

Huang CH, Peng J (2005) Transfus Med Rev 19(2):93–104. https://doi.org/10.1016/j.tmrv.2004.10.002

Huang CH, Liu P, Cheng JG (2009) Blood 114(24):4916–4924. https://doi.org/10.1182/blood-2009-06-227140

ISBT Working Party (2023) ISBT Sci Ser 18(1):1–6. https://doi.org/10.1111/voxs.12863

Izetbegovic S (2013) Occurrence of ABO and RhD incompatibility with Rh negative mothers. Mater Sociomed 25(4):255–258

Kaňková Š, Šulc J, Křivohlavá R, Kuběna A, Flegr J (2012) Slower postnatal motor development in infants of mothers with latent toxoplasmosis during the first 18 months of life. Early Hum Dev 88(11):879–884
Kelly RJ et al (1995) J Biol Chem 270(9):4640–4649. https://doi.org/10.1074/jbc.270.9.4640
Koda Y, Kimura H, Mekada E, Soejima M (1997) Expression cloning of a human α(1,2)-fucosyltransferase gene (FUT1) involved in the H blood group antigen synthesis. J Biol Chem 272(23):14114–14119
Knowles S, Regan F (2006) Br J Haematol 135(1):4–18. https://doi.org/10.1111/j.1365-2141.2006.06291.x
Kustu S, Inwood W (2006) Trends Biochem Sci 31(9):527–533. https://doi.org/10.1016/j.tibs.2006.07.004
Marini AM et al (2000) Nat Genet 26(3):341–344. https://doi.org/10.1038/81632
McGann PT et al (2012) Blood 119(10):2394–2401. https://doi.org/10.1182/blood-2011-09-381921
Moise KJ (2008) Obstet Gynecol 112(1):164–176. https://doi.org/10.1097/AOG.0b013e31817d2018
Montgomery RA, Locke JE, King KE, Segev DL, Warren DS, Kraus ES, Cooper M, Simpkins CE, Singer AL, Stewart ZA, Melancon JK, Ratner L, Zachary AA, Haas M (2009) ABO incompatible renal transplantation: a paradigm ready for broad implementation. Transplantation 87(8):1246–1255
Moulds JM et al (2011) Transfus Med Rev 25(1):36–45. https://doi.org/10.1016/j.tmrv.2010.08.002
O'Donnell J, Laffan MA (2001) The relationship between ABO histo-blood group, factor VIII and von Willebrand factor. Transfus Med 11(4):343–351
Qi L, Cornelis MC, Kraft P, Jensen M, van Dam RM, Sun Q, Girman CJ, Laurie CC, Mirel DB, Hunter DJ, Rimm EB, Hu FB, Meigs JB (2010) Genetic variants in ABO blood group region, plasma soluble E-selectin levels and risk of type 2 diabetes. Hum Mol Genet 19(9):1856–1862
Okuda H et al (1997) Transfusion 37(9):951–954. https://doi.org/10.1046/j.1537-2995.1997.37997414112.x
Olsson ML, Chester MA (2001) Transfusion 41(1):3–12. https://doi.org/10.1046/j.1537-2995.2001.41010003.x
Oriol R et al (1981) Hum Genet 58(1):53–58. https://doi.org/10.1007/BF00286440
Paré G et al (2021) Arterioscler Thromb Vasc Biol 41(2):825–836. https://doi.org/10.1161/ATVBAHA.120.315625
Patenaude SI et al (2002) Nat Struct Biol 9(9):685–690. https://doi.org/10.1038/nsb837
Patnaik MM, Spitalnik SL (2024) Blood Rev 57:101005. https://doi.org/10.1016/j.blre.2023.101005
Ramsey G (2025) The Rh blood group system: RHD update. Immunohematology 41(2):31–48. https://doi.org/10.2478/immunohematology-2025-007
Reid ME, Lomas-Francis C (2004) The blood group antigen factsbook, 2nd edn. Academic Press
Reid ME, Lomas-Francis C (2019) The blood group antigen factsbook, 3rd edn. Academic Press
Rodrigues MMO, Höher G, Waskow G, Hutz MH, Lindenau JD, Petzl-Erler ML, Callegari-Jacques SM, Almeida S, Fiegenbaum M (2021) Blood groups in Native Americans: a look beyond ABO and Rh. Genet Mol Biol 44(2):e20200255
Rowe JA et al (2007) Proc Natl Acad Sci 104(44):17471–17476. https://doi.org/10.1073/pnas.0708251104
Roychoudhuri AK, Nei M (1988) Human polymorphic genes world distribution. Oxford University Press, Oxford
Saitou N, Yamamoto F (1997) Evolution of primate ABO blood group genes and their homologous genes. Mol Biol Evol 14(4):399–411
Seto NOL et al (1999) J Biol Chem 274(49):34481–34488. https://doi.org/10.1074/jbc.274.49.34481
Shahshahani HJ, Vahidfar MR, Khodaie SA (2013) Transfusion reaction in a case with the rare Bombay blood group. Asian J Transfus Sci 7(1):86–87
Sigamani K, Gajulapalli S (2022) An insight into the distribution of allele frequency of ABO and Rh (D) blood grouping system among blood donors in a tertiary Care Hospital in Chengalpattu District of South India. Cureus 14(4):e24207

Singleton BK, Green CA, Avent ND, Martin PG, Smart E, Daka A, Narter-Olaga EG, Hawthorne LM, Daniels G (2000) The presence of an RHD pseudogene containing a 37 base pair duplication and a nonsense mutation in africans with the Rh D-negative blood group phenotype. Blood 95(1):12–18
Springer GF, Horton RE (1969) Blood group isoantibody stimulation in man by feeding blood group-active bacteria. J Clin Invest 48(7):1280–1291
Storry JR et al (2023) ISBT Sci Ser 18(2):95–115. https://doi.org/10.1111/voxs.12910
Tan M, Jiang X (2005) Trends Microbiol 13(6):285–293. https://doi.org/10.1016/j.tim.2005.04.005
van't Oever RM et al (2025) Ultrasound Obstet Gynecol 65(5):589–596. https://doi.org/10.1002/uog.29201
van der Schoot CE et al (2018) Transfusion 58(2):435–442. https://doi.org/10.1111/trf.14435
Villanea FA et al (2022) Mol Biol Evol 39(4):msac048. https://doi.org/10.1093/molbev/msac048
Vuhahula EAM, Yahaya J, Morgan ED, Othieno E, Mollel E, Mremi A (2023) Frequency and distribution of ABO and Rh blood group systems among blood donors at the northern zone blood transfusion Center in Kilimanjaro, Tanzania: a retrospective cross-sectional study. BMJ Open 13(2):e068984. https://doi.org/10.1136/bmjopen-2022-068984
Wagner FF, Flegel WA (2000) RHD gene deletion occurred in the rhesus box. Blood 95:3662–3668
Wagner FF, Frohmajer A, Ladewig B (2018) Transfusion 58(7):1520–1528. https://doi.org/10.1111/trf.14649
Ward SE, O'Sullivan JM, O'Donnell JS, Preston RJS (2020) The relationship between ABO blood group, von Willebrand factor, and primary hemostasis. Blood 136(25):2864–2874
Watkins WM (2011) Transfus Med 21(3):155–164. https://doi.org/10.1111/j.1365-3148.2011.01076.x
Westhoff CM (2004) Semin Hematol 41(1):3–9. https://doi.org/10.1053/j.seminhematol.2003.10.002
Westhoff CM (2022) Blood 139(3):342–350. https://doi.org/10.1182/blood.2021011998
Westhoff CM et al (2019) Transfusion 59(3):918–926. https://doi.org/10.1111/trf.15066
Wolpin BM, Kraft P, Gross M, Helzlsouer K, Bueno-de-Mesquita HB, Steplowski E, Stolzenberg-Solomon RZ, Arslan AA, Jacobs EJ, Lacroix A, Petersen G, Zheng W, Albanes D, Allen NE, Amundadottir L, Anderson G, Boutron-Ruault MC, Buring JE, Canzian F, Chanock SJ, Clipp S, Gaziano JM, Giovannucci EL, Hallmans G, Hankinson SE, Hoover RN, Hunter DJ, Hutchinson A, Jacobs K, Kooperberg C, Lynch SM, Mendelsohn JB, Michaud DS, Overvad K, Patel AV, Rajkovic A, Sanchéz MJ, Shu XO, Slimani N, Thomas G, Tobias GS, Trichopoulos D, Vineis P, Virtamo J, Wactawski-Wende J, Yu K, Zeleniuch-Jacquotte A, Hartge P, Fuchs CS (2010) Pancreatic cancer risk and ABO blood group alleles: results from the pancreatic cancer cohort consortium. Cancer Res 70(3):1015–1023
Yamamoto F (2021) Comparative assessment of dithiothreitol-treated ABO isoagglutinin titres performed by two methods. ISBT Sci Ser 16(2):155–163. https://doi.org/10.1111/voxs.12641
Yamamoto F, McNeill PD, Hakomori S (1990) Molecular genetic basis of the histo-blood group ABO system. Nature 345(6272):229–233
Yamamoto F, McNeill PD, Hakomori S (1992) Human histo-blood group A2 transferase coded by A2 allele, one of the A subtypes, is characterized by a single base deletion in the coding sequence, which results in an additional domain at the carboxyl terminal. Biochem Biophys Res Commun 187(1):366–374
Yamamoto F, Cid E, Yamamoto M, Blancher A (2012) ABO research in the modern era of genomics. Transfus Med Rev 26(2):103–118
Zeng J et al (2025) Enzymatic conversion of blood group B kidney prevents hyperacute antibody-mediated injuries in ABO-incompatible transplantation. Nat Commun 16(1):1506. https://doi.org/10.1038/s41467-025-56563-w
Zhao J, Yang Y, Huang H, Li D, Dongfeng G, Xiangfeng L, Zhang Z, Liu L, Liu T, Liu Y, He Y, Sun B, Wei M, Yang G, Wang X, Zhang L, Zhou X, Xing M, Wang PG (2021) Relationship between the ABO blood group and the coronavirus disease 2019 (COVID-19) susceptibility. Clin Infect Dis 73(2):328–331
Zhou J et al (2020) Differential effects of drinking water quality on phagocyte responses of broiler chickens against fungal and bacterial challenges. Front Immunol 11:584. https://doi.org/10.3389/fimmu.2020.00584
Zipursky A, Bhutani V, Odame I (2018) Rhesus disease: a global prevention strategy. Lancet Child Adolesc Health 2:536–542

Species Identification and Blood Typing

4

Abstract

Species identification has gained importance due to its role in wildlife forensics, particularly in illicit trade, criminal use of wild animals, monitoring environmental changes, and the study of disease epidemiology in the animal kingdom and the biodiversity conservation concerns. Based on conventional morphological characteristics, geographical distribution, or behavioral traits, the science is facing multiple challenges due to climate change, which in turn induced morphological and behavioral evolutions. The taxonomic identification via blood typing is a complementary approach to the conventional ones rooted in immunological principles, leveraging antigen–antibody interactions to discriminate between biological taxa. Special emphasis is placed on the forensic and wildlife applications of blood typing, including its integration with DNA-based species identification in wildlife crime investigations and biodiversity monitoring.

This chapter is focused on the major concepts of blood typing, deciphering evolutionary aspects in major taxa as primates, canines and felines, and mice, and will give a brief overview of recent advancements in the field.

Keywords

Species determination · Wildlife forensics · Blood group systems · ABO · DEA · CMAH · Molecular blood typing · Xenotransplantation · Transfusion medicine

1 Introduction

The identification of biological species is a fundamental task in taxonomy, ecology, conservation biology, and forensic science. Traditional methods of species identification—based on morphological characteristics, geographical distribution, or behavioral traits—often face limitations due to phenotypic plasticity, cryptic species, or incomplete taxonomic descriptions. In such cases, molecular and biochemical approaches offer more objective and reliable alternatives.

H. R. Dash et al., *Forensic Investigation of Blood Evidence*,
https://doi.org/10.1007/978-981-95-9183-1_4

Among these approaches, blood typing has historically played a significant role in species identification and differentiation using DNA analysis or serologically (Swart and Wilks 1982). Blood typing relies on the detection of species-specific antigens, antibodies present in blood serum, or red blood cells. These markers are genetically determined and tend to be highly conserved within a species, making them valuable tools for discriminating between closely related taxa (Ségurel et al. 2013).

This chapter will principally focus on the species antigen variability between major taxa exhibiting economic and social importance, blood-typing methods, and their particularities in species identification and will give an overview of the novel techniques gaining importance in the field.

2 Blood Origin and Antigen Evolution: Species Identification

Blood itself is a result of evolution, which appeared in the first multicellular organisms. Blood evolutionary origins trace back to a primitive phagocytic cell in an animal ancestor. Before the advent of circulatory systems, small invertebrates were capable of tissue oxygenation only through simple diffusion. This mechanism is still observed in primitive animals such as sponges and *Capsaspora owczarzaki* (Nagahata et al. 2022). The formation of a larger body structure in the latter period required more efficient oxygen delivery mechanisms. This led to the evolution of increasingly complex circulatory systems and the emergence of various pigments that were able to carry oxygen (Monahan-Earley et al. 2013).

With regard to the blood group antigen evolution across species, multiple hypotheses are known:

2.1 Pathogen-Driven Selection Hypothesis

Researchers proposed that blood group antigens may have evolved as a result of genomic reprogramming by pathogens, which insert DNA sequences into host genomes to enhance their survival. According to this hypothesis, pathogens act as a primary evolutionary pressure, utilizing blood group antigens as receptors for attachment and invasion. The most well-known example is the selective advantage conferred by the O allele of the ABO system against severe malaria, as *Plasmodium falciparum* infections are associated with the formation of rosettes that are more prevalent in non-O blood types (Rowe et al. 2007). This mechanism extends beyond humans, with similar selective pressures from diverse pathogens hypothesized to drive the evolution of the complex blood group systems observed in other primates and mammals (Garratty 2000).

2.2 Selection for Immune Diversity

Gene polymorphisms coding enzyme protein sequences and consequently their conformation, responsible for the diversity of blood group antigens, could provide a survival advantage by diversifying immune responses to pathogens. Blood group antigen genes are reported to be subjected to balancing selection. A single example is reported by Yamamoto and colleagues, mentioning the existence of ABO haplotypes in *Mus musculus,* formed by a combination of cis-AB alleles located on the same parental chromosome encoding the AB phenotype (Yamamoto et al. 2001). In humans, the presence of cis-AB polymorphism has been reported in Japanese and Koreans, highlighting the clinical importance of the variant in the transfusion process (Chun et al. 2019).

3 ABO Blood Typing in Primates: Structure and Forensic Relevance

Primates are among the most reliable animal models for laboratory use, yet they are also among the most financially expensive. Therefore, it is best to minimize losses of laboratory animals outside of experimental procedures. This can be achieved, among other things, by implementing a reliable and efficient transfusion system. Also, some nonhuman primate species are endangered due to illegal trafficking.

3.1 Structure of ABO Blood Typing

The ABO blood typing in primates is based on the expression of the H antigen, a precursor molecule present on red blood cells that serves as the foundation for the AandBantigensandtwoenzymes,thealpha-(133)-N-acetylgalactosaminyltransferase (GTA) and the alpha-(133)-galactosyltransferase, responsible for N-acetyl-D-galactosamine and D-galactose to the H acceptor, respectively (Ségurel et al. 2012). The enzymatic specificity of A and B blood group alleles is conditioned by two amino acids at positions 266 and 268 in exon 7 and is surrounded by a multitude of polymorphisms. For example, the diversity level in exon 7 is 0.0058 in a Yoruba sample, suggesting a long-term balancing selection has acted on this locus in humans (Ségurel et al. 2012).

Enzymatic sequences share structural similarities with other primates, suggesting a convergent evolution in many lineages of an ancestral polymorphism stably maintained for millions of years and inherited across species. The current consensus indicates that the A allele is ancestral, and the B allele had evolved independently at least six times in primates (in human, gorilla, orangutan, gibbons, baboon, and chimpanzee), as illustrated in Fig. 4.1.

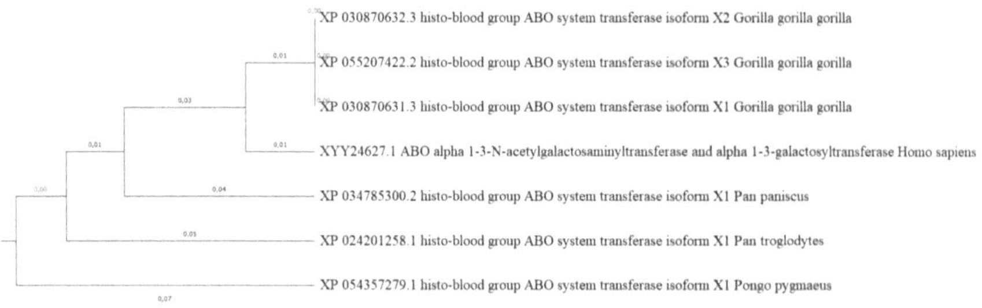

Fig. 4.1 Phylogenetic distribution of ABO transferase isoform A performed by the MEGA 11.0 software

3.2 Forensic and Wildlife Relevance of ABO Blood Typing

To distinguish nonhuman primate blood from humans, ancient blood-typing techniques such as the precipitin test are used, although considered not capable of discriminating between human and monkey blood in a monkey bite case. The work of Sivaram and colleagues had demonstrated that a quantitative precipitin test can, however, demonstrate it (Sivaram et al. 1975).

Due to the fact that apes and humans have ABO antigens on their RBC, ABO blood typing can discriminate only between the latter ones, contrary to New- and Old-World monkeys (Ségurel et al. 2013).

Nonhuman primate blood samples are used for transfusion purposes, especially for nonhuman primates living in captivity in zoos. For this case, biobanks are considered the optimal solution for the conservation of blood. To ensure the best laboratory practice of transfusion and to ensure the proper sorting of the samples, Gamble and colleagues had developed a protocol adapted to the conditions of Lincoln Park Zoo, which could be used to categorize blood groups for great ape species, information that would then be useful to develop donor databases and reduce the search time for critical patient care. The incorporation of the obtained information into the breeding management program of the zoo could potentially minimize maternal–fetal blood interactions (Gamble et al. 2011). These findings can also be applied to animal biobank management for similar purposes.

With regard to xenotransplantation feasibility, the ABO blood group compatibility was found not to play a significant role in heart transplantation compatibility between Old-World monkeys (Cooper et al. 1989).

4 Non-ABO Blood Groups in Primates

4.1 Overview of Non-ABO Blood Systems

Nonhuman primates have significantly contributed to the development of modern hematology, especially to the discovery of the Rhesus system. Indeed, the Rhesus antigens were for the first time identified in Rhesus macaques by Drs. Philip Levine and Rufus Stetson who helped to elucidate hemolytic reactions in infants. The Rhesus system, consisting of D, C, c, E, and e antigens and found in about 85% of humans, has intrigued researchers to understand the molecular bases of structural homologies of the human and primate Rhesus factors. According to Salvignol and colleagues, full-length Rh transcript amplification by RT-PCR encoded 417 amino acid-long proteins homologous to Rh polypeptides. Western blot analysis had revealed an Rh-like protein length of 30,000–40,000 M(r), which was immunologically related to human Rh proteins, indicating that Rh-like transcripts are translated into membrane proteins. Alignments of the obtained proteins to human protein sequences had a pivotal role in the understanding of the molecular basis of the blood group C/c polymorphism. In fact, the Pro103Ser substitution was correlated with the latter. Additionally, a proline residue at position 102 was critical in the expression of C and c epitopes, most likely by providing an appropriate conformation of Rh polypeptides. With the obtained data, a phylogenetic reconstruction of the RH locus evolution has been calculated, suggesting that African ape Rh-like genes would be closer to the human RhD gene than to the human RhCE gene. Also, the amplification by RT-PCR of RH-related transcripts from samples of several species of nonhuman primates showed a high degree of similarity to human blood group Rh sequences, suggesting a great conservation of the RH genes throughout evolution (Salvignol et al. 1995).

Beyond the Rhesus system, other non-ABO blood types are used in hematology, such as Kell, Kidd, Duffy, MNS, Lewis, Lutheran, and P, each having multiple phenotypes and variable prevalence in humans. Studies have reported the prevalence of the non-ABO blood groups in nonhuman primates (Table 4.1). Ramis et al.'s report encompassed 48 olive baboons, 9 chacma baboons, 14 Guinea baboons, 38 Rhesus macaque, and 30 squirrel monkeys. For all enrolled species, the highest prevalence (100%–97.4%) was observed for the Jk^a Kidd antigen, followed by the Duffy Fy^b antigen (100%–85.4%); the Kp^a Kell antigen exhibited an intermediate prevalence (100%–22.2%); the Lutheran Lu^b antigen prevalence was also intermediate (66.7%–11.1%); Le^b Lewis antigens were prevalent in all analyzed species except the squirrel monkey (100%–53.5%). All analyzed species had exhibited a null prevalence of the M, N, s, and JK^b antigens. The Rhesus c and E antigens were present only for squirrel monkeys (Ramis et al. 2013), as indicated in Table 4.1.

4.2 Forensic/Wildlife Relevance of Non-ABO Blood Systems

To date, the non-ABO blood group typing constitutes a great importance in wildlife conservation procedures in frame of the management of inbreeding nHP in situ populations.

As reported by Terao and colleagues, simian-type M had a role in the management of breeding colonies of African green monkeys (Terao et al. 1988).

To date, no other studies have been performed on the influence of other non-ABO groups on colonies' inbreeding procedures other than nHP. However, Ramis and colleagues had highlighted the role of Lewis antigens in organ rejection,

Table 4.1 Frequencies of different antigens of non-ABO blood groups in nonhuman primates

Group	Antigen	Frequencies (%) Olive baboon (*n* = 48)	Chacma baboon (*n* = 9)	Guinea baboon (*n* = 14)	Rhesus macaque (*n* = 38)	Squirrel monkey (*n* = 30)
Rh	C	0	0	0	0	100
	E	0	0	0	0	3.3
Kell	K	0	0	0	0	100
	K	29.2	0	21.4	0	56.6
	Kp[a]	62.5	22.2	85.7	73.7	100
	Kp[b]	31.3	0	57.1	2.6	56.7
P	P	2.1	0	0	0	16.7
Lewis	Le[a]	20.8	0	0	2.6	0
	Le[b]	91.7	77.8	100	55.3	0
Lutheran	Lu[a]	27.1	0	21.4	0	53.3
	Lu[b]	33.3	11.1	35.7	34.2	66.7
Kidd	Jk[a]	100	100	100	97.4	100
	Jk[b]	0	0	0	0	0
Duffy	Fy[a]	0	0	0	18.4	66.7
	Fy[b]	85.4	88.9	100	92.1	100
MNSs	M	0	0	0	0	0
	N	0	0	0	0	0
	S	0	0	0	5.3	93.3
	Ŝ	0	0	0	0	0

particularly in the case of kidney transplantation, emphasizing that any research involving transplantation or xenotransplantation with NHPs should include typing for these antigens, taking into account the fact that Lewis antigens were absent in squirrel monkeys (Ramis et al. 2013).

In the forensic context, non-ABO blood grouping can help in a more detailed classification and traceability of nHP, particularly for chimpanzees, possessing simian antigens that are counterparts of human ones, particularly R-C-E-F and V-A-B-D systems that are counterparts of the human Rh system and MNS systems, respectively. Macaques have their proper simian Drh system, G, I, J, K, Q antigens that have been linked to the Rhesus genome and the Duffy antigen, conferring resistance to *Plasmodium vivax* is particular to Old-World Primates (Socha 1997).

In practice, DNA analysis is preferred and widely used for discrimination between closely related taxa for primate species identification, offering a far greater specificity and discriminatory power. Blood grouping procedures, whether ABO or non-ABO, are now used for initial screening or excluding possibilities rather than for a definitive identification of species. The DNA's stability and its availability even in degraded samples also make it a superior tool in the forensic field compared to less durable proteins and antigens, analyzed within the blood-typing context. The major characteristics of primates' blood groups are given in Table 4.2.

Table 4.2 Main characteristics of primates' blood groups

	Prevalent blood groups								
Species	*ABO*	*Rh*	Kell	*P*	Lewis	Lutheran	Kidd	Duffy	MNS
Humans	A, B, AB, O	D,e [1]	K [1] K:-6, 7 [3]	P [2]	Le^{b} [2]	Lu^{b} [2]	JK^{a}/JK^{b} [2]	Fy^{a} -Asians[1] Fy^{b}- Caucasians[1]	M, N, S, and s in various proportions[2]
Chimpanzees	A	Polymorphic: Expresses "simian Rh" antigens (R-C-E-F system). C, c, E, e counterparts are present. R-C-E-F system	K:6, -7 [3]		Prevalent in secretions				Polymorphic: Possesses human-like M and N; complex V-A-B-D system is the simian counterpart to MNSs
Macaca mulata	B, O (rare A, AB)	Also have unique systems	Kp^{a}	–	Le^{b}	Lu^{b}	Jk^{a}	Fy^{b}	S
Macaca fascicularis	A, B, AB, O	Drh system: Possesses simian-type Rh antigens distinct from human proteins						Fy(a-b-)	Simian-type M-like antigens are common but distinct from human MNS
Papio	A, B, AB	–	k	P	Le^{b}	Lu^{a} / Lu^{b}	Jk^{a}	Fy^{b}	–
Chlorocabus sabaeus	A, O, and B	Possess simian-type antigens; lack human D-antigen protein.							M System: Polymorphic; governed by dominant M and recessive m alleles.
Gorilla gorilla	B	Rh+ expresses human-like Rh-Hr antigens			Positive in secretions				Primarily type M; lacks the complex polymorphism found in chimpanzees.

(continued)

Table 4.2 (continued)

	Prevalent blood groups								
Pan Paniscus	A	Rh_0 (D)-positive simian system differs from *P. troglodytes*.					Absent	Absent	Primarily M positive; simian-type V, CCef, and g antigens are common
Pan troglodytes	A, O	Polymorphic: Expresses "simian Rh" antigens (R-C-E-F system). C, c, E, e counterparts are present.			Prevalent in secretions				Polymorphic: Possesses human-like M and N; complex V-A-B-D system is the simian counterpart to MNSs
Pongo pigmaeus	A, B, O	Variably Rh-positive							Primarily M positive; some N-like reactivity depending on subspecies
Saimiri sciureus	A, B	Rhc	K and Kp^a	P	-	Lu^a/Lu^b	Jk^a	Fy^b	S

[1]Baruah et al. (2023)
[2]Redman et al. (1989)
[3]Owaidah et al. (2020)

Kermarrec and colleagues investigated the genetic basis of the ABO blood group system in humans and nonhuman primates, focusing on O allele sequences, analyzing not only blood but also bone marrow and saliva samples from chimpanzees, rhesus monkeys, crab-eating macaques, and humans using PCR, reverse transcription, and DNA sequencing to examine exons 6 and 7 of the ABO gene. The obtained results denoted species-specific O alleles with unique silencing mutations in each primate group, indicating that chimpanzees possess only A and O blood types, while macaques exhibit all four ABO phenotypes (Kermarrec et al. 1999).

In the context of blood typing, the principle of exploiting stable, species-specific antigen patterns for serological distinction is well illustrated by the evolutionary divergence of blood groups, such as ABO, using the immunodiffusion assay. Molecular structures defining blood group antigens on RBC and in secretions (for New-World monkeys) are genetically determined and are fixed or not, over evolutionary time, creating characteristic variations between species.

Purified antisera specific to a human blood group (e.g., human anti-A) can be used against another RBC species and will strongly agglutinate RBC from the immunizing species. When tested against blood from other species, the presence and strength of agglutination reveal shared antigenic structures. A species possessing a nearly identical antigenic determinant will show strong cross-reactivity, while one where the gene for that antigen is absent or has mutated significantly will show little to no reaction. This allows for the construction of phylogenetic relationships based on shared blood group antigens.

This method is highly effective to distinguish between primates and nonprimates' mammals, taking into account that the A-like antigen is found in humans and in apes and Old-World monkeys, even in more distantly related species like rodents. A similar antigenic structure in different species can be interpreted as strong evidence of a shared, ancient genetic lineage.

The method is challenging with closely related species such as great apes (chimpanzees, gorillas, orangutans), where variants of the A, B, and O antigens are present, but cross-reactivity with human reagents can be complex and inconsistent, making precise species identification difficult without highly refined reagents. Furthermore, the independent evolution (convergence) of similar carbohydrate structures in unrelated species can cause false-positive cross-reactions, misleading phylogenetic inferences.

Although serological methods for phylogenic identification are now rarely used, according to the literature, the immunodiffusion method is still used to identify dog breeds in veterinary transfusion medicine. In forensic science, the method was used in 1980s to identify the species of origin of meat samples (Swart and Wilks 1982).

The current chapter aims also to provide an overview of the blood group systems existing in certain species of biomedical and social interest such as Felines and Canines in the following sections.

5 Canine Systems: Rebalance Transfusion vs Species-ID Focus

Forensic Identification of Species and Subspecies of the Canidae Family by Cross-Species PCR and Real-Time PCR

Blood grouping in canines is an essential part of veterinary medicine, used for life-saving treatments and critical diseases. Canine erythrocytes, as in all placental mammals, have on their surfaces glycolipids and glycoproteins that allow for classification into blood groups. Canine blood grouping is outstanding for the multitude of blood groups discovered. Among them, five blood groups are internationally recognized and most well characterized, known under the acronym DEA (dog erythrocyte antigen), listed in Table 4.3.

5.1 DEA Blood System

Table 4.3 recapitulates the main properties of the DEA blood system.

All antigens cause a variable degree of antibody formation during the first blood transfusion (Hale 1995). If the same blood group is transfused again, the produced antibodies will affect the injected red blood cells. This property is known as immunogenicity. When antibodies interact with red blood cells, they can cause them to clump together (agglutination) or break down (hemolysis).

5.1.1 DEA1 Subtype

The DEA1 antigen has the strongest immunogenicity. DEA1+ and DEA1− include approximately 50% of dogs (Goy-Thollot et al. 2017). The DEA1+ blood type is more commonly found for DEA1+; the highest positivity rates were observed in Golden Retrievers (80.8%), Malteses (80%), mixed-breed dogs (80%), and St. Bernards. The breeds with the highest DEA1 negativity were Doberman Pinschers (75.0%), Old English Sheepdogs (66.7%), Labrador Retrievers (40.7%), Boxers,

Table 4.3 DEA system blood group repartition and main characteristics

Blood group	Prevalence	Consequences of antigen–antibody interaction
DEA 1	40–60%, depending on the region	Acute hemolytic transfusion reaction (AHTR) induction
DEA 3	5–20%	Reduced transfusion efficiency due to the gradual destruction of donor red blood cells
DEA 4	85–100%	Acute hemolytic transfusion reaction (AHTR) induction
DEA 5	10–25%	Reduced transfusion efficiency due to the gradual destruction of donor red blood cells
DEA 6	98–99%	Not studied
DEA 7	10–50%	Transfusion incompatibilities observed (Novais and Magron 2018)
DEA 8	40%	Not studied

Greyhounds, Dobermans, Irish Wolfhounds, German Shepherds, and Pit Bulls (Kim and Han 2025).

A recent meta-analysis encompassing studies on Posavaz hounds, Tornjak hounds, Galgos dogs, mixed-breed dogs, Ibizan hounds, greyhounds, Nigerian mixed-breed dogs, and indigenous dogs had reported DEA1 positivity prevalence between 39.89% and 91.30% (Mangiaterra et al. 2021).

When it is re-entered into the body of a dog with blood group DEA1, acute haemolysis occurs. This is the rapid destruction of donor red blood cells, in which there is a serious deterioration in the animal's condition. DEA1 blood group figures among the best studied ones due to its particularity to have three subtypes, DEA-1.1, DEA-1.2, DEA-1.3 and a null type (negative). Each of the subtypes is subdivided into two categories, a positive one and a negative. Only one of the four phenotypes occurs in a dog. DEA 1.1 antibodies can be produced by DEA1-negative females during pregnancy in case of a DEA1.1-positive status in puppies (Hale 1995; Corato et al. 1997). Acute hemolysis involving the blood type 1.1 can be life-threatening for the patient as it causes the destruction of incompatible blood cells within 12 h following the transfusion, causing adverse effects as hyperbilirubinemia and hemoglobinuria (Giger et al. 1995).

Also, Hale reported agglutination reactions involving DEA1.2-negative dogs receiving successive blood transfusions with DEA1.1/1.2 blood within the 24 h, causing extravascular hemolysis (Hale 1995).

5.1.2 DEA-3 Subtype

The DEA-3 blood group is known to have a low prevalence and therefore has a lesser clinical significance, unlike the DEA-1 group (Zaremba et al. 2019). According to an American study, it represents about 6% (Novais and Magron 2018); however, other authors have found a frequency of 20% in Greyhounds. Two studies on mixed-breed dogs reported a prevalence between 10.6% and 23.2%. The existence of alloantibodies against the DEA-3 antigen is found in 20% of dogs (Novais and Magron 2018; Mangiaterra et al. 2021).

Among the described clinical cases, delayed transfusion reactions such as shortened RBC lifespans have been documented (Mangiaterra et al. 2021; Zaremba et al. 2019), represented by a sequestration and destruction of red blood cells in the spleen within 72 h. However, Weiss and Wardrop (2011) observed that repeated exposure to DEA-3 cells results in antisera that are strongly reactive with all three antigen subtypes of the DEA-1 group (Weiss and Wardrop 2011). Thus, DEA-3-positive dogs should not be used as donors of blood, except for DEA-3-positive dogs (Hale 1995). Due to the low probability to find DEA-3 sera, detailed classification of this subgroup remains an open question.

5.1.3 DEA-4 Subtype

The DEA-4 subgroup is a highly prevalent one in canines representing 98% (Melzer et al. 2003), studied mainly in mixed-breed dogs and Ibizan hounds, with prevalence between 98.8% and 100% (Mangiaterra et al. 2021), and therefore has a lower probability of AHTR induction. In fact, to observe a DEA-induced AHTR, the transfused dog should have a DEA-4-negative status that represents 2% (Zaremba et al. 2019). When occurring, the adverse effects are not associated with a severe hemolysis in vivo despite an isolated case report.

5.1.4 DEA-5 Subtype

The DEA-5 antigen is reported to have a prevalence from 5% to 25% in dog population. Delayed transfusion incompatibility reactions as well as a decreased erythrocyte viability were reported by Ebelt and colleagues (2020). DEA-5 dogs are not defined as the best donors (Hale 1995). Occurrence of alloantibodies had been reported (Mangiaterra et al. 2021) causing a late transfusion reaction, similar to the anti-DEA-3 antibody. Consequently, DEA-5-positive dogs are not considered as good candidates for blood donation (Hale 1995).

5.1.5 DEA-6 Subtype

The prevalence of the DEA-6 subtypes had been described in only two studies: In 1961, Switcher and Young performed DEA blood typing of 165 dogs and found a prevalence of 99.4% of DEA-6-positive dogs, with one dog with a DEA-6-negative status (Swisher and Young 1961).

Reported prevalence in mixed-breed dogs, Ibizan hounds, and Spanish greyhounds had ranged between 4.7% and 71.7% (Mangiaterra et al. 2021).

5.1.6 DEA-7 Subtype

The DEA-7 subtype is formed of two factors and three phenotypes. It has been stipulated that DEA-7 is secreted by lymphocytes and is therefore not considered as a true antigen (Bull 1992). DEA-7 shows variable prevalence and expression (sometimes as an "acquired" expression); naturally occurring anti-DEA-7 antibodies are reported and can cause delayed transfusion incompatibilities detectable by cross-match methods and have been associated with RBC clearance/delayed reactions (Novais and Magron 2018).

In their works, Eva Spada and colleagues reported the presence of anti-DEA-7 antibodies in 38.1% of DEA-7-negative dogs not previously transfused. The DEA-7 antigen seems to be unique as it is acquired onto the RBC surface and thus varies in expression (Spada et al. 2016, 2017).

5.1.7 DEA-8 Subtype

The DEA-8 subtype has not been extensively studied in dogs due to its rarity and the consequent lack of serum to obtain adequate results. When DEA antigens-3, 5, 6, 7, and 8 interact with their antibodies, less serious adverse events occur (see Table 4.1 for details).

In addition to the official DEA blood system, antigens other than DEA had been identified.

5.2 Dal Antigen System

The Dal antigen system is considered one of the major blood systems in dogs, named after the Dalmatian breed, for the reason of the high prevalence of the Dal antigen. The Dal antigen was first identified in an anemic Dalmatian dog by a first transfusion from a Dal-positive canine blood donor (Blais et al. 2007). In the United States, the prevalence of the Dal+ blood type varied between 85.6% and 100% in Dalmatians and 43.3–78.6% in Doberman Pinschers. Dal− dogs were identified mostly in Dalmatians (11.7%), Doberman Pinschers (42.4%), and Shih Tzus (57.1%) and sporadically in mixed-breed dogs (2.5%), Lhasa Apsos (17%), and Bichon Frises (33%). Only 2.4% blood donors were found to be Dal−, including 1.38% of Doberman Pinschers. In Doberman Pinschers, a case report had described the phenomenon of Dal-induced RBC incompatibilities in a dog with von Willebrand factor deficiency, highlighting the value of cross-matching and the recognition of von Willebrand factor and Dal statuses in order to prevent complications (Goulet et al. 2017; Conti-Patara et al. 2021).

Korean authors reported a 77.14% prevalence of Dal+ dogs, and agglutination tests were negative for dog breeds not previously associated with Dal− status as Labrador Retrievers, Golden Retrievers, Shepherd, Siberian Husky, American Bully, Miniature Poodle, and Pungsan dog (Kim and Han 2025). The Cane Corso breed had been reported to have Dal-negative cases according to Proberbio and colleagues (Proverbio et al. 2020).

5.3 Kai Blood Group

Another alternative blood system used to discriminate between dog breeds in the context of transfusion is the Kai blood group. This blood group was discovered in 2017 using monoclonal antibodies produced by mouse hybridoma techniques. It was suggested that Kai 1 (200 and 50 kDa) and Kai 2 (80 kDa) antigens do not coexist in a dog from the observed prevalence. In fact, Kai 1+/Kai 2− phenotypes were found with proportions of 0.42 in Korea and 0.94% in North America, while the Kai 1−/Kai 2+ prevalence was 37% in Korea and 1% in USA; Kai 1−/Kai 2− phenotypes were 20% in Korea and 5% in North America, and no Kai 1+/Kai 2+ phenotypes were observed (Lee et al. 2017). This hypothesis had been confirmed by two studies by Ebelt and colleagues who also had not found Kai 1+/Kai 2+ during the analysis for their cohort, suggesting that Kai 1 and Kai 2 antigens could be genetically related, although being structurally very different (Ebelt et al. 2020; Euler et al. 2016; Mangiaterra et al. 2021).

DEA (Dog Erythrocyte Antigen), Dal, and Kai blood groups are primarily used to identify donor compatibility within domestic dogs. The latter blood groups are mainly used to identify patterns among specific dog breeds rather than species. The Dal-negative status, highly prevalent in Dalmatians, Doberman Pinschers, and Shih Tzus, illustrates this example; Kai 1+/Kai 2− is the most common pattern found in North American dogs.

The DEA-1 antigen was found to be in various nondomesticated canids, such as fennec foxes, red foxes, and arctic foxes. Because DEA-1 is for now known to be common to the Canidae family, discrimination cannot be made between one species and another using DEA-1 antigen alone. The prevalence of the Dal and Kai blood systems has not yet been studied in nondomesticated canids. To ensure Canidae species identification by serological methods, discoveries of additional biomarkers are necessary, but due to the uncertainty of the findings, molecular methods are the most used due to the reliability of the obtained result.

Implications for Species Identification and Forensic Applications
Dog erythrocyte antigen system is not only used for the determination of different blood groups between various dog breeds but also can be applied for the study of proximal species (Table 4.4). In their study, Charpentier and colleagues had explored DEA system's blood compatibilities between domestic and nondomestic canids. DEA-1 blood group was present in both groups however, its distribution varied, with 72.9% of nondomestic canids tested DEA-1+, compared to varying rates among domestic breeds. Cross-matching results had shown that 80% of nondomestic canids had incompatibilities with DEA-1+ sensitized dog plasma, indicating antigenic differences. However, 76% of them had shown compatibility with domestic dog red blood cells. For domestic dogs, only 51% plasma was compatible with nondomestic canid red blood cells, suggesting that xenotransfusion could be a potential option for in emergency situations (Charpentier et al. 2020).

6 Blood Typing in Felines

In the 1950s, British scientists first identified feline blood types and performed a detailed classification. Significant advances were made in the 1980s when this research gained significant attention and was implemented into practice. Unlike the complex canine blood-typing groups, the recognized feline one has an almost similar nomenclature but is structurally different from primates with A, B, and AB antigens, each one having a specific combination of syalic acids on the RBC surface. Type A RBC expresses N-glycolylneuraminic acid (Neu5Gc), and type B RBC expresses N-acetylneuraminic acid (Neu5Ac). The AB blood type expresses the two sialic acids on the erythrocyte surfaces (Table 4.5). The enzyme cytidine monophosphate-N-acetylneuraminic acid hydroxylase is encoded by the *CMAH* gene and catalyzes the conversion of Neu5Ac to Neu5Gc determining the blood groups in felines. The CMAH gene is located on the chromosome B2 and comprises 16 exons. Blood types are inherited according to Mendelian law, with A being dominant over B (Kehl et al. 2018, 2019; Tasker et al. 2014), as recapitulated in Table 4.5.

The knowledge of the AB feline blood system greatly helps in veterinary critical care during surgeries, particularly in transfusions, as the AB system is also defined to have anti-A and anti-B alloantibodies (NOAb anti-A and anti-B) in cats lacking the antigen (e.g., the A/A cat will have the NOAb anti-B and vice versa) (Silvestre-Ferreira et al. 2004).

Table 4.4 Summary of the DEA, Dal, and Kai blood groups

Blood group	Prevalence	Clinical/forensic implications	Known breed/region patterns	Usefulness for species discrimination
DEA-1	40–60% (world); up to 71% in Asia[a]	Highly antigenic. Causes acute hemolytic transfusion reactions (AHTR). No naturally occurring antibodies[a]	High in Labradors (>72%); low (<20%) in Boxers[b]	*Low. Used for breed transfusion compatibility and profiling*
DEA-3	~23.2[c]	It can induce antibodies after a mismatched transfusion; clinical significance is lower than DEA-1[c]	High in greyhounds (~23%) and Staffordshire Bull Terriers	
DEA-4	92–100%[a]	Highly prevalent, compatible with most dogs. Sensitization can still occur in rare negative dogs	Nearly 100% in most breeds; rare negatives in Samoyeds (33%) and Dobermans (25%)	
DEA-5	~11%[b]	Suspected naturally occurring antibodies exist in some dogs	Higher prevalence in beagles and harriers (~23%)	
DEA-6 & 8	73,1[c]	Limited clinical data due to a lack of available testing reagents	Breed-specific data is limited	
DEA-7	~8% to 77%[a]	Naturally occurring antibodies may cause delayed hemolytic reactions in some dogs	High in argentine Dogo (up to 54.5%) and boxers (52%)[d]	
Dal	~80% to 90% in most breeds. ~77% in some Asian surveys[a]	It can cause delayed or acute hemolysis in sensitized dogs	High frequency of Dal-negative in Dalmatians (~14%), Dobermans (~23%), and Shih Tzus (~57%)[b,d,e]	
Kai (1 & 2)	Kai 1+: ~94% to 96% Kai 2+: Rare (~1% to 3%)[b]	Can induce alloantibodies after a mismatched transfusion; no naturally occurring antibodies	Kai 1+/Kai 2– is the dominant global pattern[b]. Heavily studied in South Korea	

[a]Kim and Han (2025)
[b]Ebelt et al. (2020)
[c]Swisher and Young (1961)
[d]Proverbio et al. (2020)
[e]Goulet et al. (2017)

Table 4.5 Genetic transmission of the blood groups and associated enzymatic properties (Vieira et al. 2017)

Phenotype	Genotype	Type of sialic acid on erythrocytes	CMAH activity	Breeds
A	*A*/*A*	NeuGc, minor amount of NeuAc and two intermediate forms	Active	Siamese, Burmese, and Oriental breeds
	A/*b*			
	A/a^c			
B	*b*/*b*	NeuAc	Inactive	British Shorthairs, Devon Rex, and Cornish Rex
AB	a^c/a^c	NeuGc and NeuAc at low levels	Reduced activity	Rarely found
	a^c/b			

Clinically, the relevance of these alloantibodies is presented by the manifestation of severe, acute hemolytic transfusion reactions in type B cats after the reception of type A blood and during the development of neonatal isoerythrolysis in type A or AB kittens born to a type B mother due to the presence of the anti-A antibodies present in colostrum. Also, reduced survival of transfused RBC in type A cats receiving type B blood is reported (Gurkan et al. 2005). Even if it was reported that the type AB cats lack the alloantibodies, transfusion reactions when receiving blood types other than AB, due to anti-A or anti-B alloantibodies in the donor plasma, can occur (Giger and Bücheler 1991). That is why, AB typing of feline blood donors and recipients is critical before a first transfusion.

In addition to the abovementioned groups, other non-AB blood systems are now discovered: In 2007, another alloantibody, named Mik, was identified in a cat following a type A-matched RBC transfusion that resulted in an acute hemolytic reaction in a Mik-negative cat with anti-Mik alloantibodies that received blood from a Mik-positive donor. Anti-Mik alloantibodies are naturally occurring but could develop after an RBC transfusion in a Mik-negative cat (Weinstein et al. 2007). The prevalence of Mik-positive cats is reported to be about 100% in diverse studied cat populations, and Mik-negative subjects are rare.

In 2021, novel Feline Erythrocyte Antigens (FEA 1–7) were identified, but according to the authors, the identified group of antigens may correspond to the Mik antigen, as Mik-typed cats are rare and plasma for testing is unavailable. The newly identified antigens are not yet standardized for clinical veterinary practice (FEA blood group).

Evaluating the possibilities of transfusions between cats and nondomestic felids, a recent study conducted by Parkinson and colleagues on 162 blood samples from 18 different nondomestic felid species had concluded that only 7 (27%) were compatible, highlighting the fact that CHMA genotyping and cross-matching are not sufficient to ensure a successful transfusion in cats. This suggests the presence of unknown alloantibodies in nondomestic felids to be discovered (Parkinson et al. 2024).

However, in emergency cases, transfusion protocols between nondomestic felids and cats can be implemented, as shown in a study conducted on 30 tigers. Agglutination tests showed a 100% agglutination with type B cat erythrocytes and 76.7% agglutination with type A cat erythrocytes. Three out of ten tiger red blood cell samples were fully compatible with all cat plasmas, and one out of ten tiger plasma samples was fully compatible with type A red cells of domestic cats. Cross-matching results showed 80% major compatibility and 60% minor compatibility, indicating possible variations in tiger blood antigens (Thengchaisri et al. 2017). Blood-typing antigens are mostly studied to monitor transfusion efficacy in order to prevent adverse reactions at an intra-species and, in a minority of cases, at an inter-species level. In wild felids, the traceability is now mainly performed using DNA barcoding. As an example, the TigerBase, a DNA registry system employing 60 SNP profiles using barcoding, enhances enforcement of tiger trafficking cases and contributes to combat the illegal tiger trade (TigerBase article).

7 Key Methods in Blood Typing

Key methods include serological techniques, molecular methods, and emerging technologies like microfluidics and spectroscopy.

Blood typing exploits the antigen–antibody reactions occurring between specific blood group antigens and their corresponding antisera. These antigens are species-specific glycoproteins expressed on the surface of erythrocytes. When exposed to species-specific antisera, agglutination or precipitation occurs, which can be visually or instrumentally detected. Traditional serological methods rely on hemagglutination reactions between RBCs and specific antibodies.

Recent innovations focus on reducing the pretreatment steps and improving efficiency.

7.1 Conventional Blood-Typing Methods

Agglutination is one of the most common and conventional techniques implemented in hematology practice and species identification aiming the detection of specific antigens or antibodies in blood samples. Agglutination is the basis of the direct antiglobulin test (Coombs direct test) used for the detection of adverse reactions to blood transfusions by cross-match techniques (Garratty 2000; Weinstein et al. 2007; Spada et al. 2018). Most recent advances in this field involve the use of test cards, gel columns, slides, and immunochromatographic strips (Blais et al. 2007; Kim and Han 2025; Ebelt et al. 2020).

7.2 Advanced Methodologies of Blood Typing

With the advancement of microfluidics used for species identification (Yugovich et al. 2025) and molecular techniques, new detection tools emerged in the blood-typing field, improving the efficacy of detection. These include the use of erythrocyte aggregometer for hemagglutination discrimination, lab-on-disc platforms (Chang et al. 2018), and molecular typing methods that analyze DNA polymorphisms using PCR (Kehl et al. 2019) and NGS, enabling accurate prediction of blood groups and mutations influencing the development of hematopathologies (Westhoff and Floch 2025).

The IH-500 system introduced by Biorad showed reliable results for ABO-RhD blood typing and unexpected antibody screening and thus can be potentially implemented in a clinic for pre-transfusion compatibility and blood bank needs (Park et al. 2017).

Microfluidic technologies, such as 3D-μSTADs, offer cost-effective solutions for blood typing, classifying 105 ABO and RhD blood samples with a minimization of environmental light interference and reducing potential personal bias used to assess transfusion compatibility (Zhan et al. 2025).

7.3 Emerging Technologies

Raman spectroscopy combined with AI was reported to be able to identify blood groups for species identification without traditional antibody-based methods, offering rapid results and reduced risk of contamination. Huang and colleagues had collected human and 19 kinds of animal blood and measured them using the Renishaw inVia confocal Raman spectrometer at time points of 8, 24, 48, and 72 h and then performed data pre-processing and training of the obtained dataset by a deep learning method, obtaining a recognition performance to distinguish between species by blood analysis. Despite an existing, low probability of confusion (no more than 14%), the observed recognition accuracy was as high as 97.33%. Discrepancies and confusions occurred for various reasons, principally of the fact of the analysis of similar Raman spectra of close-related species, second because the obtained quality of data may fluctuate during measurement, and third due to the mutual influence of the obtained data on the training model and the training mechanism on the model (Huang et al. 2019).

AI-assisted spectral analysis is likely to grow and may complement DNA, where rapid, nondestructive triage is needed, as demonstrated by Jensen et al. (2023), where Raman spectroscopy label-free blood typing is demonstrated by training an artificial intelligence (AI) model on 271 blood-typed donor whole blood samples. As a result, the average area-under-the-curve value of the ABO classification is 0.91 ± 0.03, indicating a very good diagnostic performance. Post-hoc scalability analysis of the models shows the potential of RS and AI for future applications in transfusion medicine and blood banking (Jensen et al. 2023).

Another efficient tool for discrimination between species is ATR FT-IR (Attenuated Total Reflectance to perform Fourier Transform Infrared) spectroscopy, a rapid analytical method used in tandem with advance statistical tools that showed excellent potential for the rapid and nondestructive discrimination of blood collected from Asian Elephant, Indian leopard, royal Bengal tiger, and human, showing a statistically confident discrimination among selected species from dry blood traces supported with predictive models (Sharma et al. 2023).

Correct quantification of analytical parameters is essential to ensure the good reputation of a forensic/wildlife laboratory. There is an essential need to validate spectrophotometric methods to establish their correct use in the laboratories and to ensure the traceability of the data obtained. Legal governmental agencies are created to verify analytical laboratories' performances and correct work and to explain analytical methods' validation guidelines (Huber 2007). Thermo Fisher company employs ISO/IEC 17025-accredited employees, and published guidelines on the use of ATR FTIR spectroscopy for the forensic tracing of blood stain aging are available (Thermofisher 2025). Wildlife forensic laboratories can follow these guidelines in order to validate and implement the ATR-FTIR blood-typing method. Few laboratories have validated the method, and concrete examples of its implementation are not available. Laboratories working in various domains such as blood lipid quantification, alcohol concentration analysis, or *Klebsiella pneumoniae* typing (Novais et al. 2024; Gau et al. 2025; Dini et al. 2021) follow their own validation and implementation procedures that cannot be directly applied to ATR-FTIR blood typing and should be tailored as a proper process.

7.4 Validation and Quality Assurance in Forensic and Wildlife Laboratories

Before the implementation of new blood-typing methods complemented with genomic analysis methods, a validation step is imperative to ensure the accurate measurement of the analyte, confirming that the obtained results are meaningful for their purpose. According to the FBI Quality Assurance Standards for DNA Databasing Laboratories (2025), validation studies should be conducted for any new typing test kit. These studies should include the characterization of the genetic marker, species specificity, sensitivity studies, stability studies, database-type sample analysis, population studies, mixture studies, and precision and accuracy studies. PCR-based studies are also required, including the optimization of reaction conditions, assessment of differential and preferential amplification, effects of multiplexing, assessment of appropriate controls, and product detection studies. A summary of different tests used for species determination is provided in Table 4.6. Validation studies shall be conducted by each laboratory with the appropriate sample number and type to demonstrate the reliability and potential limitations of each method. Validation data shall be used to establish quality assurance parameters and interpretation guidelines. Internal validation studies should include, as applicable, evaluation using known database-type samples, precision and accuracy studies,

Table 4.6 Summary of main methods used in species identification

Method	Principle	Application field	Sample requirements	Strengths	Limitations
ABO-RhD hemagglutination	Antigen–antibody interaction	Transfusion compatibility	Fresh/EDTA blood	Rapid, standardized, clinically validated	Limited to the identification of humans and primates
Precipitin test	Immune precipitation	Wildlife species ID	Serum/ antisera, extracted protein	Simple	Low specificity, reagent scarcity
Hemagglutination (lectin-based)	Serologic	Preliminary species screening	RBC	Inexpensive	High cross-reactivity
Raman spectroscopy	Spectroscopic	Forensic species ID	Dried bloodstain	Nondestructive, portable	Requires reference library
ATR-FTIR	Molecular bonds	Species discrimination from bloodstains	Dried/solid sample	Necessities a minimal preparation, rapid in use	Overlapping spectra in close species
Microfluidic immunoassay	Serologic/microfluidic	Field-based wildlife ID	Volumes of microliter order	Low waste, high sensitivity	Device dependency
MALDI-TOF MS	Protein profiling	Veterinary/forensic ID	Purified protein	High throughput, discriminative	Database-dependent
NGS (COI barcoding)	DNA sequencing	Gold-standard species ID	Extracted DNA	Highly informative, portable	Expensive

sensitivity and stochastic studies, and contamination assessment. All results must be documented, summarized, and approved by the person responsible for the procedure prior to implementation in database applications. The typing test kit used to generate a DNA type shall be checked against an appropriate and available certified reference material (or a sample made traceable to the certified reference material) prior to the implementation of the method for database analysis. The performance of a modified method should be evaluated in comparison to the original method using similar DNA samples; the evaluation should be documented, reviewed, and approved by a technical leader prior to the implementation of the modified method into database applications (SWGDAM 2025).

ISO/IEC 17025 is another standard used to test and calibrate laboratories, setting requirements for their competence, impartiality, and consistent operation to ensure they produce valid, reliable results. ISO/IEC 17025 accreditation is held by diverse forensic laboratories, including those specialized in the identification of meat, such as ICAR-NMRI (ICAR 2025).

8 Conclusion

Accuracy of species identification by blood-typing methods, now enhanced by molecular methods, is essential when providing a more adopted animal health assessment, ecological monitoring, wildlife preservation, and prevention of illegal trafficking. Blood typing has proven to be of critical importance in the prevention of illegal trafficking and in the control of bushmeat markets. Recent technological advances in the field by the introduction of additional modern DNA-based methods, such as whole genome sequencing, barcoding, and metabarcoding contribute to a better understanding of the speed and accuracy of species identification. These approaches help overcome some limitations of other analytical techniques, such as Raman spectroscopy and ATR-FTIR analysis, which may require necessary material and financial investment, species-specific antibody production, and can face challenges such as a lack of discriminatory power for closely related taxa and cross-reactivity issues.

References

Baruah D, Bharali A, Barua A, Devi G (2023) Distribution of Rh, Kell and Duffy blood group antigens and their phenotypic characteristics in blood donors of Barpeta District of Assam: a hospital based study from North East India. Int J Pharm Clin Res 15(1):581–587. https://ijpcr.com/volume15issue1/

Blais MC, Berman L, Oakley DA, Giger U (2007) Canine Dal blood type: a red cell antigen lacking in some Dalmatians. J Vet Intern Med 21(2):281–286. https://doi.org/10.1892/0891-6640(2007)21[281:cdbtar]2.0.co;2

Bull RW (1992) Inmunohematología. In: Halliwell REW, Gorman NT (eds) Inmunologia clinica veterinaria. Editorial Acribia S. A, Zaragoza

Chang YJ, Fan YH, Chen SC, Lee KH, Lou LY (2018) An automatic lab-on-disc system for blood typing. SLAS Technol 23(2):172–178. https://doi.org/10.1177/2472630317744732

Charpentier T, Petit T, Guidetti M, Goy-Thollot I (2020) The dog erythrocyte antigen 1 blood group in nondomesticated canids and compatibility testing between domestic dog and nondomesticated canid blood. J Vet Intern Med 34(6):2365–2373. https://doi.org/10.1111/jvim.15950

Chun S, Choi S, Yu H, Cho D (2019) *Cis*-AB, the blood group of many faces, is a conundrum to the novice eye. Ann Lab Med 39(2):115–120. https://doi.org/10.3343/alm.2019.39.2.115

Conti-Patara A, Ngwenyama TR, Martin LG, Wardrop KJ (2021) Dal-induced red blood cell incompatibilities in a Doberman Pinscher with von Willebrand factor deficiency and ehrlichiosis. J Vet Emerg Crit Care (San Antonio, Tex: 2001) 31(2):274–278. https://doi.org/10.1111/vec.13015

Cooper DK, Human PA, Rose AG, Rees J, Keraan M, Reichart B, Du Toit E, Oriol R (1989) The role of ABO blood group compatibility in heart transplantation between closely related animal species. An experimental study using the vervet monkey to baboon cardiac xenograft model. J Thorac Cardiovasc Surg 97(3):447–455

Corato A, Mazza G, Hale AS, Barker RN, Day MJ (1997) Biochemical characterization of canine blood group antigens: immunoprecipitation of DEA 1.2, 4 and 7 and identification of a dog erythrocyte membrane antigen homologous to human Rhesus. Vet Immunol Immunopathol 59(3–4):213–223

Dini I, Di Lorenzo R, Senatore A, Coppola D, Laneri S (2021) Comparison between mid-infrared (ATR-FTIR) spectroscopy and official analysis methods for determination of the concentrations of alcohol, SO2, and Total acids in wine. Separations 8(10):191. https://doi.org/10.3390/separations8100191

Ebelt AK, Fuchs S, Weber C, Müller E, Giger U (2020) Survey of blood groups *DEA 1, DEA 4, DEA 5, Dal*, and *Kai 1/Kai 2* in different canine breeds from a diagnostic laboratory in Germany. Front Vet Sci 7:85. https://doi.org/10.3389/fvets.2020.00085

Euler CC, Lee JH, Kim HY, Raj K, Mizukami K, Giger U (2016) Survey of two new (Kai 1 and Kai 2) and other blood groups in dogs of North America. J Vet Intern Med 30(5):1642–1647. https://doi.org/10.1111/jvim.14572

Gamble KC, Moyse JA, Lovstad JN, Ober CB, Thompson EE (2011) Blood groups in the Species Survival Plan®, European endangered species program, and managed in situ populations of bonobo (Pan paniscus), common chimpanzee (Pan troglodytes), gorilla (Gorilla ssp.), and orangutan (Pongo pygmaeus ssp.). Zoo Biol 30(4):427–444

Garratty G (2000) Blood groups and disease: a historical perspective. Transfus Med Rev 14(4):291–301

Gau TP, Wen JH, Lu IW, Huang PY, Lee YC, Lee WP, Lee HC (2025) Application of attenuated total reflection-Fourier transform infrared spectroscopy in semi-quantification of blood lipids and characterization of the metabolic syndrome. PLoS One 20(1):e0316522. https://doi.org/10.1371/journal.pone.0316522

Giger U, Bücheler J (1991) Transfusion of type-A and type-B blood to cats. J Am Vet Med Assoc 198(3):411–418

Giger U, Gelens CJ, Callan MB, Oakley DA (1995) An acute hemolytic transfusion reaction caused by dog erythrocyte antigen 1.1 incompatibility in a previously sensitized dog. J Am Vet Med Assoc 206:1358–1362. https://doi.org/10.2460/javma.1995.206.09.1358

Goulet S, Giger U, Arsenault J, Abrams-Ogg A, Euler CC, Blais MC (2017) Prevalence and mode of inheritance of the Dal blood Group in Dogs in North America. J Vet Intern Med 31(3):751–758. https://doi.org/10.1111/jvim.14693

Goy-Thollot I, Giger U, Boisvineau C, Perrin R, Guidetti M, Chaprier B, Barthélemy A, Pouzot-Nevoret C, Canard B (2017) Pre- and post-transfusion alloimmunization in dogs characterized by 2 antiglobulin-enhanced cross-match tests. J Vet Intern Med 31(5):1420–1429. https://doi.org/10.1111/jvim.14801

Gurkan M, Arikan S, Ozaytekin E, Dodurka T (2005) Titres of alloantibodies against A and B blood types in non-pedigree domestic cats in Turkey: assessing the transfusion reaction risk. J Feline Med Surg 7(5):301–305. https://doi.org/10.1016/j.jfms.2005.03.003

Hale AS (1995) Canine blood groups and their importance in veterinary transfusion medicine. Vet Clin North Am Small Anim Pract 25(6):1323–1332. https://doi.org/10.1016/s0195-5616(95)50157-3

Huang S, Wang P, Tian Y, Bai P, Chen D, Wang C, Chen J, Liu Z, Zheng J, Yao W, Li J, Gao J (2019) Blood species identification based on deep learning analysis of Raman spectra. Biomed Opt Express 10(12):6129–6144. https://doi.org/10.1364/BOE.10.006129

Huber L (2007) Validation and qualification in analytical laboratories, 2nd edn. CRC Press. https://doi.org/10.3109/9780849382680

ICAR (2025) https://www.icar.org.in/en/icar-nmri-meat-species-identification-laboratory-beacon-integrity-meat-authentication-and-food. Accessed on 11/12/2025

Jensen EA, Serhatlioglu M, Uyanik C, Hansen AT, Puthusserypady S, Dziegiel MH, Kristensen A (2023) Label-free blood typing by Raman spectroscopy and artificial intelligence. Adv Mater Technol 9(2). https://doi.org/10.1002/admt.202301462

Kehl A, Heimberger K, Langbein-Detsch I, Boehmer S, Raj K, Mueller E, Giger U (2018) Molecular characterization of blood type A, B, and C (AB) in domestic cats and a CMAH genotyping scheme. PLoS One 13(9):e0204287. https://doi.org/10.1371/journal.pone.0204287

Kehl A, Mueller E, Giger U (2019) CMAH genotyping survey for blood types A, B, and C (AB) in purpose-bred cats. Anim Genet 50(3):303–306. https://doi.org/10.1111/age.12778

Kermarrec N, Roubinet F, Apoil PA et al (1999) Comparison of allele O sequences of the human and non-human primate ABO system. Immunogenetics 49:517–526. https://doi.org/10.1007/s002510050529

Kim H, Han HJ (2025) Prevalence of dal blood type and dog erythrocyte antigens 1 and 4 among canine blood donors and recipients in Seoul, South Korea. BMC Vet Res 21(1):500. https://doi.org/10.1186/s12917-025-04927-1

Lee JH, Giger U, Kim HY (2017) Kai 1 and Kai 2: characterization of these dog erythrocyte antigens by monoclonal antibodies. PLoS One 12(6):e0179932. https://doi.org/10.1371/journal.pone.0179932

Mangiaterra S, Rossi G, Antognoni MT, Cerquetella M, Marchegiani A, Miglio A, Gavazza A (2021) Canine blood group prevalence and geographical distribution around the world: an updated systematic review. Animals 11(2):342. https://doi.org/10.3390/ani11020342

Melzer KJ, Wardrop KJ, Hale AS, Wong VM (2003) A hemolytic transfusion reaction due to DEA 4 alloantibodies in a dog. J Vet Intern Med 17(6):931–933

Monahan-Earley R, Dvorak AM, Aird WC (2013) Evolutionary origins of the blood vascular system and endothelium. J Thromb Haemost 11(Suppl 1):46–66. https://doi.org/10.1111/jth.12253

Nagahata Y, Masuda K, Nishimura Y, Ikawa T, Kawaoka S, Kitawaki T, Nannya Y, Ogawa S, Suga H, Satou Y, Takaori-Kondo A, Kawamoto H (2022) Tracing the evolutionary history of blood cells to the unicellular ancestor of animals. Blood 140(24):2611–2625. https://doi.org/10.1182/blood.2022016286

Novais AA, Magron HF (2018) Canine blood groups: a review. Sci Elec Arch 11(5):183–187. http://www.seasinop.com.br/revista/index.php?journal=SEA&page=ar ticle&op=view&path%5B%5D=761&path%5B%5D=pdf

Novais Â, Gonçalves AB, Ribeiro TG, Freitas AR, Méndez G, Mancera L, Read A, Alves V, López-Cerero L, Rodríguez-Baño J, Pascual Á, Peixe L (2024) Development and validation of a quick, automated, and reproducible ATR FT-IR spectroscopy machine-learning model for Klebsiella pneumoniae typing. J Clin Microbiol 62(2):e0121123. https://doi.org/10.1128/jcm.01211-23

Owaidah AY, Naffaa NM, Alumran A, Alzahrani F (2020) Phenotype frequencies of major blood group systems (Rh, Kell, Kidd, Duffy, MNS, P, Lewis, and Lutheran) among blood donors in the eastern region of Saudi Arabia. J Blood Med 11:59–65. https://doi.org/10.2147/JBM.S236834

Park Y, Kim SY, Koo SH, Lim J, Kim JM, Lim YA, Kwon GC (2017) Evaluation of the automated blood bank systems IH-500 and VISION max for ABO-RhD blood typing and unexpected antibody screening. Lab Med Online 7(4):170–175

Parkinson L, Kuder H, Kehl A, Koenig A, Good J, Hausmann J, Giger U (2024) Nondomestic felid *ABC* blood phenotyping, genotyping, and crossmatching. J Zoo Wildl Med 55(1):143–154. https://doi.org/10.1638/2023-0025

Proverbio D, Lubas G, Spada E, Medina Valentin AA, Viñals Florez LM, Del Rosario Perlado Chamizo M, Perego R, Pennisi MG, Ferro E, Baggiani L, Gavazza A, Blais MC (2020) Prevalence of Dal blood type and dog erythrocyte antigens (DEA) 1, 4, and 7 in canine blood donors in Italy and Spain. BMC Vet Res 16(1):126. https://doi.org/10.1186/s12917-020-02351-1

Ramis G, Martínez-Alarcon L, Quereda JJ et al (2013) Non-ABO blood group systems phenotyping in non-human primates for blood banking laboratory and xenotransplantation. Lab Anim 47(2):100–105. https://doi.org/10.1177/0023677213475439

Redman CM, Lee S, ten Huinink D, Rabin BI, Johnson CL, Oyen R, Marsh WL (1989) Comparison of human and chimpanzee Kell blood group systems. Transfusion 29(6):486–490. https://doi.org/10.1046/j.1537-2995.1989.2968931844

Rowe JA, Handel IG, Thera MA, Deans AM, Lyke KE, Koné A et al (2007) Blood group O protects against severe Plasmodium falciparum malaria through the mechanism of reduced rosetting. Proc Natl Acad Sci 104(44):17471–17476

Salvignol I, Calvas P, Socha WW, Colin Y, Le Van Kim C, Bailly P, Ruffié J, Cartron JP, Blancher A (1995) Structural analysis of the RH-like blood group gene products in nonhuman primates. Immunogenetics 41(5):271–281. https://doi.org/10.1007/BF00172151

Ségurel L, Thompson EE, Flutre T, Lovstad J, Venkat A, Margulis SW, Moyse J, Ross S, Gamble K, Sella G, Ober C, Przeworski M (2012) The ABO blood group is a trans-species polymorphism in primates. Proc Natl Acad Sci USA 109(45):18493–18498. https://doi.org/10.1073/pnas.1210603109

Ségurel L, Gao Z, Przeworski M (2013) Ancestry runs deeper than blood: the evolutionary history of ABO points to cryptic variation of functional importance. BioEssays 35:862–867. https://doi.org/10.1002/bies.201300030

Sharma CP, Sharma S, Singh R (2023) Species discrimination from blood traces using ATR FT-IR spectroscopy and chemometrics: application in wildlife forensics. For Sci Int Anim Environ 3(100060):2666–9374. https://doi.org/10.1016/j.fsiae.2022.100060

Silvestre-Ferreira AC, Pastor J, Almeida O, Montoya A (2004) Frequencies of feline blood types in Northern Portugal. Vet Clin Pathol 33(4):240–243. https://doi.org/10.1111/j.1939-165x.2004.tb00380.x

Sivaram S, Bhatnagar RK, Bhandari SK, Sijher JS (1975) Differentiation between stains of human blood and blood of monkey. Forensic Sci 6(3):145–152

Socha WW (1997) Simian-type blood groups of nonhuman primates. In: Molecular biology and evolution of blood group and MHC antigens in primates. Springer Berlin Heidelberg, Berlin/Heidelberg, pp 254–267

Spada E, Proverbio D, Viñals Flórez LM, Del Rosario Perlado Chamizo M, Serra Y, Gómez de la Serna B, Perego R, Baggiani L (2016) Prevalence of naturally occurring antibodies against dog erythrocyte antigen 7 in a population of dog erythrocyte antigen 7-negative dogs from Spain and Italy. Am J Vet Res 77(8):877–881. https://doi.org/10.2460/ajvr.77.8.877

Spada E, Proverbio D, Priolo V, Ippolito D, Baggiani L, Perego R, Pennisi MG (2017) Dog erythrocyte antigens (DEA) 1, 4, 7 and suspected naturally occurring anti-DEA 7 antibodies in Italian Corso dogs. Vet J (London, England: 1997) *222*:17–21. https://doi.org/10.1016/j.tvjl.2017.02.008

Spada E, Perego R, Viñals Flórez LM, Del Rosario Perlado Chamizo M, Baggiani L, Dall'Ara P, Proverbio D (2018) Comparison of cross-matching method for detection of DEA 7 blood incompatibility. J Vet Diagn Invest 30(6):911–916. https://doi.org/10.1177/1040638718806023

Swart KS, Wilks CR (1982) An immunodiffusion method for the identification of the species of origin of meat samples. Aust Vet J 59(1):21–22. https://doi.org/10.1111/j.1751-0813.1982.tb02705.x

SWGDAM (2025) https://www.swgdam.org/publications. Accessed on 11/12/2025

Swisher SN, Young LE (1961) The blood grouping systems of dogs. Physiol Rev 41:495–520. https://doi.org/10.1152/physrev.1961.41.3.495

Tasker S, Barker EN, Day MJ, Helps CR (2014) Feline blood genotyping versus phenotyping, and detection of non-AB blood type incompatibilities in UK cats. J Small Anim Pract 55(4):185–189. https://doi.org/10.1111/jsap.12180

Terao K, Hiyaoka A, Cho F, Honjo S (1988) The simian-type M and the human-type ABO blood groups in the African green monkey (Cercopithecus aethiops): their inheritance, distribution and significance for the management of a breeding colony. Lab Anim 22(4):347–354. https://doi.org/10.1258/002367788780746151

Thengchaisri N, Sinthusingha C, Arthitwong S, Sattasathuchana P (2017) Comparative serological investigation between cat and tiger blood for transfusion. J Vet Med Sci 79(6):1081–1085. https://doi.org/10.1292/jvms.16-0630

Thermofisher (2025) https://documents.thermofisher.com/TFS-Assets/MSD/Application-Notes/atr-ftir-spectroscopy-forensic-tracing-blood-stain-age-an56372.pdf. Accessed on 11/12/2025

Vieira SM, Ferreira RRF, de Matos AJ et al (2017) Distribution of feline AB blood types: a review of frequencies and its implications in the Iberian Peninsula. J Feline Med Surg Open Rep 3(2). https://doi.org/10.1177/2055116917727693

Weinstein NM, Blais MC, Harris K, Oakley DA, Aronson LR, Giger U (2007) A newly recognized blood group in domestic shorthair cats: the Mik red cell antigen. J Vet Intern Med 21(2):287–292. https://doi.org/10.1892/0891-6640(2007)21[287:anrbgi]2.0.co;2

Weiss DJ, Wardrop KJ (eds.) (2011) Schalm's veterinary hematology. John Wiley & Sons

Westhoff CM, Floch A (2025) Blood group genotype matching for transfusion. Br J Haematol 206(1):18–32. https://doi.org/10.1111/bjh.19664

Yamamoto M, Lin XH, Kominato Y, Hata Y, Noda R, Saitou N, Yamamoto F (2001) Murine equivalent of the human histo-blood group ABO gene is a cis-AB gene and encodes a glycosyltransferase with both A and B transferase activity. J Biol Chem 276(17):13701–13708

Yugovich O, Bunce M, Harbison SA (2025) Point-of-need species identification using non-PCR DNA-based approaches to combat wildlife crime. Forensic Sci Int: Genet 78:103278. https://doi.org/10.1016/j.fsigen.2025.103278

Zaremba R, Brooks A, Thomovsky E (2019) Transfusion medicine: an update on antigens, antibodies and serologic testing in dogs and cats. Top Companion Anim Med 34:36–46. https://doi.org/10.1053/j.tcam.2018.12.005

Zhan T, Niu H, Huang Y, Min S, Wu X, Wang H, Xu B (2025) 3D microfluidic analytical device on a single thread for smart point-of-care blood typing. Lab Chip 25(11):2769–2779

Forensic Investigation of Blood Transfusion: Medical Errors, Detection, and Legal Accountability

5

Abstract

Blood transfusion is a vital part of modern medicine, widely used in emergency care, surgery, and chronic illness management. But despite its lifesaving potential, transfusions can lead to serious harm if safety protocols are not strictly followed. Errors—ranging from mismatched blood types to administrative oversights—can cause severe complications, including hemolytic reactions and organ failure. This chapter explores how forensic investigations help uncover the causes behind such events by combining clinical records, autopsy results, and legal analyses. It also examines the role of regulatory bodies, ethical challenges, and emerging technologies like AI and blockchain in improving transfusion safety. Using real-world case studies and policy comparisons, the review offers a comprehensive look at the forensic, clinical, and legal dimensions of blood transfusion errors.

Keywords

Forensic hematology · Blood transfusion errors · Medical malpractice · Hemovigilance · Medico-legal investigation

1 Introduction

Blood transfusion remains one of the most essential interventions in clinical medicine, playing a pivotal role in a variety of settings, from emergency trauma care and complex surgical procedures to the long-term management of chronic conditions such as anemia and cancer (Stainsby 2005). Its therapeutic potential, while well established, comes with significant risks when standard protocols are not strictly observed. Mistakes such as mismatched blood types, incorrect sample labeling, or patient misidentification can result in grave outcomes, including acute hemolytic reactions, the transmission of infectious agents, multi-organ failure, or even death (Linden and Kaplan 1994; Wecht 2007). These risks are amplified in environments

H. R. Dash et al., *Forensic Investigation of Blood Evidence*,
https://doi.org/10.1007/978-981-95-9183-1_5

characterized by urgency and complexity—emergency departments and operating rooms, for example—where medical decisions are made rapidly and under pressure.

Given the high stakes, any unexpected deterioration or death following a transfusion often prompts a forensic investigation. These investigations are designed to uncover the precise series of events that led to the adverse outcome, evaluate compliance with clinical standards, and assess whether the incident was a consequence of medical negligence or systemic error (Madea 2022). Typically, such investigations are interdisciplinary, drawing on expertise from forensic pathologists, hematologists, legal consultants, and ethical review bodies. One of the primary forensic tools used is the autopsy, which may reveal physical signs of transfusion-related injury—such as hemolysis, clotting abnormalities, or inflammatory responses—that can confirm or refute clinical suspicions (Sobh and Zaki 2024).

In addition to pathological findings, documentary evidence plays a key role. Investigators will review transfusion records to determine whether the correct blood product was ordered, how it was matched, who approved its use, and whether bedside verification was completed properly. Laboratory crossmatch logs, nursing documentation, and incident reports all contribute to a full reconstruction of the clinical timeline. A failure at any one of these checkpoints could signal a lapse in safety practices. For example, the absence of dual verification at the bedside or errors in sample identification could be critical factors leading to a patient's decline (Nitsche et al. 2023).

The legal implications of these findings can be profound. Civil litigation is common, especially when the patient or their family can demonstrate a deviation from the standard of care. In such cases, plaintiffs typically argue that the hospital or individual provider breached their duty of care and that this breach was directly responsible for the harm suffered (Pal et al. 2020). These claims fall under the broader category of medical malpractice and often hinge on detailed forensic evidence. In some situations—particularly when the error was egregious or involved deliberate concealment—criminal charges may also be filed. The threshold for criminal liability is higher, often requiring proof of gross negligence or willful disregard for patient safety (Kazarian 2020).

The intersection between forensic analysis and legal responsibility is crucial. On one hand, forensic science provides an evidence-based foundation to determine whether a transfusion error occurred and, if so, who was responsible. On the other hand, it informs legal proceedings and often influences policy reforms aimed at preventing recurrence. Forensic findings, especially when supported by autopsy and hematologic evidence, can be central to court cases and disciplinary hearings. The ripple effects of a single case can lead to changes in national guidelines, institutional protocols, and staff training procedures.

Errors in blood transfusion are generally grouped into three primary categories: clinical, technical, and administrative. Clinical errors include decisions to transfuse when they are not medically indicated or when important patient history such as previous transfusion reactions is omitted. Technical errors occur in laboratory settings and often involve mistakes in blood typing, cross-matching, or component storage. Administrative errors—arguably the most preventable—stem from issues like mislabeling, patient misidentification, or improper handling and tracking of blood units (Linden and Kaplan 1994). These human factors, whether isolated or systemic, are frequently at the root of transfusion-related complications.

One of the most dangerous yet under-recognized scenarios involves the transfusion of ABO-incompatible blood. This type of mismatch can cause an acute hemolytic transfusion reaction, where the recipient's immune system attacks the transfused red blood cells. Clinical signs may include fever, back pain, hypotension, and hemoglobinuria, with possible progression to renal failure, shock, or death within hours (Wecht 2007). Unfortunately, such symptoms are often overlooked or attributed to the patient's underlying condition, especially in complex cases involving trauma or critical illness. As a result, the true cause may only be revealed during postmortem analysis.

When such events lead to a forensic inquiry, the first step is often a detailed review of the clinical and procedural sequence. Investigators work to determine who was involved at each stage—from ordering the blood to administering it—and whether proper checks were followed. They examine bedside identification practices, protocol adherence, and equipment functionality. Blood samples are often re-analyzed for cross-match compatibility, and autopsy findings provide insight into whether hemolysis or thrombotic damage occurred (Zaami et al. 2021).

However, isolated clinical actions do not tell the whole story. Forensic investigators must also consider systemic issues that may have contributed to the error. These include environmental stressors such as understaffing, shift changes, or communication breakdowns between departments. Deviations from standard operating procedures—either due to oversight or intentional shortcutting—are especially important in establishing institutional accountability. In jurisdictions with established hemovigilance frameworks, such events are automatically reported and investigated, offering a layer of quality assurance. However, the effectiveness of these systems varies greatly by country and hospital policy (Sakai et al. 2010; De Vries et al. 2011).

The consequences of transfusion-related errors extend beyond the immediate medical outcomes. From a legal perspective, institutions and individuals may face sanctions ranging from mandatory retraining and financial penalties to license suspension or imprisonment. The distinction between civil and criminal responsibilities often rests on the intent and severity of the breach. In civil law, negligence is sufficient grounds for liability. In criminal cases, prosecutors must often prove recklessness or intent to deceive—a far higher bar (Kazarian 2020).

One illustrative example comes from Japan, where a now-infamous case—the so-called "Yamamoto case"—brought national attention to transfusion practices. In that instance, a patient received an excessive volume of transfused blood due to poor communication among medical staff. The resulting cardiac overload led to the patient's death. Forensic investigation confirmed that inadequate monitoring and failure to adhere to transfusion thresholds were contributing factors. The legal outcome included disciplinary measures against the healthcare providers and a subsequent overhaul of hospital transfusion policies (Van Grunsven 1997).

Cases like these underscore the high stakes of blood transfusion management. Though errors are relatively rare given the volume of transfusions performed globally, their consequences can be devastating. Forensic investigations help prevent such tragedies by revealing not only what went wrong but also why and how future systems can be strengthened. The lessons drawn from each case contribute to a growing body of knowledge that drives policy reform, enhances clinical vigilance, and ultimately protects patient safety.

2 Overview of Blood Transfusion Practices

Blood transfusion is a cornerstone of clinical care, widely used across surgical operations, trauma interventions, cancer therapy, obstetric emergencies, and the management of hematological disorders. Its primary objective is to restore or maintain essential blood components—such as red blood cells (RBCs), plasma, platelets, and clotting factors—to support oxygen delivery, coagulation, and circulatory stability. While the process may appear routine, it is a highly structured and sensitive medical intervention that demands strict adherence to protocols at every step.

The transfusion process begins with donor recruitment and screening. Donors, usually volunteers, undergo thorough health checks to eliminate risks of transmissible diseases such as HIV, Hepatitis B and C, syphilis, and malaria. Once collected, the blood is separated into its components and stored under precisely controlled conditions: red cells are kept at 1–6 °C, platelets at room temperature with constant agitation, and plasma is frozen to maintain the activity of clotting factors. Any deviation in these conditions can compromise the safety or therapeutic effect of the product (Stainsby 2005).

Before any transfusion, blood typing and cross-matching are conducted to ensure ABO and Rh compatibility. Cross-matching involves combining a patient's plasma with donor red cells to check for immune reactions. Even a minor mismatch—such as administering type A blood to a type B recipient—can provoke an acute hemolytic reaction, potentially leading to kidney failure or fatal shock (Linden and Kaplan 1994).

Equally critical is bedside identification. Clinical staff are required to verify patient identity through barcode scanning, verbal confirmation, and documentation. Many institutions mandate a dual verification process involving two trained professionals. Despite these systems, administrative errors—particularly patient misidentification—remain the most frequent cause of transfusion-related harm (Wecht 2007).

Hospitals rely on hemovigilance systems to monitor adverse reactions and improve safety through structured reporting (de Jonge et al. 2022). These programs are supported by transfusion committees composed of physicians, nurses, pathologists, and quality control staff, who assess compliance and investigate deviations (Mansfield et al. 2009).

Ensuring transfusion safety depends on robust clinical governance, including ongoing training, digital tracking systems, periodic audits, and decision-support

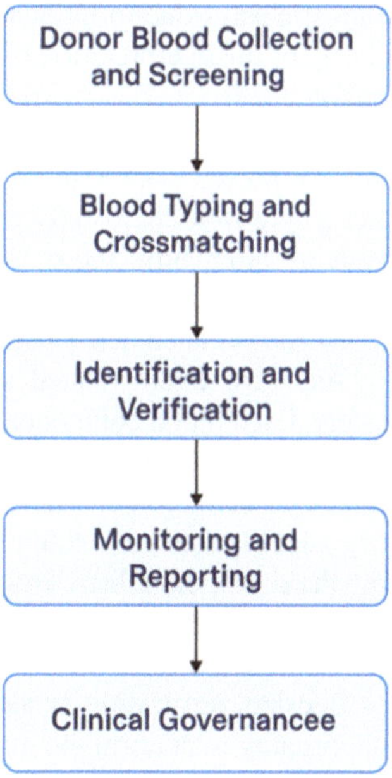

Fig. 5.1 Key stages in blood transfusion practice

technologies. High-performing centers often employ simulation-based education and recurring staff evaluations to minimize errors.

Given its high frequency and potential for harm, transfusion must be treated as a critical procedure—not a routine one. Every stage, from donor screening to post-transfusion monitoring, plays a vital role in safeguarding patients and minimizing risk. In Fig. 5.1, the flowchart illustrates the critical steps in safe blood transfusion, from donor screening through clinical governance. Each stage—collection, cross-matching, verification, monitoring, and oversight—plays a vital role in preventing errors and ensuring patient safety.

3 Types and Causes of Transfusion Errors

Despite being a lifesaving intervention, blood transfusion is susceptible to a range of preventable errors. These mistakes can occur at any point in the process—from collecting and labeling samples to storing, cross-matching, or administering blood at the bedside. Forensic investigations often focus on understanding the nature and cause of these errors to establish clinical accountability and potential malpractice.

One of the most serious and well-documented mistakes is the administration of mismatched blood. An incompatibility in ABO blood types can result in an acute hemolytic transfusion reaction (AHTR), a condition marked by fever, chills, hypotension, and renal failure. Such reactions are typically fatal if not quickly identified and treated. The underlying mechanism involves the destruction of donor red cells by the recipient's immune system. Research indicates that ABO mismatches are almost always due to human error—such as incorrect labeling, failure to double-check, or misinterpretation of blood-typing results (Stainsby et al. 2005; Rossi and Simon 2022).

Clerical mistakes, including misidentification of patients, incorrect sample labeling, or flawed data transcription, represent another significant source of risk. According to a 2023 study by Nitsche, Dreßler, and Henschler, passive reporting systems often miss these "handling errors," leading to chronic underreporting. Administrative mistakes are so widespread that they account for more than half of transfusion-related near-misses in certain hospitals (Barnhard 2020).

Technical errors related to storage and handling can also jeopardize patient safety. Each blood component has specific storage criteria: red cells require refrigeration, platelets need room temperature and constant agitation, and plasma must be frozen. If these conditions are not met, the quality and safety of the product may degrade. Platelets, for example, may clump when agitation is insufficient, increasing the risk of embolism. Bolcato et al. (2020) emphasize that poor inventory management, outdated equipment, and untrained personnel all contribute to these technical failures.

Bedside verification is another critical point of failure. This step is the final opportunity to confirm the match between the patient and the blood unit. However, staff under pressure may bypass protocols. Even with digital safeguards in place, scanning errors, verbal confirmation lapses, or procedural fatigue can result in transfusing the wrong patient. Phou and Klein (2025) report that such oversights are among the most frequent—and preventable—errors, particularly in high-stakes clinical environments.

In emergencies, clinicians may initiate uncross-matched transfusions using O-negative blood to save time. Although this is generally safe, it carries risks, particularly for patients with rare antibodies or a history of alloimmunization. Sakai et al. (2010) and De Vries et al. (2011) warn that delayed hemolytic reactions can develop days after transfusion and are often only recognized through forensic review or hemovigilance monitoring.

Systemic underreporting further compounds the issue. Many institutions fail to document near-misses and nonfatal events, impeding institutional learning. Although hemovigilance programs aim to detect and mitigate these errors, their effectiveness is inconsistent globally. Regulatory gaps, staff fear of blame, and cultural resistance often deter full transparency (Orkuma and Ayla 2014; Jones and Copplestone 2009).

In summary, transfusion errors span clinical, technical, and administrative domains. Most of them are avoidable with proper safeguards, yet when failures occur, forensic examination becomes essential for determining the cause, responsibility, and reform.

Table 5.1 outlines the major types of transfusion errors, their underlying causes, associated clinical consequences, and their relevance in forensic investigations. It highlights how preventable mishandlings—such as clerical mistakes or skipped verification steps—can result in severe patient harm and legal accountability.

Table 5.1 Summary of common blood transfusion errors

Error type	Primary cause	Clinical consequences	Forensic relevance
Mismatched blood type	ABO/Rh incompatibility due to mislabeling or lab error	Acute hemolytic reaction, renal failure, death	High—Confirms negligence or fatal mismatch
Clerical errors	Patient misidentification or documentation errors	Wrong transfusion, delayed treatment, adverse reactions	Moderate-to-high—traceable via logs and witness
Improper storage	Deviation from temperature or handling protocols	Sepsis, clotting, reduced efficacy	Moderate—may involve equipment or staff failures
Lack of bedside verification	Skipped ID check or barcode scan	Transfusion to wrong patient, fatal outcomes	High—patient safety breach, liability likely
Emergency uncross-matched transfusion	Use of O-negative blood without antibody screen	Delayed hemolysis, alloimmunization	Moderate—context-specific, sometimes justified

4 Forensic Role in Detecting Transfusion Errors

The forensic examination of deaths or adverse events linked to blood transfusions serves a dual purpose. It not only aims to determine the medical cause of death but also seeks to uncover any lapses in clinical practice that may carry legal or institutional implications. When a patient experiences sudden or unexplained deterioration following a transfusion, particularly where a transfusion reaction is suspected but not clearly documented, forensic evaluation becomes essential.

A foundational step in any such investigation involves the preservation and review of all transfusion-related records. Investigators usually begin by collecting the complete chain-of-custody documentation, including blood requisition forms, cross-matching results, nursing notes, and blood bank logs. These records help reconstruct the procedural timeline: Was the correct blood type issued? Was it stored appropriately? Were identification protocols strictly followed at the bedside? Discrepancies in these materials often point to clerical mistakes or procedural failures, such as mislabeled units or incompatible blood administration (Nitsche et al. 2023).

In cases where documentation proves inconclusive or incomplete, autopsy findings become critical. Forensic pathologists examine the body for hallmarks of acute hemolytic transfusion reactions (AHTRs). These include dark-red or cola-colored urine, generalized jaundice, enlargement of the spleen, and histological signs such as renal tubular necrosis or systemic inflammation. In some instances, the only irrefutable evidence of a transfusion-related fatality comes from pathological findings, particularly when the medical records are ambiguous or contradictory (Madea 2022).

Beyond the gross and histopathological examinations, specialized subfields like forensic toxicology and immunohematology offer key insights. These disciplines may identify elevated levels of free hemoglobin, bilirubin, or previously undetected alloantibodies—substances that may have been missed in the original compatibility testing. Advanced forensic analysis can also match DNA from transfused red blood cells with donor records, allowing investigators to verify whether the patient received blood from the intended source. This is especially valuable in cases of suspected unit mislabeling or accidental switch (Zaami et al. 2021).

Forensic teams also examine contextual and procedural variables within the hospital setting. Key questions include: Was the transfusion performed under emergency conditions where standard protocols were bypassed? Was the medical staff adequately trained and supervised? Did verification steps fail under time constraints or due to fatigue? Understanding the environment and human factors surrounding the event helps distinguish between individual error, systemic failure, and potential negligence. Investigators may also conduct interviews with hospital personnel, examine shift schedules, and, in rare cases, review surveillance footage to construct a minute-by-minute chronology.

Modern forensic investigations now benefit from a wide array of diagnostic and analytical tools that significantly enhance the accuracy and reliability of conclusions drawn in transfusion-related cases. Among the most impactful are the following:

- *Hemovigilance systems*, which are structured frameworks designed to log and analyze transfusion-related incidents. These databases help forensic professionals identify broader patterns and systemic risks by providing historical data on similar cases within an institution or across a national registry (De Vries et al. 2011).
- *DNA profiling*, used to detect mismatches between transfused blood and intended recipients. This tool is especially useful in cases where labeling errors are suspected or chain-of-custody documentation is unreliable (Zaami et al. 2021).
- *Immunohematological assays*, which can identify unexpected antibodies that may have triggered delayed hemolytic reactions. These reactions can develop over several days and are often missed without specialized testing.
- *Cross-match incompatibility reviews*, which revisit the laboratory data produced before transfusion. Often, errors in this phase—such as technical oversights or misinterpreted test results—go unnoticed until forensic analysis reveals the mismatch.
- *Autopsy with histopathology*, which remains the gold standard for forensic examination. Detailed microscopic analysis of the kidneys, liver, spleen, and vascular tissues can uncover evidence of immune-mediated damage such as hemolysis, thrombosis, or widespread inflammation (Zaami et al. 2021; Perper et al. 1993; Sobh and Zaki 2024).

When used in tandem, these tools offer a powerful framework for confirming or ruling out transfusion-related complications. They also enhance the legal defensibility of findings, ensuring that investigative conclusions are not only medically sound but also admissible in court or institutional proceedings.

An increasingly valuable resource for investigators is hemovigilance data. These systems allow cross-referencing of past incidents, revealing whether certain staff members or departments have histories of repeated transfusion errors. They can also identify recurring patterns of the same error across institutions. This type of meta-data plays a vital role in informing both policy reform and litigation (De Vries et al. 2011).

Ultimately, the forensic process is designed to answer three central questions. First, did the transfusion contribute directly to the adverse outcome or death? Second, if so, could that outcome have been prevented through adherence to established clinical protocols? And third, is there sufficient evidence to hold an individual practitioner or institution accountable?

By addressing these questions, forensic investigations serve more than a diagnostic role—they promote systemic learning, support regulatory oversight, and offer closure to affected families. In this way, the fusion of forensic medicine and transfusion science becomes an essential safeguard in modern healthcare.

5 Medico-Legal Implications

When a blood transfusion error results in patient injury or death, the consequences extend far beyond clinical concerns. These incidents often become matters of legal scrutiny, particularly when evidence suggests the error could have been avoided. Whether the issue involves mismatched blood, missing documentation, or insufficient oversight, such events frequently lead to medico-legal investigation. At the heart of these inquiries lies a pivotal question: Did the care provided meet acceptable medical standards?

In civil law, transfusion-related incidents commonly lead to tort claims for medical negligence. For such claims to succeed, plaintiffs generally must prove three elements: that a duty of care existed, that the duty was breached, and that this breach caused harm. This framework is rooted in the Bolam test, a legal standard used in many common law jurisdictions to assess medical negligence (Pal et al. 2020). For example, if a healthcare provider failed to verify patient identity before administering blood, or disregarded standard cross-matching protocols, they may be found legally liable for the resulting harm.

Criminal responsibility, while less frequent, may apply in cases of gross negligence or willful disregard for patient safety. Acts such as knowingly administering unverified blood, falsifying records, or repeatedly ignoring safety procedures may constitute criminal negligence or misconduct. In some jurisdictions, this can lead to charges of manslaughter or professional negligence (Kazarian 2020). The line between civil and criminal liability is typically drawn based on intent and the severity of deviation from standard practices.

Legal interpretations and enforcement differ globally. In the United Kingdom, the General Medical Council (GMC) may suspend or revoke licenses even in the absence of criminal prosecution. In the United States, malpractice suits often result in financial compensation and may include corrective measures such as retraining. By contrast, in many developing nations, legal protection for patients is limited, and accountability for transfusion-related malpractice may be rare due to insufficient regulatory infrastructure (Orkuma and Ayla 2014).

Increasingly, legal accountability extends to healthcare institutions. Hospitals can be sued for systemic issues, including inadequate training, poor equipment maintenance, or failure to enforce verification protocols. Courts have acknowledged that transfusion errors are frequently the result of institutional shortcomings, not just individual oversight (Mansfield et al. 2009). In such cases, institutions may be compelled to implement new policies, improve monitoring systems, or publicly report outcomes.

Legal systems also play a preventive role by promoting risk mitigation strategies. Many hospitals now operate transfusion committees, not only to ensure clinical standards but also to manage legal exposure. These groups routinely review adverse events and near-misses in consultation with legal and forensic professionals. In some regions, reporting near-miss events to national hemovigilance systems is mandatory, supporting policy development and broader public health monitoring (De Vries et al. 2011).

However, this intersection of medicine and law also raises ethical concerns. Fear of litigation may lead clinicians to practice defensive medicine—ordering unnecessary tests or avoiding high-risk procedures—to protect themselves legally. This can escalate healthcare costs and sometimes compromise patient care. Moreover, fear of legal consequences may deter staff from reporting errors, hindering institutional learning and safety improvements (Stainsby et al. 2005).

In conclusion, the medico-legal implications of transfusion errors are multifaceted and depend on the specifics of each case. Forensic investigation plays a crucial role in bridging clinical facts with legal standards, supporting both justice and patient safety reform.

6 Case Studies in Transfusion Malpractice

Real-life cases offer some of the clearest illustrations of how seemingly minor errors—whether administrative, procedural, or technical—can lead to tragic and preventable outcomes. These events not only highlight individual lapses but often uncover deeper systemic issues, making them important references for both legal accountability and clinical reform.

One widely cited case is the *Yamamoto case* in Japan, which brought national attention to the risks of over-transfusion and inadequate communication among healthcare teams. In this incident, an elderly patient was given multiple units of blood in rapid succession, far exceeding what the patient's body could safely handle. The transfusion was administered without proper physiological monitoring or reassessment between units. Shortly afterward, the patient developed acute pulmonary edema and suffered cardiac failure. Forensic analysis confirmed that volume overload had played a direct role in the death. The legal investigation led to disciplinary actions against several staff members and prompted the hospital to adopt new safeguards—including tighter transfusion thresholds and more rigorous fluid monitoring during emergency care (Van Grunsven 1997).

Another notable case involved a pathology lab that mislabeled a unit of blood. Although the patient's name on the label was correct, the recorded blood type was wrong. A patient with blood group A was mistakenly transfused with group B red cells. The result was a rapid-onset hemolytic reaction, which progressed to multi-organ failure. Forensic investigation revealed that the mistake occurred due to multiple breakdowns: no barcode scanning was used, and verification steps were skipped. Legal proceedings determined that institutional processes had failed at several critical points, and the hospital was held liable. As a result, nationwide reforms were introduced in labeling practices and bedside verification procedures (Van Grunsven 1997).

What these examples show is that transfusion errors often stem not from malice or incompetence but from gaps in communication, insufficient checks, and system-level oversights. Yet, the consequences can be catastrophic. That is why such cases are frequently used as teaching points in both clinical education and legal training—they underscore how even routine procedures, when handled carelessly, can end in irreversible harm.

They also have wider ripple effects. These forensic findings influence not only the outcome of individual malpractice suits but also help shape national and institutional policies on transfusion safety. Policies on staff training, unit verification, and documentation practices have all evolved as a result of these landmark cases. In this way, forensic casework becomes a feedback loop: it exposes vulnerabilities and helps ensure that the same mistakes are less likely to be repeated.

7 Ethical Considerations

Although blood transfusion is considered a routine medical procedure, it brings with it a range of ethical questions—especially when things go wrong. These concerns intersect with law, personal belief systems, informed consent, and broader public health responsibilities. In forensic contexts, ethical missteps—such as failure to disclose risks or obtain proper consent—can deepen the consequences of clinical error and become key issues in investigations.

One of the core ethical principles in transfusion medicine is the *requirement for informed consent*. Patients must be clearly told why a transfusion is needed, what the potential risks and benefits are, and whether any alternatives exist. In emergencies, when patients are unconscious or unable to communicate, clinicians may rely on implied consent. But if documentation of this decision is missing—especially when complications arise—it can lead to ethical and legal scrutiny during a forensic review (Pal et al. 2020).

Another ethical obligation is *transparency about risks*. Even though transfusions today are generally safe, patients still face potential complications such as transfusion-transmitted infections (TTIs), allergic reactions, alloimmunization, or acute hemolytic events. Ethically, healthcare providers must fully disclose these possibilities, even if the chances are low. Respecting a patient's autonomy also includes honoring their right to refuse blood products, provided they understand the consequences and are competent to decide.

Donor screening and selection practices raise additional ethical challenges. Blood banks must strike a balance between respecting donor privacy and protecting recipients. For example, donors from high-risk groups—such as those who recently traveled to malaria-endemic regions or have a history of intravenous drug use—must be carefully assessed. If TTIs are transmitted because of oversight or falsified donor histories, this can carry both ethical and forensic implications (Mansfield et al. 2009).

Religious objections to transfusion present one of the most complex ethical areas. Some adult patients, particularly Jehovah's Witnesses, refuse blood products on religious grounds. These choices are typically respected under ethical principles of autonomy. However, complications arise when parents refuse transfusions for their children. In such cases, many legal systems invoke the doctrine of *parens patriae*, allowing courts to override parental authority in order to protect the child's life. When a child dies after a refused transfusion, forensic investigators must weigh whether the medical team did enough to intervene while also considering the family's rights and beliefs (Kazarian 2020).

Equity also poses ethical concerns. In rural or low-resource healthcare settings, patients may receive poorly screened or mismatched blood due to limited access or substandard infrastructure. If a fatal event occurs in such circumstances, forensic experts must examine whether the outcome was a result of unavoidable limitations—or preventable negligence stemming from poor planning or policy gaps.

Ultimately, practicing ethical transfusion medicine goes beyond simply following written protocols. It requires thoughtful engagement with patients, transparent communication, and institutional cultures that prioritize informed decision-making. In difficult cases—especially those that result in harm—ethics boards, national guidelines, and clear documentation become essential tools for navigating the balance between patient rights, provider responsibility, and public health.

8 Global Regulatory Frameworks and Policies

Blood transfusion practices around the world are governed by different regulatory frameworks, each with the shared goal of maximizing safety and minimizing risk. However, the specifics of how these frameworks are implemented—and how effective they are—vary greatly depending on the region, resources, and governing bodies involved.

In the *United States*, the Food and Drug Administration (FDA) and the American Association of Blood Banks (AABB) play key roles in enforcing safety standards. Their regulations are legally binding and cover everything from donor screening and labeling procedures to adverse event reporting and facility accreditation. Institutions must comply with these standards not just to ensure patient safety but also to maintain licensure. Hemovigilance systems are mandatory, and reporting of adverse reactions is required at the national level. While this approach promotes consistency and traceability, it can be costly and demanding—especially for smaller hospitals, where meeting every compliance metric may pose logistical challenges (Mansfield et al. 2009).

In the *European Union*, transfusion practices are regulated under the *EU Blood Directive*, which aims to harmonize safety standards across all member states. The directive requires traceability, proper documentation, and standardized procedures for blood collection and usage. However, enforcement is left to each member state, and this has led to significant variation in how thoroughly the rules are applied. Some countries have well-established monitoring systems and excellent compliance, while others struggle due to resource constraints or fragmented healthcare structures. Still, the EU framework offers a useful model for regional coordination, especially when it comes to data sharing across borders (De Vries et al. 2011).

For *low- and middle-income countries*, the *World Health Organization (WHO)* offers a set of guidelines designed to promote transfusion safety on a global scale. These are not legally binding but rather advisory in nature, aiming to help countries develop their own systems based on best practices. The WHO emphasizes voluntary blood donation, adequate testing, safe storage, and proper training of healthcare workers. However, without the legal authority to enforce compliance, adoption depends largely on local political will, funding, and infrastructure. In many regions, these remain significant barriers to full implementation (Mansfield et al. 2009).

Each framework includes some form of *hemovigilance*—a system for detecting, reporting, and analyzing transfusion-related incidents. In high-income countries, these systems are typically well integrated into hospital operations and national reporting databases. In contrast, in lower resource settings, hemovigilance often lacks the technical infrastructure, staff training, or legal backing needed to be fully effective. This gap can make it harder to track and prevent errors, and it also weakens the forensic process in the event of adverse outcomes.

To better understand how these frameworks compare, Table 5.2 outlines their focus areas, enforcement models, hemovigilance requirements, and key challenges. While all three prioritize safety, their success depends heavily on how well they are enforced and supported within the healthcare system.

In summary, the global effort to ensure safe transfusion practices reflects a common objective—but the path toward that goal differs widely depending on geography and resources. Strong regulatory frameworks, when enforced consistently, play a crucial role in preventing errors, supporting forensic investigations, and promoting accountability across healthcare systems.

Table 5.2 Comparative review of international blood transfusion policy frameworks

Policy framework	Core focus	Enforcement mechanism	Hemovigilance mandate	Challenges
FDA and AABB (USA)	Standardized procedures, product labeling, donor screening, hemovigilance, and licensing	Legally binding through FDA regulation; AABB standards used for accreditation	Mandatory reporting of adverse events and near-misses to national systems	High compliance costs; variation in adherence among institutions
EU blood directive	Harmonization of standards across EU states, traceability, and safety reporting	Binding on member states; enforcement varies depending on national implementation	Mandatory across EU with national and EU-level data sharing	Uneven implementation across EU states; resource constraints
WHO guidelines	Global best practices for low- and middle-income countries, safety promotion, and donor care	Nonbinding; recommendations to be adopted voluntarily by countries	Encouraged but not enforced; relies on member state initiative	Limited adoption in low-resource settings; lack of infrastructure

9 Challenges and Strategic Solutions in Transfusion Error Investigation

Investigating errors in blood transfusion is rarely straightforward. These cases often sit at the intersection of medicine, law, and administration, making them complex to untangle—especially when medical complications, multiple providers, or institutional gaps are involved.

One of the major hurdles is *proving causation*. Even when a patient experiences harm following a transfusion, it is not always easy to determine whether the transfusion was directly responsible. Many patients receiving blood are already critically ill, and overlapping treatments can obscure the clinical picture. This makes forensic conclusions difficult and can weaken legal cases, particularly in civil or criminal court proceedings (Zaheer 2012).

Another key issue is the *lack of forensic infrastructure*—especially in smaller or rural hospitals. These facilities may not have access to trained forensic pathologists, immunohematology labs, or standardized audit systems. As a result, important evidence can be missed or improperly preserved, limiting the accuracy of post-event investigations (Jacobs et al. 2025).

Documentation problems are also widespread. Transfusion records, chain-of-custody logs, and incident reports are sometimes incomplete or missing altogether. In high-pressure environments like emergency rooms, record-keeping often takes a back seat. This becomes a serious challenge during retrospective reviews, where missing details can make it difficult to determine if a mistake occurred and who was involved.

There is also the problem of *underreporting*, often tied to a fear of punishment. In many hospitals, staff may avoid fully documenting near-miss events or minor errors because they are concerned about disciplinary action or legal exposure. This defensive behavior—though understandable—undermines transparency and limits opportunities for learning and prevention (Orkuma and Ayla 2014).

Emergency situations present another layer of complexity. During trauma care, for example, clinicians may bypass normal verification procedures to save time—using uncross-matched O-negative blood as a lifesaving measure. While sometimes justified, this increases the risk of transfusion-related reactions, particularly in patients with rare antibodies. If something goes wrong, it is not always clear whether the decision was negligent or clinically appropriate.

Legal knowledge is yet another gap. Many frontline staff are unaware of how their actions (or omissions) could lead to civil or criminal liability. This lack of awareness can result in unintentional violations of both clinical and legal standards (Thrakul et al. 2023).

To address these issues, a number of *strategic solutions* have been developed and implemented across different healthcare systems. One approach is to use *root cause analysis (RCA)* and structured audits after any adverse transfusion event. These tools help trace the sequence of errors and provide a roadmap for process improvements.

To combat the shortage of forensic experts, some regions are now building *regional training centers* or offering *remote consultation services* that smaller hospitals can access when needed.

Improving *digital record-keeping* is another priority. Using electronic systems with built-in safety checks—such as barcode or RFID verification—can improve tracking and reduce errors. These technologies also help maintain complete and searchable logs that support forensic reviews.

Protecting *whistleblowers* and creating anonymous reporting systems can also make a big difference. When staff feel safe reporting mistakes, hospitals can learn from incidents before they escalate into harm (Orkuma and Ayla 2014).

Educational programs that combine *clinical training with legal literacy* are becoming more common. These sessions help medical personnel understand not only the importance of protocol compliance but also the legal implications of failing to follow them.

To summarize these challenges and responses, Table 5.3 outlines key investigation barriers and the tools used to address them. These strategies emphasize a shift from blame to prevention—supporting both clinical safety and legal defensibility.

Ultimately, tackling transfusion errors requires more than just good intentions. It demands coordinated effort, technological support, and a culture that values both patient safety and professional accountability.

Table 5.3 Challenges in transfusion error investigation and strategic solutions

Challenges in investigation/litigation	Strategic solutions and prevention tools
Difficulty proving causation in adverse events	Implement structured root cause analysis (RCA) and audit pathways
Lack of forensic expertise in rural hospitals	Develop regional forensic training centers and remote consultation
Incomplete transfusion records or logs	Enforce digital record-keeping and chain-of-custody logging
Defensive documentation or underreporting	Protect whistleblowers and anonymize hemovigilance reporting systems
Emergency transfusions bypassing verification	Install barcode and RFID-based unit verification before release
Limited awareness of legal standards	Conduct legal education programs for transfusion personnel

10 Future Directions in Forensic Hematology

As transfusion medicine continues to evolve, so too does the field of forensic hematology. New technologies are transforming how errors are detected, investigated, and prevented. These innovations are not just improving clinical accuracy—they are also strengthening the evidence base for forensic and legal investigations.

One of the most promising developments is the use of *artificial intelligence (AI)* in transfusion monitoring. Machine learning algorithms can now analyze vast amounts of data in real time, flagging irregularities such as mismatched entries, skipped verification steps, or unusual patterns in vital signs during transfusion. These alerts can prompt early intervention, reducing harm and making forensic reviews more precise when adverse events do occur.

Blockchain technology is also emerging as a valuable tool in traceability. By providing a secure, tamper-proof digital ledger, blockchain ensures that each step in the donor-to-recipient chain is documented. This includes donor screening, unit testing, storage conditions, cross-match results, and transfusion timing. In forensic settings, this level of traceability can help investigators confirm whether protocols were followed and pinpoint exactly where a breakdown occurred (Zaami et al. 2021).

Another area of innovation is *real-time alert systems*, which can immediately notify clinicians of potential reactions or discrepancies during the transfusion process. These systems integrate data from bedside monitoring, electronic health records, and blood bank databases to create a dynamic safety net. When paired with forensic tools, they help ensure that any post-event analysis starts with complete and time-stamped evidence.

Efforts are also underway to improve the *integration of clinical and forensic databases*. When a transfusion-related incident is reported, having access to shared digital platforms enables pathologists, hospital administrators, and legal teams to coordinate their reviews more efficiently. This also improves transparency and accountability across departments and institutions.

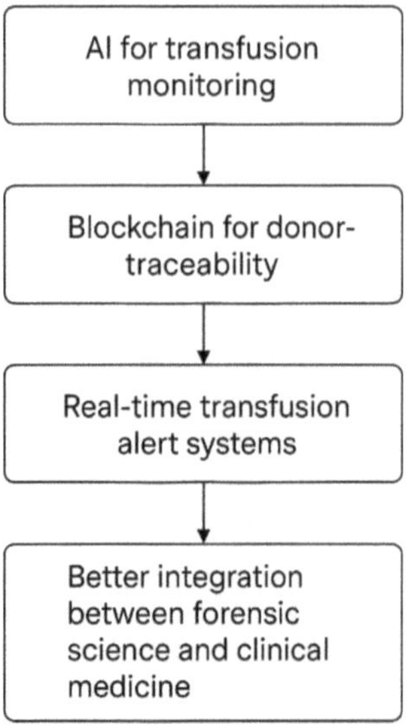

Fig. 5.2 Future Directions in Forensic Hematology (Zaami et al. 2021)

Figure 5.2 presents a flowchart summarizing these emerging directions—highlighting how the combination of digital innovation and forensic science is reshaping transfusion safety. These technologies are not just futuristic concepts; they are already being piloted or implemented in hospitals worldwide.

As these tools become more widely adopted, they offer an opportunity to shift forensic hematology from a reactive field to a more proactive one—helping healthcare systems identify risks earlier, respond faster, and create stronger safeguards for patients.

11 Conclusions

Blood transfusion is one of the most critical and lifesaving interventions in modern medicine. Yet, as this review has shown, when carried out improperly, it can have devastating consequences. From clinical missteps to administrative errors, the risks are real—and the need for rigorous oversight is essential.

Forensic investigation plays a central role in bridging the gap between clinical outcomes and accountability. Through tools like autopsies, DNA profiling, crossmatch reviews, and documentation audits, forensic teams help determine whether an error occurred, how it happened, and whether harm could have been prevented. These investigations inform not just legal cases but also hospital protocols, national policies, and training standards.

As global regulatory bodies like the *FDA*, *EU Blood Directive*, and *WHO* continue to shape safety protocols, the consistency of enforcement and available resources remains a challenge—particularly in low-resource settings. Despite these gaps, the field is advancing rapidly.

Emerging technologies like *AI-powered monitoring*, *blockchain-based traceability*, and *real-time alerts* are bringing new levels of precision to both clinical and forensic processes. When paired with stronger regulatory systems and ethical frameworks, these tools offer real hope for reducing preventable harm.

Ultimately, transfusion safety depends on more than equipment or policies—it depends on a culture of vigilance, transparency, and accountability. By integrating forensic insight with clinical practice and legal standards, healthcare systems can reduce risk, promote justice, and uphold patient trust.

References

Barnhard S (2020) "This is the wrong patient's blood!": evaluating a near-miss wrong transfusion event. AHRQ. https://psnet.ahrq.gov/web-mm/wrong-patients-blood-evaluating-near-miss-wrong-transfusion-event

Bolcato M, Russo M, Trentino K, Isbister J (2020) Patient blood management: the best approach to transfusion medicine risk management. Transfus Apher Sci 59(4):102776. https://doi.org/10.1016/j.transci.2020.102776

de Jonge LL, Wiersum-Osselton JC, Bokhorst AG (2022) Haemovigilance: current practices and future developments. University of Groningen. https://research.rug.nl/en/publications/haemovigilance-current-practices-and-future-developments

De Vries RRP, Faber JC, Strengers PFW, Board of the International Haemovigilance Network (2011) Haemovigilance: an effective tool for improving transfusion practice. Vox Sang 100(1):60–67

Jacobs JW et al (2025) Blood Borders: state laws, vaccine misinformation, and the threat to the United States blood supply. Wiley Online Library. https://onlinelibrary.wiley.com/doi/abs/10.1111/trf.18312

Jones J, Copplestone A (2009) Regulatory aspects of blood transfusion. In: Practical transfusion medicine. Wiley, pp 189–200

Kazarian M (2020) Criminalising medical malpractice: a comparative perspective. Routledge

Linden JV, Kaplan HS (1994) Transfusion errors: causes and effects. Transfus Med Rev 8(3):169–183

Madea B (2022) Chapter 32: Medical malpractice. In: Madea B (ed) Handbook of forensic medicine, 2nd edn. Wiley-Blackwell. https://doi.org/10.1002/9781119648628.ch32

Mansfield EM, Berg TK, Leifheit KA, Sweeney JP, Rigney PR (2009) Current legal issues in transfusion medicine. In: Simon TL, Snyder EL, Solheim BG, Stowell CO, Strauss RG, Petrides M (eds) Rossi's principles of transfusion medicine, vol 4, pp 993–1009

Nitsche E, Dreßler J, Henschler R (2023) Systematic workup of transfusion reactions reveals passive co-reporting of handling errors. J Blood Med 14:103–112. https://doi.org/10.2147/JBM.S411188

Orkuma JA, Ayla ON (2014) Ethico-legal aspects of hospital-based blood transfusion. Int J Med Biomed Res 3(3):137–143

Pal R, Ghosh A, Sinha D, Pal S (2020) Medicolegal issues in blood transfusion practice. In: Legal issues in medical practice. Jaypee Brothers Medical Publishers, pp 294–310

Perper JA et al (1993) Detection of fatal therapeutic misadventures by an urban medico-legal system. ASME 38:327

Phou S, Klein NJ (2025) Patient verification. In: Pediatric patient blood management and transfusion. Springer, pp 315–324. https://doi.org/10.1007/978-3-031-81666-6_19

Rossi EC, Simon TL (2022) Rossi's principles of transfusion medicine, 5th edn. Wiley. https://doi.org/10.1002/9781119719809

Sakai K, Takatsu A, Shigeta A, Fukui K, Sekikawa A (2010) Potential medical adverse events associated with death: a forensic pathology perspective. Int J Qual Health Care 22(1):9–13. https://doi.org/10.1093/intqhc/mzp058

Sobh ZK, Zaki MK (2024) Forensic medical examiners' role in resolving malpractice claims. Naif Arab University

Stainsby D (2005) Errors in transfusion medicine. Anesthesiol Clin North Am 23(3):553–567

Stainsby D, Russell J, Cohen H (2005) Reducing adverse events in blood transfusion. Br J Haematol 132(1):3–12. https://doi.org/10.1111/j.1365-2141.2005.05702.x

Thrakul S, Insorn T, Chokprajakchat S, Jarupan M (2023) Preventive measure of criminal prosecution for nursing and midwifery errors: case studies of public hospitals. Int J Crim Justice Sci 18(1):33–46. https://ijcjs.com/menu-script/index.php/ijcjs/article/view/622

Van Grunsven PR (1997) Medical malpractice or criminal mistake?-an analysis of past and current criminal prosecutions for clinical mistakes and fatal errors. DePaul J Health Care L 2:1

Wecht CH (2007) Clinical and forensic pathology. In: The medical malpractice survival handbook. Mosby, Philadelphia

Zaami S et al (2021) Twin-to-twin transfusion syndrome and forensic diagnostic standards. Diagnostics 11(3):445. https://www.mdpi.com/2075-4418/11/3/445

Zaheer HA (2012) Establishing good practice in difficult settings. Blood. https://www.academia.edu/download/92166053/j.1423-0410.2012.01615_1.pdf

Role of Blood Evidence in Criminal Investigation

6

Abstract

Blood is one of the most informative and probative forms of biological evidence in forensic science, owing to its distinct biochemical composition, predictable physical behavior, and high individualizing potential. Its presence at crime scenes supports event reconstruction, establishes spatial–temporal relationships, and provides robust links between victims, suspects, and locations. This chapter presents a comprehensive overview of the scientific foundations of blood evidence and its evolving role in criminal investigation. A structured integrative review approach was adopted to examine classical serological techniques—such as ABO blood grouping, enzyme polymorphism, and confirmatory crystal tests (Takayama and Teichmann)—alongside modern molecular and analytical advancements. Contemporary laboratory methodologies, including DNA profiling, STR analysis, mitochondrial DNA sequencing, and bloodstain pattern analysis (BPA), are critically evaluated for their forensic utility. Emerging innovations such as molecular and proteomic profiling, RNA-based tissue identification, epigenetic markers, and microbiome signatures are explored for their potential to enhance specificity and improve temporal estimations. Despite substantial advancements, the interpretation of blood evidence continues to face challenges, including contamination, environmental degradation, mixed-profile complexities, and investigator bias. The chapter also examines ethical and legal implications related to evidence admissibility and chain-of-custody integrity. Future directions emphasize the integration of omics technologies, machine learning, artificial intelligence, and computational modeling to improve accuracy, reduce subjectivity, and enhance real-time crime-scene applications. Overall, the chapter underscores the central role of blood evidence in forensic investigations and highlights the scientific, technological, and operational developments that are shaping the next generation of forensic blood analysis.

Keywords

Blood evidence · Forensic serology · DNA profiling · Bloodstain pattern analysis · Forensic investigation · Molecular forensics

H. R. Dash et al., *Forensic Investigation of Blood Evidence*,
https://doi.org/10.1007/978-981-95-9183-1_6

1 Introduction

Blood is one of the most significant and informative forms of biological evidence encountered at crime scenes. Its detection not only indicates that a violent incident may have occurred but also provides critical insights into the sequence of events, the individuals involved, and the manner in which the crime was committed. The forensic value of blood was first recognized in the late nineteenth century, when researchers identified human-specific reactions to blood proteins, allowing investigators for the first time to distinguish human blood from animal blood. Since then, advances in analytical science have transformed forensic serology from basic presumptive testing into a sophisticated molecular discipline capable of yielding extensive biological, behavioral, and temporal information (Singh et al. 2021).

Historically, forensic serology relied heavily on ABO blood grouping, enzyme polymorphism, and protein-based assays to establish biological connections between samples and potential sources. Although these classical methods contributed significantly to early forensic investigations, they lacked the discriminatory power required for individual-level identification. The introduction of DNA profiling in the 1980s marked a paradigm shift, enabling scientists to identify individuals with high precision based on unique genetic markers. Modern methodologies—including short tandem repeat (STR) analysis and mitochondrial DNA sequencing—have further enhanced the capacity to generate reliable results from minute, aged, or degraded blood samples (Yadav and Sharma 2024).

Beyond identification, blood evidence also provides valuable information about the physical dynamics of a crime. Bloodstain pattern analysis (BPA) allows investigators to reconstruct events by examining the size, shape, distribution, and spatial arrangement of bloodstains. Parameters such as angle of impact, velocity of force, and the nature of the target surface help determine whether injuries were caused by blunt force, sharp instruments, or firearms. When integrated with DNA profiling, BPA contributes to a detailed reconstruction of both the biological and physical aspects of a crime scene.

Recent advances in molecular biology, forensic genomics, and proteomics have significantly expanded the scope of blood analysis. Techniques such as mRNA and microRNA profiling can identify the specific tissue origin of a biological stain, while proteomic and metabolomic approaches enable the detection of drugs, toxins, and biomarkers related to physiological states. Emerging research into blood-associated microbial communities (the blood microbiome) offers novel possibilities for estimating the postmortem interval and differentiating individuals. Additionally, the application of artificial intelligence and machine learning is revolutionizing pattern recognition and enhancing the interpretation of complex analytical datasets.

Despite these remarkable advancements, challenges remain in the interpretation and preservation of blood evidence. Environmental degradation—caused by heat, humidity, sunlight, or microbial activity—can affect the quality of biological material. Inadequate collection, improper storage, and contamination may further compromise results. Human interpretive bias and the complexities associated with mixed DNA samples also continue to pose difficulties. Nevertheless, the integration of omics technologies, AI-driven analytics, and computational modeling paves the way for faster, more accurate, and field-deployable forensic solutions, reinforcing the critical role of blood evidence in modern criminal investigation.

2 Nature and Composition of Blood

Blood is a uniquely complex biological fluid that plays an essential role in human physiology, supporting processes ranging from oxygen and nutrient transport to waste removal and homeostatic regulation. In forensic science, these same biochemical and structural properties make blood one of the most valuable forms of evidence encountered at crime scenes. Its composition carries a wealth of genetic, molecular, and physiological information that can link individuals to crime scenes, reconstruct events, and assist in both investigative and legal decision-making. A thorough understanding of the nature and composition of blood forms the scientific foundation for its analysis and forensic interpretation.

Blood consists of two major components: the cellular fraction and the plasma fraction. Approximately 45% of total blood volume comprises the cellular elements—erythrocytes (red blood cells), leukocytes (white blood cells), and thrombocytes (platelets)—while the remaining 55% is plasma, a straw-colored fluid responsible for transporting metabolic solutes, proteins, and signaling molecules (Krebs 1950). Together, these components create a dynamic biological system that preserves critical genetic and biochemical signatures central to forensic investigations.

Red blood cells (RBCs) are the most abundant cellular components and are primarily responsible for oxygen transport via the iron-containing protein hemoglobin. Mature human RBCs lack nuclei, making them unsuitable for nuclear DNA analysis. However, hemoglobin variants—such as HbA, HbS, and HbC—can serve as population-specific biochemical markers and may provide supplementary investigative clues, especially in certain genetic or epidemiological contexts.

White blood cells (WBCs), although far fewer in number, are of far greater forensic significance. Unlike RBCs, leukocytes contain nuclei that house the full complement of nuclear DNA, making them the primary source of genetic material for forensic DNA profiling. Even microscopic or degraded bloodstains can yield sufficient DNA for analysis through amplification of short tandem repeats (STRs), enabling highly specific comparisons between evidentiary samples and reference profiles. In addition, emerging molecular analyses—such as epigenetic profiling, microRNA expression, and other RNA-based assays—can now offer insights into the tissue origin of the stain, estimate the time since deposition, and infer physiological states of the donor (Zagorodniy and Zagorodniy 2024).

Platelets, or thrombocytes, are small anucleate cell fragments involved in blood clotting and wound repair. Although they do not contain nuclear DNA, their protein and enzyme composition carries forensic value. Prior to the widespread adoption of DNA profiling, polymorphic enzyme markers—such as phosphoglucomutase (PGM) and esterase D—play a significant role in forensic serology. These enzymatic systems remain useful in cases where DNA is severely degraded or absent, such as in environmentally exposed samples.

Plasma, constituting more than half of total blood volume, is composed of approximately 90% water and contains a rich mixture of solutes, including electrolytes, hormones, glucose, amino acids, enzymes, antibodies, and major plasma proteins such as albumin, globulins, and fibrinogen. Plasma components often retain clues regarding an individual's physiological condition at or near the time of the incident. Forensic toxicologists analyze plasma for the presence of drugs, alcohol, and toxic agents, while forensic pathologists may examine specific protein or hormone levels to infer the cause of death, timing of injury, or pathological states (Vandewoestyne et al. 2015).

Blood also carries highly discriminating antigenic markers. The ABO and Rh blood group systems were among the earliest serological tools available to forensic scientists, aiding in the exclusion of suspects and comparison of bloodstains. Although less specific than modern DNA analysis, blood group testing retains value, particularly when DNA is degraded, absent, or insufficient. Additional antigen systems—such as MN, Kell, and Duffy—further enhance the discriminatory capacity of serological profiling.

Beyond genetic markers, the molecular and proteomic diversity of blood provides additional layers of forensic information. Proteins, peptides, and enzymatic markers often remain stable even when DNA has deteriorated due to environmental exposure. Current research in proteomics and metabolomics is exploring personalized molecular signatures and environmental influences preserved within bloodstains, offering new avenues for identification, ageing of stains, and contextual interpretation.

In summary, blood is a biologically rich and forensically powerful fluid. Its cellular, genetic, biochemical, and molecular constituents collectively form a comprehensive source of evidence for human identification, crime-scene reconstruction, and investigative interpretation. Advances in molecular biology, proteomics, and forensic genomics continue to expand the analytical potential of blood, affirming its central role in modern forensic science.

3 Collection and Preservation of Blood Evidence

The proper collection and preservation of blood evidence is critical in forensic investigations, as the reliability of any analytical, serological, or DNA result depends directly on the integrity of the sample. From the moment a suspected bloodstain is located, investigators must adhere to strict scientific and procedural standards to ensure that the evidence remains uncontaminated, interpretable, and legally admissible.

Initial Documentation All visible blood evidence must first be documented through high-resolution photography, sketches, and comprehensive notes. Photographs should include overall, mid-range, and close-up views, each incorporating measurement scales to depict the exact size and spatial relationships of stains. Such documentation supports later reconstruction of events—including the positions of individuals, trajectories of impact, and potential weapon use (Castro and Coyle 2013).

Collection Procedures To prevent contamination and biochemical degradation, appropriate collection techniques must be used. Fresh or liquid blood should be collected with sterile swabs or pipettes. The collected swabs must be air-dried at room temperature and stored in breathable paper containers, since plastic packaging can trap moisture and promote microbial growth. Because degradation begins rapidly upon environmental exposure, samples intended for DNA or toxicological analysis should be transferred to the laboratory without delay. If immediate transport is not feasible, temporary refrigeration at 4 °C is advised; long-term preservation requires freezing at −20 °C or preferably −80 °C.

Dried Bloodstains Dried stains may be recovered by scraping the surface, excising the stained portion of fabric, or using a moistened sterile swab to resuspend the material. The method selected should depend on the substrate and analytical requirements—for instance, scraping is suitable for smooth surfaces, while cutting is recommended for fabrics. All samples must be stored in clean, labeled paper envelopes or boxes to prevent moisture accumulation and contamination (Singh 2021).

Chain of Custody Maintaining an unbroken chain of custody is essential for evidential integrity. Every transfer—whether from the scene to storage or from storage to the laboratory—must be logged with the collector's name, date, time, and purpose. Sample containers should bear tamper-evident seals and labels containing the case number, sample type, anatomical origin (e.g., femoral vs. cardiac blood), and the collector's signature.

Sampling Sites and Containers In postmortem investigations, peripheral blood (commonly from the femoral vein) is preferred due to its reduced susceptibility to postmortem redistribution. Preservatives such as sodium or potassium fluoride (1–5%) are added to inhibit enzymatic degradation and microbial fermentation, particularly when assessing volatile substances like ethanol. Samples must be stored in airtight polypropylene containers or, for volatile analytes, in glass vials with Teflon-lined caps to minimize diffusion losses.

Environmental Influences Environmental factors—such as heat, humidity, and direct sunlight—can accelerate hemolysis, stimulate microbial proliferation, fragment DNA, and alter drug/metabolite concentrations. Therefore, blood collection should be prioritized in warm or tropical settings. Portable coolers or insulated containers should be used to stabilize samples until controlled storage conditions are available.

Use of Forensic Collection Kits Standardized blood collection kits typically include sterile syringes, swabs, labeled containers, absorbent materials, and tamper-proof seals. All containers should be precleaned and sterilized to avoid introducing contaminants, particularly in cases requiring toxicological or trace element analysis. Containers must not be overfilled to reduce headspace and limit evaporation of volatile compounds.

Laboratory Handling Within the laboratory, strict contamination-control procedures are followed. Analysts must wear gloves, masks, and protective clothing, changing them between samples. Workspaces are routinely disinfected with bleach or UV irradiation.

Storage and Preservation For short-term preservation, blood samples should be kept at 4 °C; for long-term storage, freezing at −20 °C or below is essential. Ethanol and other analytes may change in concentration if samples are improperly stored, as microbial activity can lead to increases or decreases even in vitro (Dinis-Oliveira et al.). Samples requiring serum or plasma separation must be processed prior to freezing, as repeated freeze–thaw cycles compromise cellular and protein integrity.

Documentation and Quality Assurance Each specimen must be accompanied by a detailed laboratory request form specifying the required analyses (e.g., DNA, toxicology, and serology) and any environmental factors that could affect interpretation. Samples should be sealed in opaque, tamper-evident evidence bags and transported promptly. Modern forensic practice also requires periodic auditing of collection procedures, ensuring compliance with best-practice standards. Since no universal protocol can account for all environmental and physiological variations, each case must be evaluated individually.

Ultimately, the evidential value of blood depends on its scientific reliability and contextual interpretation. Properly collected and preserved samples can support DNA profiling, toxicological assessment, and event reconstruction, while poorly managed samples may compromise investigations or result in inadmissible evidence. Therefore, the use of standardized procedures—such as those recommended by the European Council of Legal Medicine—is essential to uphold scientific rigor and judicial integrity (Ip and Yu 2021).

4 Preliminary Tests for Blood Detection

The presumptive tests serve as rapid screening tools to determine whether a stain or trace material may contain blood. Although these tests cannot confirm the presence of human blood, they are essential for locating otherwise undetectable traces and for guiding investigators toward subsequent confirmatory analyses.

One of the most established chemical presumptive tests is the Kastle–Meyer (KM) test, developed in the early twentieth century. The test operates on the peroxidase-like activity of hemoglobin, the principal oxygen-binding protein in red blood cells. The reagent contains reduced phenolphthalein, a colorless compound. When hydrogen peroxide is added to a blood-containing sample, the heme group catalyzes the breakdown of peroxide, releasing oxygen that oxidizes phenolphthalein into its vivid pink form. This rapid color change indicates a positive presumptive reaction. Experimental evidence shows that the KM test is highly sensitive, detecting blood diluted to as low as one part in several hundred thousand. Notably, hemoglobin's catalytic activity persists even in aged samples or stains exposed to moderate heat, allowing detection long after visible coloration has faded.

The procedure itself is simple: a moistened swab of the suspected stain is treated with a few drops of KM reagent, followed by hydrogen peroxide. A bright pink color appearing within seconds suggests probable blood. The test's speed, ease of use, and minimal sample requirement make it superior to earlier methods that required prolonged incubation or soaking. However, excessive peroxide must be avoided, as it can produce misleading foaming or transient hues. When performed with precision, the KM test is stable, highly sensitive, and produces very few false positives (Glaister 1926).

Extensive testing of common environmental and household contaminants—such as rust, plant material, detergents, and various salts—shows that these substances rarely generate a full positive reaction. Although certain chemicals, such as iodine or nitrate compounds, may produce faint pink tones, these are short-lived and easily distinguished from the intense magenta produced by genuine hemoglobin. For these reasons, the Kastle–Meyer test remains a preferred rapid prescreening method in forensic laboratories.

Beyond colorimetric tests, chemiluminescent reagents such as luminol and Bluestar® have transformed the detection of latent bloodstains. These reagents are particularly valuable for visualizing blood patterns that are invisible to the naked eye. Their mechanism is similar: in the presence of hemoglobin, hydrogen peroxide decomposes in alkaline conditions, releasing energy that excites the luminol molecule, causing a characteristic blue glow. This luminescence allows detection of diluted, cleaned, or wiped blood on floors, walls, fabrics, and outdoor surfaces.

Bluestar® produces brighter, longer-lasting luminescence than conventional luminol and is more resistant to temperature fluctuations. It can detect blood at very high dilutions—sometimes exceeding one part per hundred thousand—and often remains reactive on older stains or surfaces that have been partially cleaned. However, both luminol and Bluestar® are susceptible to interference from oxidizing agents such as bleach, which can produce sparkling false positives. The addition of glycine has been shown to reduce such interference by enhancing the contrast between true blood signals and cleaning residues (Barni et al. 2007).

Despite their exceptional sensitivity, chemiluminescent tests are strictly presumptive. They detect peroxidase-like activity rather than blood specifically, and certainly not human blood. Metals, certain plant peroxidases, and various chemical oxidants can produce similar luminescent or color reactions. Therefore, all positive results—whether obtained via KM reagent, luminol, or Bluestar®—must be followed by confirmatory tests such as microscopic identification, crystal tests, or DNA profiling.

In forensic practice, the true value of presumptive tests lies in their ability to guide investigators during scene processing. By revealing hidden, diluted, or cleaned blood patterns, they assist in reconstructing events and determining where to collect samples for definitive identification. When applied judiciously, these tests provide critical direction for subsequent detailed analysis and help investigators uncover the narrative behind violent incidents.

5 Confirmatory Tests for Blood Identification

Confirmatory tests play a crucial role in forensic serology because they provide definitive evidence of the presence of blood and help distinguish it from substances that may yield false positives in presumptive testing. These assays target hemoglobin, its derivatives, or species-specific proteins. The most widely used confirmatory approaches include microcrystal tests, immunological assays, and advanced spectroscopic and chromatographic techniques.

5.1 Microcrystal Tests: Teichmann and Takayama Reactions

Teichmann Test (Hemin Crystal Test)

First introduced in 1853, the Teichmann test is one of the earliest confirmatory methods for detecting blood. The reaction involves treating a suspected blood sample with glacial acetic acid and a halide salt—typically sodium chloride. Upon gentle heating, characteristic brown rhomboid crystals (Fig. 6.1a) of hemin (ferriprotoporphyrin chloride) form and can be observed under a microscope.

While highly specific, the test has notable limitations:

- It requires intact heme and may fail with old or degraded stains.
- Overheating or insufficient heating may lead to false negatives.
- Sensitivity is lower compared to modern techniques (Abid et al. 2024).

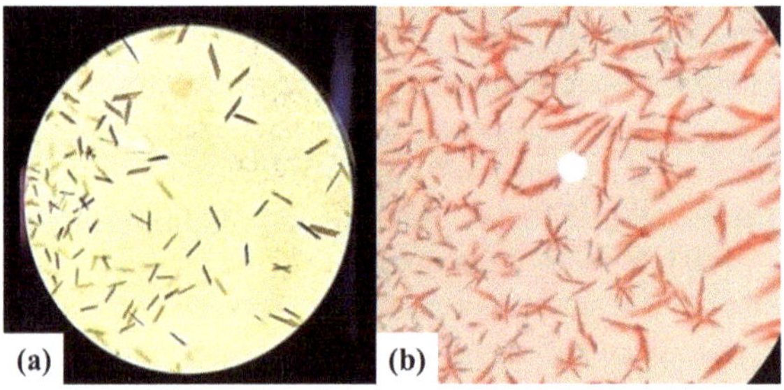

Fig. 6.1 Crystal tests for the confirmatory detection of blood: (**a**) Teichmann brownish rhomboid crystals of hemin; (**b**) Takayama hemochromogen crystals under a microscope

Takayama Test (Hemochromogen Crystal Test)
Developed in 1912, the Takayama test improves upon the Teichmann method. In this reaction, a pyridine-based reagent converts hemoglobin into pink, feathery crystals of pyridine ferroprotoporphyrin (Fig. 6.1b) (hemochromogen). These crystals form more readily, under milder conditions, and often persist even in:

- Old stains
- Partially washed materials or
- Diluted samples

Research shows that Takayama crystals remain detectable even after repeated washing, indicating the method's high sensitivity and reliability (Dixit et al. 2023).

5.2 Immunological Techniques: Ouchterlony Double Diffusion Test

Immunological confirmatory tests determine the species origin of blood—critical for differentiating human blood from that of animals. The classic method is Ouchterlony double diffusion, developed in the 1940s.

In this assay:

- Antibody and antigen solutions diffuse through an agar gel.
- When they meet at optimal concentrations, they form a visible precipitin line.
- Each line is species-specific, enabling a clear distinction between human and nonhuman blood (Bailey 1996).

Strengths of the Ouchterlony test:

- Effective with small or moderately degraded samples
- Species-specific and visually interpretable

Limitations:

- Requires high-quality antisera.
- Diffusion conditions must be tightly controlled.

Modern immunological methods—such as ELISA and lateral-flow immunochromatographic strips—retain the same antigen–antibody principles but offer greater sensitivity, speed, and ease of use.

5.3 Spectroscopic and Chromatographic Methods

Advances in analytical science have introduced highly sensitive, nondestructive techniques capable of confirming blood and characterizing its biochemical components.

Raman Spectroscopy

Raman spectroscopy identifies blood by detecting unique vibrational spectra associated with hemoglobin and its variants. This method:

- Is reagent-free and nondestructive
- Discriminates between hemoglobin forms (e.g., HbA, methemoglobin, HbA1c)
- Can detect aged or environmentally altered bloodstains

Chemometric analysis further enhances sensitivity, allowing detection well below physiological concentrations. Raman methods also show potential for assessing post-depositional changes where traditional microcrystal tests may fail.

High-Performance Liquid Chromatography (HPLC)

HPLC is another powerful confirmatory method. It separates hemoglobin variants based on their chromatographic behavior, allowing the identification of:

- Human vs. nonhuman hemoglobin
- Modified derivatives (carboxyhemoglobin, methemoglobin)
- Degradation products

The precision and ability to detect trace analytes make HPLC valuable for both confirmatory identification and forensic interpretation of degraded blood samples.

5.4 Comparative Evaluation and Modern Integration

A comparison of classical and contemporary confirmatory protocols demonstrates that each category offers distinct yet complementary advantages. Traditional microcrystal assays—such as the Takayama and Teichmann tests—remain valuable due to their simplicity, minimal equipment requirements, and rapid formation of characteristic heme-derived crystals. Immunological assays, including Ouchterlony double diffusion, provide an additional layer of specificity by distinguishing human blood from that of other species, making them indispensable for species determination.

Modern instrumental approaches, including Raman spectroscopy and high-performance liquid chromatography (HPLC), provide molecular-level precision, high reproducibility, and strong potential for automation. Comparative analyses of presumptive and confirmatory tests have shown that results obtained through microcrystal and spectroscopic techniques remain robust across various sample dilutions and even after washing, underscoring their reliability in forensic contexts. These findings reinforce the value of maintaining a multimodal analytical framework that integrates both classical serological assays and advanced instrumentation.

Confirmatory blood identification methods form the foundation of forensic serology, spanning early microcrystal reactions to contemporary spectroscopic and immunological technologies. Together, these methods deliver not only decisive

Table 6.1 Comparison of presumptive and confirmatory tests for blood detection

Parameter	Presumptive tests (Kastle–Meyer, Luminol, Bluestar)	Confirmatory tests (Teichmann, Takayama, Immunological, Raman spectroscopy)
Purpose	Initial screening to indicate the possible presence of blood	Definitive identification of blood and often species origin
Target analyte	Peroxidase-like activity of hemoglobin	Hemoglobin derivatives, heme crystals, species-specific proteins
Sensitivity	Very high (detects highly diluted or latent stains)	Moderate to high depending on the test type
Specificity	Low-to-moderate; false positives possible	High specificity; confirms blood even in aged/degraded samples
Common reagents	Phenolphthalein, hydrogen peroxide, luminol, Bluestar	Glacial acetic acid, pyridine reagent, antisera, spectroscopic methods
Advantages	Fast; good for scene scanning; inexpensive	Highly reliable; admissible in court; species identification possible
Limitations	Cannot confirm blood; affected by oxidizing agents	Requires equipment; may fail if heme is heavily degraded
Forensic use	Screening and locating stains for collection	Court-admissible confirmation and detailed analysis

confirmation of the presence of blood but, in many cases, insight into its species of origin. Advances in optical and chromatographic technologies have significantly enhanced analytical sensitivity, enabling the detection of highly diluted, aged, or degraded blood samples previously beyond the limits of classical tests (Table 6.1). Emerging hybrid systems—combining Raman spectroscopy, HPLC, and AI-assisted spectral interpretation—are poised to become standard practice, offering unparalleled accuracy and reliability in characterizing blood evidence within criminal investigations.

6 Blood Typing and Serological Analysis

Before the advent of modern DNA profiling, forensic identification relied heavily on blood group analysis and serology to associate biological evidence with individuals. Although DNA-based techniques now dominate forensic practice due to their high discriminatory power, serological methods continue to play an essential role—particularly when DNA is degraded, insufficient, or absent.

The foundation of forensic serology lies in the ABO blood group system, discovered by Karl Landsteiner in 1901. This system classifies blood based on the presence or absence of the A and B antigens on red blood cells and their corresponding antibodies in the plasma. Individuals with type A blood possess A antigens and anti-B antibodies; type B individuals have B antigens and anti-A antibodies; type AB individuals express both antigens but lack antibodies; and type O individuals lack both antigens while possessing both antibodies. These antigen–antibody interactions form the basis of classical blood typing and provide valuable information for transfusion compatibility and forensic casework.

Historically, ABO typing was frequently used to include or exclude suspects or victims. For example, if a crime-scene bloodstain was identified as type B, a person with type A blood could be immediately excluded as its source. Although ABO typing cannot uniquely identify an individual, it effectively narrows down potential contributors and remains valuable today, especially when analyzing aged or environmentally degraded samples that do not yield viable DNA (Lv et al. 2023).

Another critical serological marker is the Rhesus (Rh) system, most notably the RhD antigen. Individuals who express the D antigen are Rh-positive, whereas those lacking it are Rh-negative. Combining the Rh factor with ABO typing increases serological discrimination. For instance, a blood sample classified as "B negative" can eliminate any individual with a "B positive" profile. Even partial antigenic profiles obtained from degraded samples may still provide meaningful exclusionary evidence.

Beyond ABO and Rh systems, forensic serologists historically relied on enzyme polymorphisms and protein markers to further differentiate individuals. Enzymes such as phosphoglucomutase and esterase D show genetically determined variants detectable through electrophoresis. These polymorphisms follow Mendelian inheritance and were valuable in criminal investigations and paternity testing. Protein markers like haptoglobin and transferrin also vary among populations, contributing to a more refined biochemical profile when used alongside blood group data.

In routine practice, forensic serology includes:

- Detection of blood
- Species determination
- Identification of blood group antigens
- Analysis of polymorphic enzymes or proteins when possible.

Methods such as absorption–elution, absorption–inhibition, and agglutination assays exploit highly specific antigen–antibody interactions to identify serological markers under controlled laboratory conditions.

Despite its value, serological typing has inherent limitations. Its discriminatory power is significantly lower than DNA profiling, as serological markers categorize individuals into broad groups shared by large segments of the population. Environmental factors—such as heat, humidity, and microbial activity—may degrade antigens, compromising test reliability. Mixed or contaminated samples can also obscure antigen–antibody reactions, leading to interpretive difficulties.

Nevertheless, serology retains practical relevance. In cases involving decade-old stains, skeletal remains, or evidence exposed to extreme environmental stress, DNA may be irretrievable, yet certain blood group antigens or protein markers can persist. Additionally, in resource-limited forensic laboratories where advanced DNA technologies are unavailable, serological testing offers an inexpensive and efficient means of preliminary biological screening. In modern practice, serology complements rather than replaces DNA analysis: it provides early guidance on the nature and possible origin of stains before conducting more sophisticated molecular testing. Moreover, decades of serological research continue to inform population genetics and the study of hereditary traits (Li 2012).

7 DNA Profiling from Blood Evidence

Before the development of modern DNA profiling, forensic investigators relied primarily on blood typing and serological analysis to narrow down suspects or establish biological relationships in criminal cases. Although contemporary forensic science depends heavily on DNA-based identification due to its exceptional accuracy, serological methods continue to play an important role—particularly when DNA is degraded, insufficient, or absent.

At the center of classical serology is the ABO blood group system, discovered by Karl Landsteiner in 1901. This system classifies blood according to the presence or absence of surface antigens—A and B—on red blood cells and their corresponding antibodies (anti-A and anti-B) in plasma.

- Type A blood possesses A antigens and anti-B antibodies.
- Type B blood carries B antigens and anti-A antibodies.
- Type AB expresses both antigens but lacks antibodies.
- Type O has neither antigen but contains both antibodies.

These antigen–antibody interactions form the basis of blood typing, which is fundamental not only for transfusion compatibility but also for preliminary forensic analysis (Singh 1991).

Historically, ABO typing was widely used to include or exclude individuals during investigations. For example, a bloodstain identified as type B immediately rules out an individual with type A blood. Although ABO typing cannot uniquely identify a person, it can significantly reduce the pool of potential contributors. This exclusionary power was critical prior to the advent of DNA profiling and remains valuable today in cases involving degraded or limited biological material from which DNA cannot be recovered.

Another major serological system is the Rhesus (Rh) group, particularly the RhD antigen. Individuals possessing the D antigen are classified as Rh-positive; those without it are Rh-negative. When combined with ABO typing, the Rh factor further increases discriminatory capability. A sample described as "B negative," for instance, can exclude any person who is "B positive." Even partial antigen profiles recovered from deteriorated specimens can provide meaningful investigative leads.

Beyond ABO and Rh systems, forensic serologists previously relied on enzyme polymorphisms (e.g., PGM, EsD) and protein markers (e.g., haptoglobin, transferrin) to help distinguish individuals. These markers show genetically inherited variants detectable through electrophoretic methods. Their Mendelian inheritance patterns made them useful in criminal investigations and paternity testing, contributing to more refined biochemical profiles before the widespread use of DNA analysis.

Practically, forensic serological analysis includes:

- Detection of blood
- Identification of species origin
- Determination of blood group antigens
- Analysis of polymorphic enzymes or proteins when available

Techniques such as absorption–elution, absorption–inhibition, and agglutination assays exploit highly specific antigen–antibody reactions that can be visually detected under controlled laboratory conditions.

Despite their usefulness, serological methods have notable limitations. Their discriminatory power is significantly lower than that of DNA profiling, which can uniquely identify individuals (excluding monozygotic twins). Serological markers place people into broad categories shared by large segments of the population. Additionally, environmental factors such as heat, humidity, and microbial activity can degrade antigens, leading to weak or ambiguous results. Mixed samples pose further challenges, as overlapping antigen–antibody reactions may obscure accurate interpretation.

Nevertheless, serology remains relevant. In cases involving severely degraded samples—such as decade-old stains, skeletal remains, or evidence exposed to environmental extremes—DNA may be unrecoverable, yet some antigens or protein markers persist. In regions with limited forensic infrastructure, where advanced DNA technologies may be unavailable, serological methods continue to provide a cost-effective and efficient means of preliminary biological screening (Zhou et al. 2011).

8 Bloodstain Pattern Analysis (BPA)

Bloodstain Pattern Analysis (BPA) is a crucial forensic discipline that interprets the physical characteristics of bloodstains—such as their size, shape, distribution, and location—to reconstruct the events that produced them. By applying principles of fluid dynamics, physics, and human biology, BPA assists investigators in understanding the sequence of actions, validating or disproving witness or suspect statements, and differentiating between accidental, suicidal, and homicidal incidents. It bridges biological evidence with physical scene reconstruction, offering a contextual narrative of violent events.

Blood behaves as a fluid subject to the laws of motion, gravity, and surface tension. When expelled from the body, blood forms spherical droplets rather than the commonly depicted teardrop shape. As these droplets travel, factors such as velocity, air resistance, and angle of impact influence their trajectory and eventual stain morphology. Upon impact, the blood's kinetic energy and the properties of the surface determine the resulting stain characteristics. With trigonometric analysis and geometric calculations, BPA experts can determine the angle of impact, direction of travel, point of convergence, and point of origin (Peschel et al. 2011). These findings allow reconstruction of movements, blows, weapon usage, and relative positions of the individuals involved.

Historically, the origins of BPA date to the late nineteenth century, when Eduard Piotrowski conducted controlled experiments on bloodstains produced by blunt force trauma, laying the foundational principles for the field. Public recognition increased in the mid-twentieth century with the Samuel Sheppard case, where Dr. Paul Kirk demonstrated how bloodstain interpretation could elucidate the sequence of events at a homicide scene. Since then, BPA has evolved into an internationally recognized forensic specialty practiced by trained analysts, law enforcement professionals, and scientists.

At its core, BPA examines how fluid blood interacts with its environment. Stain characteristics are influenced not only by blood's physical properties but also by the nature of the impacted surfaces—whether smooth, rough, absorbent, or porous. Accordingly, bloodstains are broadly categorized into passive, spatter, and altered patterns.

1. *Passive Bloodstains*

   These stains form solely under the influence of gravity. They include drops, flows, pools, and transfer or contact stains.

   - *Transfer stains* occur when a bloodied object or body part contacts another surface, often producing recognizable impressions such as fingerprints or footwear marks.
   - *Flows* indicate movement of liquid blood due to gravity and can reveal body or object repositioning.
   - *Pooling* occurs when blood accumulates in a static location, providing insight into the position of the victim at the time of bleeding and subsequent movement.

2. *Spatter Patterns*

   Spatter occurs when an external force greater than gravity impacts a blood source. It is strongly associated with violent events such as stabbings, shootings, or blunt force trauma.

   - *High-velocity spatter*, typically from gunshots, produces a fine mist.
   - *Medium-velocity spatter*, commonly from beatings, creates larger droplets.
   - *Cast-off patterns* result when blood is released from a moving, bloodied object (e.g., a swinging weapon). These stains often appear as linear arcs and can indicate the number and direction of blows. Offenders may remain relatively unstained because centrifugal force propels the blood away from the weapon rather than onto them (Attinger et al. 2013).

   Other forms include *arterial spurting*, characterized by sinusoidal or pulsing patterns due to pressurized blood from a severed artery, and *expiated patterns*, produced when blood mixed with air is expelled from the mouth or nose.

3. *Altered Bloodstains*

Altered stains have undergone physical, chemical, or biological changes after deposition. Examples include diluted stains from cleaning attempts, insect activity, and natural clotting. Clear serum separation over pooled blood, for instance, is a normal consequence of clotting—not necessarily evidence of tampering. Distinguishing genuine alterations from misinterpretations is critical to accurate reconstruction.

Correct interpretation requires careful recognition that similar-appearing stains may arise from different mechanisms (Table 6.2). For example, small droplets at a scene could represent either impact spatter or expiated blood. Therefore, BPA must be conducted at the earliest stages of an investigation, before critical patterns are disturbed. Analysts must thoroughly document not only bloodstained areas but also unstained voids, which may indicate objects or persons that blocked blood travel.

Scene documentation—particularly through high-resolution photography from multiple angles—is essential. Blood on clothing or a victim's skin must be

Table 6.2 Classification of bloodstain patterns and their forensic significance

Bloodstain type	Description	Mechanism/source	Forensic significance
Passive drops	Round, uniform drops	Gravity acting on stationary source	Indicates stationary bleeding; used to track movement
Transfer/ contact patterns	Wipes, smears, handprints, footprints	Blooded object/person contacts a surface	Shows interactions, body/ contact movement, object identification
Flow patterns	Directional trails of blood	Movement of liquid blood due to gravity	Indicates body position and changes over time
Pooling	Large accumulated blood	Prolonged bleeding in one spot	Suggests victim's position and time spent at location
Impact spatter	Small to medium droplets	Force applied to blood source (beating, stabbing)	Determines type of force/ weapon and directionality
High-velocity spatter	Fine mist-like droplets	Gunshots or explosive force	Indicates firearm use; helps locate shooter/victim positions
Cast-off patterns	Linear arcs of droplets	Blood flung from a moving object (weapon swings)	Estimates number of blows and assailant's movement
Arterial spurting	Wave-like or pulsatile patterns	Breach of major artery	Indicates severe injury; direction of victim movement
Expirated blood	Stains with bubbles or mucus	Blood expelled from mouth/nose by breathing	Suggests internal injury; mixture with saliva is diagnostic
Altered stains	Clotted, diluted, wiped, insect activity	Environmental or intentional modification	Helps determine cleaning attempts, time effects, or tampering

photographed before movement or cleaning can alter the evidence (Brodbeck 2012). Chemical and optical enhancement techniques, such as luminol, fluorescein, amido black, and leucocrystal violet, may be used to visualize faint or cleaned stains. However, these reagents are not specific to blood and must always be followed by confirmatory testing.

BPA does not operate in isolation. It works alongside DNA profiling, pathology, ballistics, and crime-scene reconstruction. While DNA can identify the individuals involved, BPA provides the dynamic context—where they were positioned, how they moved, and how the assault unfolded. This becomes especially vital in cases where a suspect's presence at the scene is undisputed, and only the sequence of actions is contested.

Ultimately, BPA allows investigators to reconstruct events that occurred in seconds, yet are preserved in bloodstain patterns. As the field advances, modern techniques such as high-speed imaging, computational modeling, and three-dimensional digital reconstruction are increasingly being incorporated, enhancing the accuracy, reproducibility, and objectivity of BPA interpretations.

9 Forensic Applications of Blood Evidence

Blood evidence is one of the most valuable forms of biological material recovered from crime scenes, and its forensic utility has remained critical across a wide spectrum of criminal investigations. Beyond simply indicating the presence of an individual, blood provides insight into the dynamics, timing, and sequence of events that occur during violent encounters. Owing to its biochemical composition, characteristic patterns, and sensitivity to environmental influences, blood enables forensic experts to draw scientifically grounded conclusions that hold strong evidentiary value in legal contexts. This section outlines the major forensic applications of blood evidence, including identification, reconstruction, timing estimation, and linkage analysis—each of which contributes to constructing factual narratives in cases of homicide, assault, and related offenses.

9.1 Identification of Individuals

A primary forensic application of blood evidence is individual identification. With the advent of DNA profiling, blood has shifted from being merely an indicator of presence to a powerful genetic identifier capable of distinguishing individuals with near-absolute certainty. Blood is rich in DNA due to the nucleated white blood cells it contains, allowing forensic laboratories to generate STR profiles that accurately differentiate one person from another. These profiles can be compared with reference samples from victims, suspects, or national DNA databases to include or exclude individuals with high reliability.

Beyond simple identification, modern genetic approaches—including SNP typing, ancestry inference, sex determination, and forensic genetic genealogy—allow deeper insights into biological traits and familial relationships. Even minute or invisible traces of blood can yield robust DNA profiles, making blood an indispensable tool in cases lacking eyewitness accounts or direct testimony.

9.2 Reconstruction of Crime Events

Blood evidence also plays a pivotal role in reconstructing the sequence and mechanics of criminal acts. Bloodstain Pattern Analysis (BPA) examines the size, shape, distribution, and spatial orientation of bloodstains to interpret the actions that produced them. Through BPA, investigators can determine the positions of the victim and assailant, direction and magnitude of force, type of weapon involved, and the likely sequence of movements.

Different types of patterns—such as impact spatter, cast-off trails, transfer stains, arterial spurts, expirated blood, and passive drips—provide crucial clues about the mechanism of injury. BPA can reveal whether the victim was standing, seated, or lying down when injured, whether the assailant changed position during the assault, or whether the weapon was swung multiple times. These findings become especially important when testimonial accounts conflict with the physical evidence, enabling investigators to corroborate or challenge statements with objective scientific data.

9.3 Estimation of Time Since Deposition

Another important forensic application of blood evidence is estimating the age of a bloodstain. Although determining an exact timestamp remains challenging, significant progress has been made in understanding chemical, physical, and molecular changes that occur as blood ages. Indicators such as hemoglobin degradation, color transitions from bright red to brown or black, clotting behavior, drying times, oxidative changes, and enzymatic decay all provide context-dependent clues about the likely time since deposition.

Advances in molecular biology have also introduced promising techniques, including RNA degradation markers, microbial succession patterns, and biomolecular decay signatures. While environmental factors—temperature, humidity, airflow, substrate characteristics—strongly influence these processes, they still enable investigators to estimate approximate timelines. These estimates can support or contradict alibis, help sequence multiple injuries, or clarify whether blood was deposited before, during, or after a particular event.

9.4 Linkage Across Scenes, Individuals, and Objects

Blood evidence often serves as a critical linking mechanism that connects suspects, victims, and multiple crime scenes. A DNA profile obtained from blood can bind different locations or objects together, demonstrating continuity in an investigation. For example, a matching profile recovered from a crime scene, a discarded weapon, and a suspect's clothing strongly supports that these items are part of the same chain of events.

This linkage is especially important in serial offenses, cases involving multiple victims, or situations where perpetrators attempt to relocate or conceal evidence. Blood on a suspect's clothing may indicate proximity or involvement in a struggle, while blood on an object can show handling, movement, or disposal. Such associations help investigators construct coherent case theories grounded in scientific evidence, thereby strengthening the investigative and legal process (Kumar and Singh 2020).

Blood evidence often remains when other forms of evidence degrade, are removed, or fail to capture critical information. Its biological uniqueness, environmental resilience, and compatibility with advanced scientific techniques make it one of the most reliable resources in forensic investigation. Through its capacity to identify individuals, reconstruct violence, estimate timelines, and link scenes or persons, blood evidence continues to be indispensable in uncovering the truth behind violent crimes.

10 Interpretation and Reporting of Blood Evidence

Interpretation and reporting represent the final—and often the most critical—stages of forensic blood analysis. While laboratory testing can identify biological profiles and characterize bloodstain patterns, these results only gain meaning when integrated into the broader context of the case. Interpretation is therefore inherently interdisciplinary, requiring not only scientific expertise but also knowledge of investigative circumstances, crime-scene reconstruction principles, and legal standards. This section outlines the key considerations, responsibilities, and ethical obligations involved in interpreting and reporting blood evidence in criminal investigations.

10.1 Distinguishing Primary and Secondary Blood Transfer

A fundamental aspect of interpretation is differentiating between primary and secondary bloodstains.

- *Primary bloodstains* originate directly from a bleeding source—typically a victim's wound—and retain information about the mechanism of injury, direction of force, and relative positions of individuals involved. Impact spatters, arterial spurts, and projected stains fall into this category and are essential for reconstructing violent events.
- *Secondary transfer stains*, in contrast, result from blood being moved from one surface to another through contact, dragging, or object movement. These patterns may reflect post-incident activity and can be easily misinterpreted if viewed

without context. Distinguishing between primary and secondary patterns helps avoid erroneous conclusions, such as mistaking a smear for evidence of an assault when it may simply indicate post-event movement.

10.2 Integrating Laboratory Findings with Crime-Scene Evidence

Meaningful interpretation requires careful synthesis of laboratory data with physical crime-scene observations. While biological testing reveals *whose blood is present*, contextual factors determine *what the presence of that blood signifies*. Critical considerations include:

- The stain's location and spatial distribution
- Pattern morphology
- Volume and directionality
- Environmental effects such as drying, heat, or disturbance

Analysts must evaluate several possibilities: whether the blood was deposited before, during, or after the crime; whether it originated from active bleeding, passive dripping, expirated blood, or mechanical transfer; and whether it reflects the actions of the victim, suspect, or another individual. A holistic and scientifically supported interpretation minimizes oversimplification and ensures that conclusions withstand investigative and legal scrutiny.

10.3 Structure and Content of Forensic Reports

Formal forensic reports should present findings in a clear, logical, and scientifically transparent manner. Essential components include the following:

(a) *Description of methods*

Reports must detail all procedures used, including:

- Presumptive and confirmatory tests for blood
- DNA extraction, quantification, and amplification methods
- Quality assurance protocols and controls

Comprehensive documentation ensures reproducibility and facilitates expert review.

(b) *Presentation of results*

Analytical findings should be reported with precision. For DNA profiling, this includes:

- Markers tested
- Whether the profile is full or partial
- Statistical interpretation of matches or exclusions

For bloodstain pattern analysis, the report should describe:

- Stain types observed
- Their distribution and location
- Conclusions regarding probable mechanisms of origin

Any limitations, anomalies, or incomplete results must be transparently noted, as these factors may influence interpretation.

10.4 Scientific Communication and Expert Testimony

Forensic experts have an ethical responsibility to communicate findings neutrally, clearly, and scientifically—without bias toward prosecution or defense. Expert testimony should:

- Explain complex concepts in terms understandable to judges and juries.
- Avoid overstating conclusions or implying certainty where uncertainty exists.
- Clearly distinguish between what the evidence supports and what it cannot determine.

For instance, a DNA match may strongly support the involvement of an individual but cannot, on its own, establish guilt or reconstruct the exact sequence of events. Similarly, bloodstain patterns can suggest likely mechanisms but cannot reproduce the entire crime with absolute accuracy.

10.5 Ethical Obligations

Transparency, impartiality, and scientific integrity are central to proper interpretation and reporting. Key duties include:

- Maintaining a complete and unbroken chain of custody
- Adhering to validated protocols and quality assurance standards
- Conducting analyses without preconceived expectations
- Fully disclosing all results, including inconclusive or exculpatory findings

Failure to meet these obligations risks flawed conclusions and jeopardizes the fairness of judicial outcomes.

Ultimately, the interpretation and reporting of blood evidence serve as the bridge between scientific data and judicial decision-making. When performed rigorously, transparently, and ethically, these processes provide courts with reliable information essential for fair and accurate adjudication (Ackerman 2006).

11 Recent Advances in Blood Evidence Analysis

Recent developments in molecular biology, biochemistry, bioinformatics, and digital technologies have significantly transformed the forensic analysis of blood evidence. These innovations have improved the precision, reliability, and interpretive power of blood analysis, allowing investigators to answer questions that were previously beyond the scope of traditional laboratory techniques. Advanced analytical platforms—such as proteomics, RNA profiling, microbiome sequencing, and artificial intelligence—have expanded the forensic toolkit, enhancing individual identification, crime-scene reconstruction, and temporal estimation of bloodstains.

11.1 Proteomic Advances in Blood Analysis

One of the most revolutionary developments in recent years has been the application of proteomics, which enables large-scale analysis of proteins and their degradation patterns. In contrast to DNA, which remains relatively stable over long periods, proteins degrade in more predictable, time-dependent ways. This makes them highly valuable for estimating the age of a bloodstain.

Specific proteins—including hemoglobin variants, fibrinogen fragments, and oxidative breakdown products—show measurable changes over time. Using mass spectrometry, forensic scientists can identify these molecular signatures and generate more accurate timelines for when blood was deposited. Proteomic analysis also aids in:

- Distinguishing human vs. nonhuman blood
- Identifying physiological conditions or disease markers
- Contributing to biological characterization when DNA is degraded or unavailable

With advances in instrumentation and expanding protein databases, proteomics is poised to become a central component of forensic interpretation, especially for timeline estimation and characterizing complex blood evidence.

11.2 RNA Profiling and Time-Since-Deposition (TSD) Estimation

Complementing proteomic methods are RNA-based techniques, which represent another major leap forward in forensic science. RNA molecules—particularly mRNA and microRNA—degrade rapidly, making them sensitive indicators for determining time since deposition.

Key applications include:

- *TSD estimation* using degradation curves that correlate RNA loss with environmental conditions

- *Body fluid identification*, as certain mRNA markers are highly specific to blood, saliva, semen, or menstrual fluid
- *Distinguishing closely related fluids*, such as venous blood from menstrual blood

MicroRNAs, being shorter and more stable than mRNA, provide strong discriminatory power in mixed or degraded samples. These techniques are especially valuable in cases with disputed timelines, mixed stains, or uncertainty about the fluid's origin.

11.3 Microbiome-Based Forensic Analysis

An emerging frontier in blood forensics involves the analysis of the microbiome associated with bloodstains. Every biological fluid contains a unique microbial community that shifts predictably as it ages and interacts with the environment.

Using high-throughput sequencing, forensic microbiologists examine microbial succession to determine:

- Approximate age of the bloodstain
- Whether the stain was deposited indoors or outdoors
- Environmental influences such as sunlight, humidity, and substrate type
- Indicators of cleaning attempts or disturbance

Although microbiome forensics is still developing, early research highlights its enormous potential, especially when integrated with proteomic and RNA-based methods. Together, these tools provide a multidimensional view of bloodstain aging and environmental exposure.

11.4 Artificial Intelligence and Digital Forensics in BPA

Advances in artificial intelligence (AI), machine learning, and computer vision have greatly enhanced bloodstain pattern analysis (BPA)—a field historically criticized for subjectivity and dependence on analyst expertise.

AI and digital tools now enable:

- Automated classification of bloodstain patterns (e.g., impact spatter, cast-off, transfer, expirated stains)
- Identification of subtle features that may not be visible to the human eye
- Reduction of expert bias and increased reproducibility

Deep learning models trained on thousands of annotated images provide consistency across analyses and help standardize BPA interpretation. Computational fluid dynamics and digital simulations allow investigators to estimate:

- Angle of impact
- Force and velocity of blood projection
- Droplet trajectories and point-of-origin estimates

These tools introduce greater objectivity and scientific rigor into BPA, addressing long-standing criticisms about variability in expert interpretation (Jithin et al. 2018).

11.5 Three-Dimensional Crime-Scene Reconstruction and Spatial Mapping

Digital forensic innovations, including 3D laser scanning and high-resolution photogrammetry, allow comprehensive recording of bloodstains along with precise spatial data. These technologies enable:

- Creation of virtual, interactive crime scenes
- Analysis of bloodstains from multiple angles
- Preservation of scene information long after cleanup or environmental changes
- Enhanced communication of complex spatial information in court

These 3D reconstructions can be integrated with AI-assisted BPA outputs, giving investigators powerful tools to visualize blood dynamics and reconstruct events with greater accuracy.

12 Case Studies and Landmark Investigations

Blood evidence has played a decisive role in many landmark criminal investigations, shaping both forensic practice and legal standards. Historical case examples illustrate not only the evidentiary power of bloodstains and DNA but also the vulnerabilities that arise when biological material is mishandled or misinterpreted (Rajesh 2024). These cases collectively demonstrate how scientific methods have evolved and how blood evidence has become central to crime reconstruction and adjudication.

One of the most widely publicized cases relying heavily on blood evidence was the 1995 O.J. Simpson homicide trial in the United States. The murders of Nicole Brown Simpson and Ron Goldman produced extensive blood evidence found at the crime scene, inside Simpson's vehicle, and within his home. DNA profiling linked these stains to both victims and to Simpson with statistically powerful certainty.

However, despite the strength of the scientific results, the defense successfully challenged the integrity of the evidence on procedural grounds. Allegations of improper packaging, possible contamination, inconsistent documentation, and weaknesses in the chain of custody cast doubt on the reliability of the tests. These criticisms ultimately shaped the jury's perception more strongly than the scientific findings themselves. The Simpson trial became a landmark reminder that even highly reliable biological techniques require meticulous collection, preservation, and documentation to withstand legal scrutiny. It also triggered worldwide discussions on laboratory quality assurance and evidence-handling standards.

In contrast, the Colin Pitchfork case (United Kingdom, 1986) showcased the transformative potential of DNA evidence when properly used. Following the murders of Lynda Mann and Dawn Ashworth, investigators lacked solid leads until forensic scientists analyzed biological material using the newly developed technique of DNA fingerprinting, pioneered by Sir Alec Jeffreys. This analysis exonerated an initially suspected teenager by proving the crime-scene DNA did not match his. Recognizing thc powcr of the emerging technology, police initiated the first-ever mass DNA screening, collecting voluntary blood samples from thousands of local men.

Pitchfork attempted to evade detection by persuading a colleague to submit a sample on his behalf, but the deception was uncovered. When his own sample was tested, it matched both crime scenes, making him the first individual in history to be convicted on the basis of DNA evidence. This case revolutionized forensic science, demonstrating that genetic profiling could simultaneously protect the innocent and identify the guilty, leading to its rapid global adoption (Vennemann 2021).

In subsequent decades, advancements in DNA extraction, amplification, and interpretation have allowed investigators to analyze even highly degraded or minimal bloodstains. Such developments have been pivotal in solving cold cases, where aged biological samples—sometimes preserved for decades—finally produced usable genetic profiles. Many long-unsolved homicides, serial offenses, and unidentified victim cases were resolved only after modern methods enabled successful analysis of old blood traces.

Blood evidence also remains fundamental to bloodstain pattern analysis (BPA), a discipline that helps reconstruct events at violent crime scenes. Characteristics such as stain size, distribution, directionality, and morphology can provide insight into the positions of individuals, the sequence of events, and the types of forces involved. When integrated with DNA results and other forensic findings, BPA contributes to a more complete reconstruction of crime dynamics. More recently, investigative tools such as familial DNA searching and forensic genetic genealogy have expanded the utility of blood evidence even further. In situations where no direct database match exists, these methods can identify biological relatives of an unknown offender, significantly narrowing suspect pools and enabling breakthroughs in otherwise stalled investigations.

Collectively, these landmark cases and evolving techniques underscore the profound impact of blood evidence on modern forensic science. When collected, analyzed, and interpreted with scientific rigor, blood remains one of the most powerful and reliable forms of forensic material available to investigators and courts.

13 Future Perspectives in Blood Forensics

Blood forensics is poised for transformative advancements as innovations in molecular biology, analytical science, and digital technologies converge. These emerging tools promise to expand the depth of information obtainable from blood evidence while addressing long-standing challenges related to sample degradation, subjectivity in interpretation, and laboratory constraints. The future of blood analysis is expected to be more rapid, precise, and integrative than ever before.

One of the most promising developments is the adoption of multi-omics approaches, which combine genomics, transcriptomics, proteomics, and microbiomics to generate comprehensive biological profiles. Traditional DNA profiling focuses primarily on genetic markers for identification; however, multi-omics has the potential to reveal the origin, age, and post-depositional changes of a bloodstain in far greater detail.

- *Genomics* continues to strengthen individual identification and kinship analysis.
- *Transcriptomics*, particularly RNA-based signatures, provides estimates of stain age and can distinguish between various biological fluids.
- *Proteomics* contributes by identifying proteins and peptides whose degradation patterns can refine timelines for bloodstain aging.
- *Microbiomics*, which examines microbial communities associated with blood, offers environmental insights that may assist in determining location or exposure history.

Together, these integrated datasets can form a powerful, layered framework that may fundamentally reshape how investigators interpret biological evidence. Equally significant is the growing role of artificial intelligence (AI) and computational modeling in bloodstain pattern analysis (BPA). Traditional BPA has often faced criticism for its reliance on expert judgment, which can introduce inconsistency or bias. AI-driven systems—including machine learning and deep learning algorithms—are capable of analyzing vast collections of digital images to classify bloodstain patterns with remarkable accuracy. These tools can detect subtle geometric and spatial features that might be overlooked by human observers, thereby improving objectivity.

Advanced computational simulations are also emerging as essential complements to BPA. These models can recreate blood dynamics under varying conditions, enabling investigators to simulate impact forces, projection angles, and spatial relationships between victims and assailants. Such digital reconstructions provide scientifically grounded scenarios that enhance the reliability of crime-scene interpretations (Rana 2018). As research and technology continue to advance, the integration of multi-omics, AI-assisted analysis, and sophisticated modeling will likely redefine the future of blood forensics. These innovations aim not only to increase accuracy and speed but also to strengthen the scientific foundations that support forensic conclusions in the justice system.

14 Conclusions

From early serological methods such as blood group typing and enzyme analysis to today's sophisticated molecular, imaging, and computational technologies, the forensic examination of blood has evolved into a highly advanced scientific discipline. Despite these advancements, the fundamental value of blood evidence remains unchanged—its ability to provide a direct and objective biological connection between individuals, objects, and crime scenes. In many investigations, it offers clarity even when other forms of evidence are limited or absent.

A defining strength of blood evidence is its remarkable versatility. At the most basic level, visible bloodstains indicate where violence occurred and who may have been involved. Beyond this, detailed interpretation of bloodstain patterns can reveal the angle and force of impact, the type of weapon used, sequence of events, or possible attempts to stage or alter a crime scene. Such analyses support accurate event reconstruction and help corroborate or contradict witness and suspect statements. When combined with modern DNA technologies, the evidentiary power of blood becomes even greater. A microscopic quantity of blood can identify or exclude individuals with exceptional precision. While STR profiling remains the cornerstone of forensic DNA identification, emerging techniques—including SNP analysis, mRNA profiling, proteomics, and epigenetic markers—allow investigators not only to determine whose blood is present but also to infer tissue origin, estimate stain age, and occasionally detect the aspects of the donor's physiological state. These advances have significantly enhanced the interpretive capabilities and scientific reliability of blood evidence.

Technological innovations are further transforming practical forensic workflows. Miniaturized analytical devices, rapid DNA systems, and integrated digital platforms enable faster, sometimes on-site, analysis of biological material. Additionally, artificial intelligence is increasingly being applied to interpret complex patterns and large datasets, reducing the subjectivity associated with traditional approaches and promoting greater consistency across laboratories. In summary, the enduring importance of blood evidence lies in its unique ability to bridge rigorous scientific inquiry with the realities of criminal behavior. When collected, preserved, and analyzed according to established standards, it provides objective and often decisive information that can guide investigations, reveal otherwise hidden narratives, and support fair judicial outcomes. As scientific advancements continue to emerge, blood evidence will remain one of the most indispensable and informative components of modern forensic practice.

References

Abid IF, Rezakola E, Hanura IY (2024) Teichmann test: assessment of hemoglobin crystals on blood spots exposed to powder detergent. Jurnal Biosains Pascasarjana 26(1):23

Ackerman MJ (2006) Forensic report writing. J Clin Psychol 62(1):59–72

Attinger D, Moore C, Donaldson A, Jafari A, Stone HA (2013) Fluid dynamics topics in bloodstain pattern analysis: comparative review and research opportunities. Forensic Sci Int 231(1–3):375–396

Bailey GS (1996) Ouchterlony double immunodiffusion. In: The protein protocols handbook. Humana Press, pp 749–752
Barni F, Lewis SW, Berti A, Miskelly GM, Lago G (2007) Forensic application of the luminol reaction as a presumptive test for latent blood detection. Talanta 72(3):896–913
Brodbeck S (2012) Introduction to bloodstain pattern analysis. J Police Sci Pract 2(51):57
Castro DM, Coyle HM (2013) Biological evidence collection and forensic blood identification. Forensic Science Department, Henry C. Lee College of Criminal Justice & Forensic Science, University of New Haven 300(3):1221–1229
Dixit A, Yadav J, Lagarkha R, Yadav A, Singhal P (2023) Tracing the clues: exploring dried bloodstain analysis on multiple objects in varied environmental conditions. Int J Multidiscip Res 5(4)
Glaister J (1926) The Kastle-Meyer test for the detection of blood: considered from the medicolegal aspect. Br Med J 1(3406):650
Ip SC, Yu EY (2021) Blood DNA preservation on various forensic swab devices. JFI 71(1)
Jithin P, Kadar N, Mohan A, Pynadath MK, Mohammed PK, Kalliath C (2018) Recent advancements in human blood identification in forensics. Oral Maxillofac Pathol J 9(2)
Krebs HA (1950) Chemical composition of blood plasma and serum. Annu Rev Biochem 19(1):409–430
Kumar A, Singh J (2020) Significance of blood and its identification: a perspective in forensic investigation. Vigyan Varta 1(7)
Li R (2012) Forensic serology. In: Forensic chemistry handbook, 1st edn. Wiley, pp 269–290
Lv YJ, Liang XF, Wu YP (2023) Clinical application of ABO blood typing. Technol Health Care 31(4):1375–1383
Peschel O, Kunz SN, Rothschild MA, Mützel E (2011) Bloodstain pattern analysis. Forensic Sci Med Pathol 7(3):257–270
Rajesh J (2024) The OJ Simpson trial: forensic evidence and narrative construction. Int J Law Manag Humanit 7(6):2250
Rana AK (2018) The future of forensic biology. J Biomed 3:13–18
Singh L (1991) DNA profiling and its applications. Curr Sci 60:580–585
Singh HN (2021) Collection, preservation and transportation of biological evidence for forensic DNA analysis. Int J All Res Educ Sci Methods 9(9):1123–1130
Singh P, Gupta N, Rathi R (2021) Blood pattern analysis—a review and new findings. Egypt J Forensic Sci 11(1):9
Vandewoestyne M, Lepez T, Van Hoofstat D, Deforce D (2015) Evaluation of a visualization assay for blood on forensic evidence. J Forensic Sci 60(3):707–711
Vennemann M (2021) New challenges in forensic DNA analysis
Yadav PK, Sharma S (2024) Application of forensic techniques to blood analysis. In: Advancements in body fluid analysis in forensics: current and emerging methods. Springer Nature Switzerland, pp 17–90
Zagorodniy DLSV, Zagorodniy DLSI (2024) Non-destructive method of blood serum composition analysis in forensic science. Technium BioChemMed 8:22–29
Zhou Y, Li S, Zhou J, Wang L, Song X, Lu X et al (2011) DNA profiling in blood, buccal swabs, and hair follicles of patients after allogeneic peripheral blood stem cell transplantation. Legal Med 13(1):47–51

Disease Biomarkers and Predictive Forensics in Blood

7

Abstract

Disease blood biomarkers are defined as measurable indicators in blood circulation reflecting physiological, pathological, or therapeutic states. Blood biomarkers are playing a dual transformative role in modern healthcare: as clinical tools for diagnosis and management and as predictive forensic evidence in legal and regulatory contexts. This chapter will examine their application across both communicable (e.g., viral hepatitis, COVID-19) and noncommunicable diseases (e.g., cardiovascular disease, diabetes, cancer), highlighting their evolving forensic utility. In communicable diseases, blood biomarkers enable early detection, severity stratification, and monitoring of viral dynamics. For hepatitis B, markers such as HBsAg, HBeAg, and HBV DNA quantify infection status and viral load, while anti-HDV IgM and HDV-RNA confirm hepatitis D co-infection. From a forensic perspective, these biomarkers can: validate occupational disease claims, substantiate medical negligence allegations (e.g., delayed myocardial infarction diagnosis); support insurance or disability determinations (e.g., chronic kidney disease progression); and inform regulatory actions regarding drug safety or environmental hazards. Ultimately, blood biomarkers represent a powerful interface between medicine and law: they translate biological signals into actionable evidence, improving patient outcomes while advancing fairness and accountability in forensic contexts.

Keywords

Blood biomarkers · Predictive medicine · Forensic medicine · Communicable diseases · Non-communicable diseases · Diagnostic evidence · Legal applications

1 Introduction

The identification and validation of disease biomarkers have revolutionized modern medicine, providing objective, quantifiable indicators of physiological states, pathological processes, and therapeutic responses. According to the U.S. Food and Drug

H. R. Dash et al., *Forensic Investigation of Blood Evidence*,
https://doi.org/10.1007/978-981-95-9183-1_7

Administration (FDA), a biomarker is defined as "a characteristic that is objectively measured and evaluated as an indicator of normal biological processes, pathogenic processes, or pharmacological responses to a therapeutic intervention" (FDA 2025) and is having critical roles across the healthcare system, going from the early diagnosis and disease monitoring to treatment personalization and outcome prediction.

In the clinical practice, biomarkers enable an early detection of diseases before symptom onset, a stratification by severity or subtype of the disease, a selection of the treatment on the molecular profiles, ensures a response monitoring through serial measurements (e.g., HbA1c in diabetes); and aids in the estimation of prognosis using risk-associated markers (e.g., troponin in acute coronary syndromes).

When turning into forensics and applying the notion to diseases, predictive forensic biomarkers are considered as bridges between medicine and justice as they constitute the use of medical knowledge to legal issues. Biomarkers in forensic prediction can objectify symptoms that are often considered as subjective symptoms as neurofilament light chain for neurological injury in long COVID, by establishing temporal sequences of disease onset and progression, linking the exposure to the harm in occupational or environmental litigation, serving to the verification or to refute of claims in insurance, disability, or personal injury cases and supporting public health investigations into outbreaks or policy failures. Sometimes, disease biomarkers can refine the investigation of criminal cases, in cases where medical records of the suspect/victim are available.

The biomarkers used in forensic science transform biological evidence into admissible data for legal proceedings, regulatory decisions, and epidemiological accountability.

The investigation of blood biomarkers spans molecular (DNA/RNA/proteins), biochemical (enzymes/metabolites), and cellular (circulating cell subsets) fields and is of forensic value when measured according to standardized collection protocols by reproducible laboratory assays, longitudinal monitoring capacity, and compliance with regulatory frameworks (CLIA/ISO).

Despite their promising potential, challenges persist, including a lack of assay standardization, biological variability across populations, and ethical concerns around data privacy. Genomics, proteomics, metabolomics, and artificial intelligence are accelerating biomarker discovery, paving the way for more precise individualized predictions and expanding forensic applications into workplace safety litigation, vaccine injury claims, environmental health disputes, and disability benefit determinations. As analytical methods improve and regulatory pathways evolve, disease biomarkers will continue to transform both patient care paradigms and the administration of justice (Hood and Flores 2012).

2 Blood Biomarkers of Cancers

Beyond being recognized as a natural cause of death, cancers are progressively gaining significance in forensic investigations, indicating exposure to environmental contamination, inadequate medical care, or occasionally, hidden criminal intent.

Alongside conventional forensic methods such as autopsies, analytical methods involving plasma and molecular blood markers in the context of the "molecular autopsy" are tending to gain ground as confirmatory methods.

2.1 Blood Biomarkers of Solid Cancers

Although the gold-standard diagnosis of solid cancers is biopsies, blood-based biomarkers represent a noninvasive alternative in forensic medicine in cases when only blood is available.

2.1.1 Prostate Cancer Biomarkers

Prostate cancer is subdivided into hormone-sensitive prostate cancer and castration-resistant prostate cancer (mCRPC). PSA is considered a gold standard in clinics but is rarely used in forensics during autopsies when the diagnosis is known. Takaso and colleagues had demonstrated that forensic autopsies are useful in determining Latent Prostate Cancer Prevalence (Takaso et al. 2020).

In the forensic context, PSA (Prostate Specific Antigen) is used to elucidate cases of sexual offenses by collecting *postmortem* forensic samples to indicate the presence of seminal fluid. A study by Müller and colleagues performed a retrospective analysis of 484 samples collected in sexual assault cases. The positive rate of positive SERATEC® PSA Semiquant tests PSA was 33% for women and 67% for men. The use of male-specific qPCR male DNA was detected in only one of the female samples, indicating a high false-positive rate for both sexes. Globally, the study demonstrated a very limited probative value of PSI as a marker for seminal fluid postmortem, independently of the degree of decomposition (Müller et al. 2025).

Jones and colleagues had demonstrated that total PSA levels were similar in ante- and postmortem sera if autopsy blood was drawn by ~24 h following the time of death. Free PSA levels, however, were increased approximately twofold in most postmortem sera analyzed (Jones et al. 2005).

The prognostic value of circulating tumor cells (CTCs) in both (mHSPC) and(mCRPC) has been highlighted. The decrease in CTC count from ≥ 5 to ≤ 5 or a 30% after 12 weeks of treatment serves as an important prognostic marker (Castro et al. 2025). In mCRPC, AR-V7 expression in CTCs was shown to be independently associated with a shorter progression-free survival (HR 1.95, 95% CI 1.1–3.3) and an overall survival (HR 4.2, 95% CI 2.1–8.5) (Castro et al. 2025; Armstrong et al. 2019). Circulating tumor DNA (ctDNA) is also of prognostic use, particularly during the late stages of the disease, taking into account that the ctDNA "conversion" occurs after 4 weeks of androgen receptor pathway inhibitor treatment and after 6 weeks of Lu-PSMA therapy, providing clinically relevant insights, with a high concordance between ctDNA and tissue analyses, enabling predictive biomarker detection. CtDNA also proved to be valuable for monitoring the resistance to PARP inhibitors. Additionally, a whole-exome study of Beltran and colleagues using bisulfite sequencing had revealed that CRPC-neuroendocrine features are detectable in circulation, deepening the understanding of disease progression.

2.1.2 Ovarian Cancer Blood Biomarkers

Ovarian cancer is characterized by a high mortality due to its asymptomatic nature in early stages. Among the most established serum markers figures, CA125, which, despite limited sensitivity in early-stage disease (detected in only 50% of cases) during the late 1980s (Bast Jr and Knapp 1985), remains a key component of diagnostic schemes, particularly in postmenopausal women in combination with other biomarkers during the assessment of the Risk of Malignancy Index (RMI) and Risk of Ovarian Malignancy Algorithm (ROMA) (Skates et al. 2004).

HE4 (human epididymis protein 4) has emerged as a promising biomarker, demonstrating high sensitivity (73%) and specificity (95%) in the detection of early-stage ovarian cancer (Diamandis et al. 2002). The combination of the latter one with CA125 significantly improves the diagnostic accuracy (Skates et al. 2004).

Among other significant conventional plasmatic biomarkers, osteopontin (OPN) shows a strong prognostic value, as its elevated levels were reported to be correlated with poor outcomes and metastasis (Diamandis et al. 2002). Notable predictive potential had been reported for Kallikreins (KLKs), as 12 out of 15 family members were reported to be upregulated in ovarian cancer, with specific correlations to prognosis and chemoresistance (Diamandis et al. 2002).

Bikunin, a Kunitz-type protease inhibitor and a heavily glycosylated protein, is associated with survival outcomes in ovarian cancer. Preoperative plasma bikunin levels below 11.5 μg/mL are linked to advanced disease and poor response to treatment, while higher levels are associated with improved survival (Petricoin et al. 2002).

The elevated levels of VEGF (vascular endothelial growth factor) were also indicative of poor survival outcomes (Diamandis et al. 2002). Among biomarkers included in multiplex panels, Prostacin (PSN) overexpression showed a specificity of 94% during ovarian cancer diagnosis (Diamandis et al. 2002). Decreased levels of Apolipoprotein A-I (apoA-I) have been reported to achieve a 98% specificity.

Among emerging blood-based biomarkers, aldehyde dehydrogenase 1 (ALDH1) was associated with a longer survival and stem cell characteristics, folate receptor alpha (FOLR1) was reported to be overexpressed in 69% of uterine serous carcinomas, glutathione S-transferase polymorphisms (GSTP) were linked to chemotherapy response, and copper isotope ratios showed potential as early disease indicators.

Multimarker panels have demonstrated superior performance over single biomarkers: combining CA125, HE4, and mesothelin, a significant enhancement of sensitivity and specificity was demonstrated. The integration of proteomic and genomic approaches has identified novel candidates, such as exosomal markers and noncoding RNAs, which show promise for future diagnostic applications.

2.1.3 Colorectal Cancer Blood Biomarkers

The recognized gold standard in colorectal cancer diagnosis is the tissue biopsy, which, however, has limitations. In the context of blood-based biomarkers, carcinoembryonic antigen (CEA) and CA-19.9 are conventional biomarkers used to diagnose colorectal cancer (Cenin et al. 2018).

During the last decade, the liquid biopsy approach emerged as a complementary method overcoming limitations such as intratumoral heterogeneity and the invasive nature of biopsy procedures.

As mentioned earlier, circulating tumor DNA (ctDNA) is actively investigated in cancer fields, and colorectal cancer is not an exception, where ctDNA represents a pivotal noninvasive biomarker for assessing recurrence risk, treatment monitoring, and therapy personalization (Vojjala et al. 2025).

Despite its promise, standardization of ctDNA analysis methods remains a challenge. Commonly used approaches include the droplet digital PCR (ddPCR), real-time PCR (qRT-PCR), and next-generation sequencing (NGS), and each method has its advantages and limitations, complicating cross-study result comparisons (Bettegowda et al. 2014).

In summary, ctDNA shows significant potential in managing colorectal cancer, yet further research is needed to standardize methods and clarify their role across diverse clinical scenarios.

2.1.4 Hepatocellular Carcinoma Blood-Based Biomarkers

The early detection of hepatocellular carcinoma is critical for improving survival rates, which currently remain below 20% with a prevalence of over 860,000 new cases annually. Traditional molecular screening includes alpha-fetoprotein (AFP) testing but is limited by a sensitivity of 40–60%, a rate of false-negative of 30–40%, and a specificity of 80–90% (Huang et al. 2022, 2023; Yang et al. 2019).

Among blood-based emerging biomarkers, proteins are still the most informative as Glypican-3 (GPC3) with a sensitivity of 50–72%, heat shock protein 70 (HSP70), glutamine synthetase (GS), and the AFP-L3 variant and des-γ-carboxyprothrombin (DCP) (Li et al. 2018). DNA-based markers are also in the trend as methylated DNA patterns, circulating tumor DNA (ctDNA) analysis, and gene mutation detection. Emerging RNA markers include microRNAs and circulating RNA analysis.

2.1.5 Pancreatic Cancer Blood-Based Biomarkers

Pancreatic ductal adenocarcinoma (PDAC) is one of the deadliest cancers (Sung et al. 2021; Siegel et al. 2020), with a mortality rate of approximately 7% of all cancer lethal cases globally (Sung et al. 2021). Early detection and effective management are critical for improving outcomes.

Blood-based biomarkers offer a promising alternative to tissue-based diagnosis. The primary established serum marker, CA 19–9, is reported to have limited utility due to issues with sensitivity and specificity (O'Neill and Stoita 2021). Other blood-based biomarkers are in development in the proteic context, utilizing techniques such as mass spectrometry, ELISA, and immunoassays to discover new candidates.

These are often investigated alongside DNA-based biomarkers, such as circulating tumor DNA (ctDNA), KRAS gene mutation screening, and the identification of DNA-methylation patterns in order to provide prognostic information (Cox et al. 2025; Bettegowda et al. 2014). Detection methods for these DNA biomarkers include digital droplet PCR (ddPCR), next-generation sequencing (NGS), and whole-exome sequencing (WES) (Bettegowda et al. 2014).

2.1.6 Blood-Based Biomarkers of Lymphoma

Lymphoma is a heterogeneous group of blood cancers originating in the lymphatic system. It includes over 50 subtypes, divided into Hodgkin lymphoma (HL) and non-Hodgkin lymphoma (NHL). The disease manifests through uncontrolled growth of lymphocytes, forming tumors in lymph nodes and other organs. In the US, approximately 8830 new Hodgkin lymphoma cases are diagnosed annually, with a 5-year survival rate of 89%. For NHL, around 80,620 new cases are expected yearly, constituting about 4% of all cancers (Siegel et al. 2020).

Predictive biomarkers in lymphoma include epigenetic markers such as DNA methylation patterns and histone modifications; the detection of genetic alterations such as translocations, mutations, and amplifications is in conventional use. MicroRNAs (miRNAs) and other noncoding RNAs are also used as biomarkers, detected by transcriptomic methods through the study of gene expression regulation (Voropaeva et al. 2022). Immunophenotypic markers such as CD15 and CD30 are also used in the context of noninvasive biomarkers (Gharbaran et al. 2013). Microenvironment factors involved in tumor–host interactions are also subjected to detection as epigenetic markers like P16/INK4A, MGMT, KLF4, MLL2, MEF2, and EZH2; genetic markers involving mutational screening of BCL2, TP53, MYC, BCL6, MYD88, and SPIB genes; microRNAs such as miR-155, the miR-17-92 cluster, and miR-15a/16-1; and immunophenotypic markers such as CD30, Bcl-6, cyclin D1, Bcl-2, and Myc (Gao et al. 2024).

2.1.7 Breast Cancer

Breast cancer is the most prevalent cancer in women, accounting for approximately 13% of all cancer diagnoses globally. In 2022, the overall mortality accounted for over 670,000 deaths worldwide (WHO 2025a, b).

Main breast cancer subtypes include the luminal A type (approximately 40% of cases), luminal B type, HER2-positive type (approximately 30% of cases), and the triple-negative type (10–15% of cases). These classifications often rely on gene expression patterns and immunohistochemistry (Ahern et al. 2014; Dai et al. 2015).

Key biomarkers in use are of a histological type, specifically the hormone receptors (oestrogen receptor [ER], progesterone receptor [PR]), human epidermal growth factor receptor 2 (HER2) status, and the Ki-67 proliferation index. Serological markers currently used for monitoring recurrence and treatment response include CA 15-3, CEA, and CA 27.29, as well as circulating tumor cells (CTCs) (Duffy 2006).

Genetic markers utilized for hereditary risk assessment, prognosis, and treatment guidance include the screening of BRCA1/BRCA2, PIK3CA, TP53, and GATA3 mutations, often detected via liquid biopsies and next-generation sequencing techniques (Al Sharhan et al. 2022).

2.1.8 Endometrial Cancer

Endometrial cancer (EC) is the most common gynecological malignancy in high-income countries, with over 420,000 new cases diagnosed annually worldwide (Sung et al. 2021). Early detection is crucial for improving patient outcomes, but current diagnostic methods, such as invasive biopsies, are invasive and lack sensitivity.

Among the novel protein biomarkers investigated, NGAL (neutrophil gelatinase-associated lipocalin) and visfatin have shown promising diagnostic value, with AUC (area under the curve) values reported up to 0.92 in some studies. The DJ-1 protein demonstrates high prognostic value, and TFPI2 (tissue factor pathway inhibitor 2) shows potential as a prognostic marker (Wang et al. 2023).

Metabolomic markers, often included in diagnostic panels, achieve high accuracy with AUC values up to 0.97. Specifically, lipid profiles show promise in early-stage detection, and plasma metabolomics has demonstrated high diagnostic accuracy in various studies (Wang et al. 2023). MicroRNA markers such as miR-29b show significant diagnostic potential with an AUC of 0.97 in specific contexts, contrasting with some microRNA panels that have reported AUC values up to 0.78 (Wang et al. 2023).

2.2 Blood Biomarkers of Non-Solid Cancers

The use of blood-based biomarkers in forensic science is not a primary focus even in the field of nonsolid cancer research but is indirectly applied in postmortem diagnosis, by identifying leukemia subtypes through the analysis of blood or bone marrow samples. Genetic profiling is also another option, in cases where DNA is available, genetic markers associated with leukemia could be analyzed to establish the presence of the disease. Protein analysis techniques can be applied, such as immunohistochemistry. The forensic application of these biomarkers is limited by several factors such as sample integrity, where postmortem degradation of tissues and DNA can complicate analysis, and specificity, taking into account that many biomarkers are not unique to leukemia and could be present in other conditions and methodological challenges as forensic labs may not have the equipment or expertise for advanced molecular testing typically used in clinical diagnosis.

2.2.1 Blood Biomarkers of Leukemia

Leukemia represents a group of blood cancers characterized by an uncontrolled proliferation of abnormal white blood cells. The four main types of leukemia are classified based on the type of cell affected (lymphoid or myeloid) and the speed of progression, either acute or chronic. Each type has distinct biomarkers that help in diagnosis, prognosis, and treatment planning and prediction.

Acute Lymphoblastic Leukemia Biomarkers

Acute lymphoblastic leukemia (ALL) is characterized by a clonal expansion of immature lymphoblasts in the bone marrow, blood, and occasionally extramedullary sites (Duffield et al. 2023). ALL is primarily classified into two major subtypes based on the cell of origin: the B cell ALL that accounts for 75–85% of the cases and the T cell ALL that represents 15–25% of the cases (Duffield et al. 2023; Inaba and Mullighan 2020). B ALL is most common in children with a peak incidence from 2 to 5 years and older adults, while T ALL frequently presents in adolescents and young adults with mediastinal involvement (Inaba and Mullighan 2020; Teachey and Pui 2019). Both subtypes are further stratified by recurrent genetic abnormalities—such as BCR: ABL1 fusions in B ALL or NOTCH1 mutations in T ALL, informing on prognosis and therapeutic approaches (Duffield et al. 2023; Al Kuwari et al. 2025). Prompt diagnosis and risk-stratified treatment, including chemotherapy, targeted agents, and stem cell transplantation, have significantly improved survival rates, particularly in pediatric patients (Teachey and Pui 2019).

Conventionally, and during autopsies, immunohistochemical markers are used. Drouot and colleagues had investigated a rare AML sudden death case that was diagnosed postmortem and had emphasized diagnostic challenges. The AML diagnosis was confirmed by immunohistochemistry using markers such as CD43+, CD15+, and MPO+ differentiating by this way myeloid leukemia from other hematologic diseases. The study underscores that a comprehensive approach, integrating autopsy findings, blood biomarker analysis, and immunohistochemical profiling, is essential for correctly diagnosing AML in sudden death cases (Krebs-Drouot et al. 2023).

Emerging AML screened mutations such as FLT3-ITD, NPM1, and CEBPA (Labaj et al. 2017; Bagacean et al. 2022) combined with cytogenetic abnormalities like the presence of the t(8;21) translocation or inv(16) (Labaj et al. 2017). Diagnostic methods include bone marrow biopsy, cytogenetic analysis, and molecular testing (Labaj et al. 2017) and immunohistochemistry (Krebs-Drouot et al. 2023).

Chronic Lymphocytic Leukemia Biomarkers

Chronic lymphocytic leukemia (CLL) is the most common leukemia in adults, characterized by clonal proliferation of mature B cells. Despite its indolent nature, treatment refractoriness remains a significant clinical challenge. The disease is diagnosed by protein markers such as ZAP-70 and CD38, which are associated with a more aggressive form of the disease (Labaj et al. 2017). The levels of BCR-ABL mRNA can be monitored to assess treatment response (Loh et al. 2020). Diagnostic methods include PCR, including the detection of the Philadelphia chromosome (BCR-ABL fusion gene), a hallmark of CML, bone marrow examination, and molecular testing (Labaj et al. 2017).

Although CLL sudden death cases are rare, a case was reported by Gito, when a sudden death of a 70-year-old man in the absence of symptoms was caused by leukostasis, characterized by a highly elevated leukemic cell count and decreased tissue perfusion symptoms that led to severe complications and death. Leukostasis symptoms are increasing blood viscosity due to large leukemic cell populations, a high metabolic activity, and cytokine production by leukemic cells, mechanisms that are responsible for local hypoxic damage that can be fatal (Gitto et al. 2021).

Genetic alterations like mutations in the TP53 gene and deletions of chromosome 13q are analyzed. Other conventional diagnostic tools are blood tests and flow cytometry (Campo et al. 2018).

In forensic science, these biomarkers can potentially be used to identify the presence of leukemia in postmortem samples, though their application in this field is less commonly discussed in the literature (Loh et al. 2020).

3 Blood Biomarkers of Infectious Diseases

3.1 Blood Biomarkers of Tuberculosis

Recent advances in tuberculosis (TB) diagnostics have significantly expanded the understanding of blood-based biomarkers, which play a crucial role in predicting disease progression and treatment outcomes (MacLean et al. 2019). These biomarkers encompass both pathogen-derived molecules and host immune responses, with immunological markers such as cytokines and chemokines showing particular promise in distinguishing active TB from latent infection.

Emerging research has demonstrated the potential of host circulating immunometabolism-associated biomarkers for early diagnosis of active TB, highlighting the value of multi-omics approaches in identifying reliable diagnostic signatures (Yang et al. 2025). Additionally, metabolomic and proteomic profiling have revealed novel biomarker candidates for tuberculous meningitis diagnosis, further expanding the diagnostic toolkit (Van Reenen et al. 2025).

A comprehensive integration of pathogen- and host-derived blood biomarkers has shown particular promise in enhancing TB diagnosis accuracy (Nogueira et al. 2023). This approach leverages both genetic susceptibility factors and immune response markers to provide a more holistic understanding of disease pathogenesis. Mass spectrometry-based proteomic and metabolomic profiling has further validated several serum biomarkers signatures with potential diagnostic utility (Gonçalves et al. 2022).

In addition to the abovementioned biomarkers, TB postmortem diagnosis is usually preceded by PCR from DNA extracted from lung tissue samples, ensuring specificity. These advances represent a significant step forward in TB diagnostics, offering the potential for more accurate and timely disease detection. However, continued research is needed to validate these biomarkers in diverse populations and clinical settings to ensure their broad applicability (Colby 2023).

3.2 Blood Biomarkers of HIV

Predictive forensic biomarkers in HIV infection encompass several key blood-based indicators that collectively enable clinicians to monitor disease progression and treatment efficacy (Conway et al. 2019). The CD4+ T-lymphocyte count serves as the primary measure of immune integrity, with levels below 200 cells/μL indicating a high risk for opportunistic infections (Pasternak et al. 2015).

HIV RNA viral load in plasma quantifies active viral replication, guiding antiretroviral therapy (ART) adjustments and predicting transmission risk. Neurofilament light chain (NfL) levels in serum act as a noninvasive marker of neuronal injury, particularly valuable for detecting HIV-associated neurocognitive disorders (HAND) (Hagberg et al. 2022).

Cell-associated HIV RNA (CA HIV RNA) reflects transcriptionally active viral reservoirs, offering insight into potential viral rebound during treatment interruptions (Rouzioux and Avettand-Fenoël 2018). Total HIV DNA in peripheral blood mononuclear cells (PBMCs) quantifies the viral reservoir size, correlating with long-term progression risk and mortality (Pasternak et al. 2015).

When interpreted together—and in context with clinical data—these biomarkers facilitate personalized treatment strategies, though their diagnostic specificity requires careful consideration due to potential confounding factors (Gisslén et al. 2015).

The postmortem detection of HIV is usually performed using microparticle enzyme immunoassay using the Abbott AxSym analyzer on HIV Ag / Ab Combo (Eriksen et al. 2009).

3.3 Blood Biomarkers of Hepatitis

Predictive blood biomarkers for hepatitis subtypes play a crucial role in diagnosis, monitoring disease progression, and assessing treatment efficacy. These markers vary depending on the specific virus (A, B, C, D, E) and the stage of infection. Table 7.1 describes the detailed properties and diagnostic interpretation and the detection timeframe for each viral subtype.

Advances in biomarker research have significantly enhanced the monitoring and prediction of chronic viral hepatitis progression. Machine learning models incorporating multiple blood biomarkers are used to forecast the progression from hepatitis B virus (HBV) infection to cirrhosis and hepatocellular carcinoma (HCC).

Immunological markers such as PD-L1 and TGF-β1 are under investigation for their potential to predict fibrosis and cirrhosis development in chronic hepatitis C patients, providing valuable insights into the immune response and disease progression pathways.

Oncomarkers, including AFP, AFP-L3, and GP73, have proven particularly useful in the early diagnosis of HCC among patients with chronic hepatitis B or C infections (Oncomarkers section). These biomarkers enable clinicians to detect malignancy at an earlier stage, thereby improving patient outcomes through timely intervention.

In forensics, the postmortem detection of HIV and hepatitis viruses is performed using licenced assays for cadaveric screening in the frame of the ISO17025 norm with a specificity of >95% (Greenwald et al. 2018), by microparticle enzyme immunoassay using the Abbott AxSym analyzer and AUSAB to detect anti-HBs, and

Table 7.1 Properties of viral subtypes of Hepatitis virus

Biomarker		Properties and diagnostic interpretation	Detection timeframe	References
Hepatitis A	Anti-HAV IgM	Indicates current acute or recent infection	Detectable 14–20 days after exposure; persists up to 6 months	Coskun et al. (2019)
	Anti-HAV IgG	Provides long-term immunity; indicates past infection or vaccination	–	
	Platelets	Potential marker of inflammation during acute infection	Higher PCT levels correlate with lower ALT levels (indicating liver inflammation)	
Hepatitis B	HBsAg	Indicates active infection; if persistent for >6 months—chronic infection	–	Hayashi et al. (2021) Nan et al. (2022)
	HBeAg	Sign of active viral replication and high infectivity	>10 weeks in acute phase may indicate chronic progression	
	Anti-HBe	Seroconversion (from HBeAg to anti-HBe) in the acute phase is a positive prognostic sign, indicating viral suppression	–	
	HBV DNA	Quantifies viral load	High levels (>2000 IU/mL) correlate with increased risk of hepatocellular carcinoma (HCC)	
	Serum HBV RNA	Emerging biomarker for enhanced diagnostic efficiency and blood product safety screening	–	
	HBcrAg (Hepatitis B core-related antigen)	New marker for monitoring treatment response and viral replication	–	
	pgRNA (pregenomic RNA)	New marker for monitoring treatment response and viral replication	–	

(continued)

Table 7.1 (continued)

Biomarker		Properties and diagnostic interpretation	Detection timeframe	References
Hepatitis C	Anti-HCV	Initial marker of infection; detectable in acute and chronic phases	–	Bagheri et al. (2024)
	HCV RNA	Confirms active infection; used to assess viral load and treatment response	Quantitative PCR	
	ALT (Alanine aminotransferase)	Elevated levels indicate liver cell damage	Fluctuates during infection	
	APRI (AST to Platelet Ratio Index)	Used to assess liver fibrosis	APRI >1.4 suggests advanced fibrosis	
	Ferritin	Elevated levels associated with hepatic steatosis and fibrosis in HCV-infected patients	–	
	Lymphocyte count, neutrophil count, and PLR (Platelet-to-lymphocyte ratio)	Strong predictors of hepatitis C infection	$p < 0.0001$, AUC = 1	
Hepatitis D	Anti-HDV IgM	Indicates acute infection	Second week of illness	Sandmann et al. (2025)
	HDV-RNA	Detects active viral replication	Confirmed via PCR	
	Quantitative anti-HBc	Lower levels independently associated with development of liver-related events	–	
	Serum cholinesterase	Predictive biomarker of hepatic reserves in chronic hepatitis D	–	
Hepatitis E	Anti-HEV IgM	Marker of acute infection	–	Taneja et al. (2009)
	Anti-HEV IgG	Indicates past infection or immunity	–	

Core was used to detect anti-HBc. HCV version 3.0 was used to detect anti-HCV (Eriksen et al. 2009).

Hepatitis B virus DNA and HBsAg from postmortem blood and bloodstains were detected by Hara and colleagues using HBV-infected blood and bloodstains stored up to 60 days at room temperature by cell culture, immunohistochemistry, and PCR by HBV-positive postmortem blood to a cell line to produce HBV to produce virions still detectable over 60 days. This approach is crucial when handling not only living material infected with HBV as well as body fluids from cadavers or medical waste (Hara et al. 2018).

3.4 Blood Biomarkers of COVID-19

COVID-19-predictive blood biomarkers encompass a range of cytokines, inflammatory markers, and vascular indicators. A recent study aimed to identify and validate predictive biomarkers for a COVID-19 patient cohort collected in Bangladesh to determine whether they will require hospitalization or can be treated as outpatients. Biomarker's validation was performed in a Japanese cohort. Fifteen key serum biomarkers were identified: BAFF, CTACK, Eotaxin, HGF, IL-6, IL-13, M-CSF, MCP-3, MMP-3, Osteocalcin, SDF-1a, TNF-R1, TNF-R2, TRAIL, and TWEAK, obtaining an 80–86% accuracy in Japanese validation for Delta and Omicron variants and a 70% accuracy for Wuhan and Alpha variants (Uno et al. 2024).

Hepatocyte growth factor (HGF) has been identified as a potential biomarker for the severe form of COVID-19 and showed high sensitivity and specificity when discriminating severe forms from nonsevere cases (Perreau et al. 2021). During the acute phase, Ferritin was strongly associated with macrophage-related markers, including IL-18 and sCD163-related factors, same as Procalcitonin (PCT) that was correlated with a wide range of cytokines, chemokines, and growth factors, including ADM, PGF, TRAILR2, and IL-6.

An upregulation of interferons β,γ,-λ, IL-6, and PTX3 was linked to post-acute sequelae of SARS-CoV-2 (PASC) (Kratzer et al. 2025). Other vascular markers include D-dimer, von Willebrand factor (VWF) (Ladikou et al. 2020), and markers of endothelial dysfunction (e.g., sVCAM-1, sVEGFR) (Vassiliou et al. 2023).

In the context of long COVID-19, elevated levels of Interleukin-6, CRP, and TNF-α were among the most frequently reported (Faraj and Jalal 2023).

Vascular markers such as ANG-1 (angiopoietin-1) and P-SEL (P-selectin) are promising biomarkers for long COVID when combined, demonstrating a 96% classification accuracy. A combination of these markers demonstrated 96% classification accuracy in distinguishing long COVID patients from healthy controls and acutely ill COVID-19 patients (Patel et al. 2022).

Lactate dehydrogenase (LDH) is associated with long COVID symptoms. Elevated LDH levels in COVID-19 patients tend to correlate with persistent symptoms (Udeh et al. 2023).

Hemoglobin levels are often decreased in long COVID patients, particularly those with vascular and neurological symptoms (Lechuga et al. 2023).

In patients with neurological symptoms of long COVID, elevated levels of neurofilament light chain (NFL) and glial fibrillary acidic protein (GFAP) have been observed (Plantone et al. 2024).

Amyloid beta 42, pTau181, and TDP-43 were found in plasma neuronal-enriched extracellular vesicles (nEVs) of long COVID patients with neurological impairment (Tang et al. 2024).

Going beyond clinical diagnosis, the forensic aspect of COVID-19 blood biomarkers lies in their potential to serve as objective, measurable evidence in legal, regulatory, or investigative contexts, to support fact-finding, attribution, and accountability.

In order to establish temporal sequence and causality, biomarkers such as IL-6, CRP, TNF-α, and LDH are useful for the reconstruction of the timeline of disease progression if serial blood samples are available. This strategy is of great help in disability claims, workplace exposure cases, or vaccine injury inquiries and may objectively document the onset and the severity of the acute infection, the transition to long COVID, and the duration of physiological disturbance (Medscape 2025).

At a molecular level, COVID-19 is usually detected by a ready-to-use RT-qPCR kit containing primers and probes for ORF, N, and S. The biological material can be either a nasopharyngeal swab during life or lung tissue obtained postmortem. A higher expression of the virus was observed in lung tissue samples, and a lower detectable level was in the nasopharyngeal swab samples, explaining the false-negative results obtained during life when analyzing swabs. Low levels of the virus in children could lead to a false-negative result, potentially making them a source of undiagnosed infection and a subsequent sudden death (Lisman et al. 2023).

The forensic aspect of COVID-19 blood biomarkers lies in their potential to serve as objective, measurable evidence in legal, regulatory, and investigative contexts, going beyond clinical diagnosis to support fact-finding, attribution, and accountability. Table 7.2 illustrates the key forensic implications of blood biomarkers of COVID-19 infection.

3.5 Biomarkers of Sepsis

Sepsis is a syndrome characterized by an excessive and dysregulated response to infections, often leading to multiorgan failure and death, whose onset and evolution over time depend both on pathogen factors and host factors (Singer et al. 2016).

The lung, representing the most impaired organ in sepsis, is the target organ for immunohistochemistry analysis. Table 7.3 shows the main immunohistochemical biomarkers used for postmortem sepsis diagnosis from lung autopsy.

Procalcitonin (PCT) levels are routinely evaluated in biological fluids both in clinical and autopsy settings, and their role is that of differentiating bacterial from nonbacterial sepsis. In their work, Maiese and colleagues had tested the PCT expression on brain, heart, lung, liver, and kidney samples from 10 sepsis cases and 5 nonsepsis control cases. The study revealed PCT negativity on samples belonging to the nonsepsis group; on the other hand, the sepsis group showed an immunopositive PCT, depending on localization (Maiese et al. 2017).

The C-reactive protein is another sepsis biomarker that functionally acts by precipitating the somatic C-polysaccharide of *Streptococcus pneumoniae* and is a

Table 7.2 Forensic use of COVID-19 infection biomarkers

Goal	Approach
To establish temporal sequence and causality	Biomarkers like IL-6, CRP, TNF-α, and LDH can help reconstruct the timeline of disease progression if serial blood samples are available. In disability claims, workplace exposure cases, or vaccine injury inquiries, rising or persistent levels may objectively document the onset and severity of an acute infection, the transition to long COVID form of the disease, and the duration of physiological disturbance (Faraj and Jalal 2023; MedScape 2025)
To objectify subjective symptoms	The long COVID often involves symptoms as fatigue, cognitive fog, or dyspnea. These are often hard to quantify. Elevated NFL (neurofilament light) and GFAP (glial fibrillary acidic protein) provide biochemical proof of neurological injury, turning patient reports into measurable pathophysiological data admissible in disability hearings or personal injury litigation (Plantone et al. 2024; MedScape 2025)
Differentiating long COVID from comorbidities	The "signature" of complement activation and endothelial markers (ANG 1, P SEL, VWF, sVCAM 1) can help distinguish long COVID from pre-existing conditions (e.g., chronic fatigue syndrome, small fiber neuropathy) (Patel et al. 2022). This is critical in: Insurance adjudication; workers' compensation claims; medical malpractice cases (e.g., delayed diagnosis) (MedScape 2025)
Linking exposure to outcome	In occupational health investigations (e.g., healthcare workers, first responders), elevated HGF, IL-6, or PCT in early-phase samples may serve as molecular footprints of high viral load exposure, supporting claims of infection acquired in the line of duty (MedScape 2025)
Verifying vaccination or infection history	While not explicitly listed above, serological markers (anti-SARS-CoV-2 IgG/IgM) combined with inflammatory profiles can retrospectively confirm prior infection. In immigration, travel, or employment disputes, this may be used to: validate or challenge claimed immunity; assess compliance with health mandates (MedScape 2025)
Detecting subclinical or asymptomatic cases	Subtle elevations in complement factors, D dimer, or sCD163 may reveal silent endothelial or coagulopathic changes in people who were never formally diagnosed. This has implications for epidemiological tracing and liability in institutional outbreaks (e.g., nursing homes, prisons) (MedScape 2025)
Standardizing eligibility for benefits or care	Forensic utility arises when biomarkers become criteria for program enrolment (e.g., long COVID clinics, disability pensions). For instance, a threshold NFL level could qualify a patient for neurorehabilitation funding; persistent CRP/IL-6 might trigger extended sick leave (MedScape 2025)
Excluding alternative diagnoses	A biomarker profile inconsistent with long COVID (e.g., normal complement, no endothelial markers) may help rule out malingering or psychosomatic explanations in contested cases—or, conversely, expose misdiagnosis (MedScape 2025)
Supporting public health accountability	At a population level, aggregated biomarker data a high HGF/IL-18 clusters in specific workplaces. It could serve as forensic evidence in negligence lawsuits against employers, policy evaluations of the adequacy of PPE during waves (MedScape 2025)

Table 7.3 Sepsis biomarkers

Biomarker	Cell type	Expression pattern	p-value	Reference
VLA-4	Intravascular, interstitial, and interalveolar leukocytes	Strong in all sepsis cases (mean = 1.8)	$p < 0.001$	Tsokos et al. (2001)
ICAM-1	Macrophages, lymphocytes	Strong in all sepsis cases (mean = 2.2)	$p < 0.001$	Tsokos et al. (2001)
Lactoferrin (LF)	Pulmonary leukocytes	Significantly higher in sepsis (mean = 9.0)	$p < 0.001$	Tsokos et al. (2001)
TNF-α	Lung macrophages	25% positive ratio in sepsis vs 13% in nonsepsis	$p < 0.05$	Miyashita et al. (2006)
CCR2	Monocytes/macrophages	41.6% positive in sepsis vs 8% in nonsepsis	$p < 0.01$	An et al. (2009)
CX3CR1	Monocytes/macrophages	36.2% positive in sepsis vs 9.2% in nonsepsis	$p < 0.01$	An et al. (2009)

systemic marker of inflammation and tissue damage, widely used in the diagnosis and management of various infectious and inflammatory diseases. The CRP molecule is composed of five polypeptide subunits, each containing 206 amino acid residues. In healthy young adults, the 99th percentile is 10 mg/L, but following an acute-phase stimulus, it could increase to 500 mg/L; hence, the plasmatic CRP is produced only by hepatocytes, predominantly under the transcriptional control of IL-6 (Samraj et al. 2013; Fujita et al. 2002; Pepys and Hirschfield 2003; Ho and Lipman 2009). In the forensic context, Laurier and colleagues had measured CRP values in postmortem serum and pericardial fluid in a series of 26 forensic autopsies that were selected based on agony duration (long and short agonies). The results of this study showed statistically significant increases in CRP levels in both postmortem serum and pericardial fluid in long agonies. The authors attributed this result to agonal pericarditis, possibly resulting from agonal myocardial necrosis, proposing a diagnostic application of CRP (Laurier et al. 1991).

Uhlin-Hansen (2001), Astrup and Thomsen (2007), and Tsokos et al. (2001) compared antemortem and postmortem CRP levels and observed increased concentrations of this marker in septic cases, reflecting the existence of ongoing inflammatory processes at the time of death. CRP levels were also found to be higher than postmortem concentrations, likely suggesting molecular proteolysis due to decompositional changes, along with irrelevant postmortem molecule release from hepatocytes in the early postmortem period. Maeda and colleagues (Maeda et al. 2008) had demonstrated that low CRP levels in some infantile and elderly cases of pneumonia were age-dependent, with low inflammatory responses. Astrup and Thomsen (2007) had denoted the stability of CRP levels in postmortem serum stored at 5 °C for several weeks.

Interleukin-6 is a critical proinflammatory 26 kDa cytokine primarily produced by leukocytes, fibroblasts, and endothelial cells and is a key regulator of the acute-phase inflammatory response, functioning as a determinant of the hepatic CRP production and driving the immunologic response development and facilitating immunoglobulin production. Clinically, IL-6 is rapidly expressed in response to trauma, infection, and stress and contributes to leukocytosis, lymphocyte activation, and acute-phase protein synthesis (Palmiere and Augsburger 2015). Serum levels up to 1000 pg/mL are indicators of a high risk of sepsis-related death. Postmortem levels may exceed antemortem values due to cell autolysis (Tsokos et al. 2001).

3.6 Blood Biomarkers in Cardiovascular Diseases

Cardiovascular blood biomarkers play a crucial role in the diagnosis, prognosis, and management processes of these categories of pathologies, which remain the leading causes of global morbidity and mortality. Blood biomarkers are also used to investigate sudden cardiac deaths in forensics.

Cardiac blood biomarkers appear in the bloodstream during the severe myocardial stress and injury caused by insufficient oxygen supply. The American College of Cardiology and the European Society of Cardiology have established a revised definition of acute myocardial infarction (AMI), which now incorporates cardiac biomarkers as key diagnostic elements (American College of Cardiology; European Society of Cardiology).

Cardiovascular blood biomarkers are subdivided into categories, depending on their structural origin, as they can be of lipidic and proteic nature. Table 7.4 illustrates the blood-based cardiovascular biomarkers by type.

Cardiac troponins are considered the gold standard for myocardial infarction diagnosis, exhibiting high specificity and sensitivity.

Hormonal markers like copeptin and endothelin-1 highlight neurohormonal activation. The inclusion of microRNAs and long noncoding RNAs represents a breakthrough in biomarker research, offering sensitive tools for early detection, risk stratification, and therapeutic targeting. These biomarkers provide valuable insights into various aspects of CVD pathophysiology, from myocardial injury detection to inflammation and oxidative stress assessment (Netala et al. 2025).

3.6.1 Ischemic Heart Disease (IHD) Blood-Based Biomarkers

Cardiovascular blood biomarkers are playing a crucial role in the diagnosis, prognosis, and management of cardiovascular diseases, which remain a leading cause of global morbidity and mortality, accounting for 9 billion lethal cases in 2021. The discovery and validation of novel biomarkers of cardiovascular diseases help to detect the onset and progression, and therapeutic responses, as well as in postmortem forensics (WHO).

Among the established, gold-standard biomarkers, cardiac troponins (cTns) are recognized as the primary indicators of ischemic heart disease, used also for the

Table 7.4 Blood-based cardiovascular biomarkers and their forensic/diagnostic use

Category	Biomarker	Function	Key characteristics	Diagnostic use	References
Lipidic	Lp-PLA2	Predicts acute coronary syndromes	Indicates inflamed plaques Sign of cardiovascular risk	Forensic predictor of acute coronary syndromes Cardiovascular death evaluation	Epps and Wilensky (2011)
	Oxylipins	Metabolic fingerprinting	Detects subtle changes in cardiac profiles Used in forensic metabolomics	Identifies cardiac profile changes Complements traditional autopsy	Le et al. (2021)
Proteic	CRP	Survival time estimation	Stable up to 18 days postmortem Differentiates acute/non-acute death	Evaluates the agonal period Systemic inflammation marker	Laurier et al. (1991)
	H-FABP & MPO	Myocardial damage indicators	Detects early myocardial injury Measures oxidative stress	Assesses early cardiac damage Oxidative stress evaluation	Gedikli et al. (2015)
	PTX3	Vascular inflammation marker	Associated with cardiovascular death	Vascular inflammation assessment Adverse event prediction	Tojo et al. (2019)
	CK-MB, LDH, HBDH	AMI diagnosis	Measured in pericardial fluid Appear earlier than in serum	Distinguishes between SCD and AMI Forensic AMI diagnosis	Aznar et al. (1983)
	AST & CA-III	Myocardial necrosis assessment	Differentiates cardiac/skeletal muscle injury	Necrosis evaluation Muscle injury distinction	Väänänen et al. (1990)

detection of potential drug-induced myocardial injury in both clinical trials and preclinical research.

Boulagnon and colleagues evaluated the diagnostic performance of cardiac troponin T (cTnT) in postmortem AMI detection: Findings revealed that although cTns demonstrated high sensitivity, biomarkers lacked sufficient specificity as a standalone analytical tool for *postmortem* AMI diagnosis (Boulagnon et al. 2011).

Sabucedo and colleagues described cardiac troponin I as a crucial regulatory protein within the ternary complex governing calcium-dependent muscle contraction, detected by the separation via sodium dodecyl sulphate–polyacrylamide gel electrophoresis (SDS PAGE) of extracted proteins, followed by visualization using western blot analysis using cTnI-specific monoclonal antibodies. Developed using a bovine model, the technique revealed distinct bands in human cadaver samples. The obtained results demonstrated a pseudo-linear correlation between the degree of cTnI degradation and the logarithm of time elapsed since death. The comparison of qualitative degradation band patterns to established standards demonstrated the ability to estimate the *postmortem* interval. This approach offers several advantages over direct temperature-based methods, including applicability across extended *postmortem* intervals, quantifiable degradation patterns, semiquantitative temporal relationships, and controlled temperature dependence (Sabucedo and Furton 2003).

Cardiac troponins remain the most widely used biomarkers due to their exceptional sensitivity and persistence in the bloodstream long after other biomarkers return to baseline levels. Two primary types are measurable: troponin T and troponin I. Notably, troponin I exhibits high cardiac specificity and remains elevated for a longer duration compared to creatine kinase MB (CK MB). The American Heart Association (AHA) now restricts the use of alternative biomarkers such as CK, CK MB, and myoglobin.

3.6.2 Emerging Biomarkers of IHD

The emergence of blood biomarkers for forensic use in IHD gained attention in the context of unexplained death without typical cardiac damage. The trendiest ones include microRNAs, and circRNAs. Serum SELENBP1 and VCL could be effective biomarkers for both the clinical and forensic diagnosis of coronary artery spasm, as demonstrated by Lin and colleagues, where in forensic samples, serum SELENBP1 alone diagnosed CAS-induced deaths at a sensitivity of 100.0% and specificity of 72.73%, and its combination with VCL yielded a diagnostic specificity of 100.0%, which was superior to the traditional biomarkers of cTnI and CK-M (Lin et al. 2022).

3.7 Diabetes Mellitus (DM) Blood-Based Biomarkers

Acute metabolic disorders of diabetes mellitus (DM), such as diabetic ketoacidosis, hyperosmolar hyperglycemic state, and hypoglycemia, are life-threatening and difficult to diagnose postmortem due to a lack of macroscopic and microscopic findings, particularly in cases when the medical history of the patient is unavailable.

According to Palmiere & Mangin, it is impossible to diagnose hyperglycemia/hypoglycemia on the basis of postmortem blood glucose levels because these are unreliable due to postmortem changes (Palmiere and Mangin 2012). To overcome this limitation, legal medicine had found an alternative to detect ketone bodies in blood in order to diagnose diabetic ketoacidosis. However, the ketone bodies seem not to be DM-specific, as they can be found in alcoholic ketoacidosis, severe starvation, and severe hypoxia (Mitchell et al. 1995).

According to the literature, hemoglobin A1c reflects stability after death and varies slightly from antemortem values (Winecker et al. 2002). Therefore, in the forensic context, HbA1c is a biomarker of choice to measure in order to evaluate the metabolic state of the patient prior to lethality, through the observation of long-term fluctuations of the average blood glucose level. This fact highlights a limitation of hemoglobin A1c, as it does not reflect short-term fluctuations that could potentially lead to fatal acute hyperglycemic/hypoglycemic states. A study by Lipska and colleagues denoted that even when the levels of HbA1c are near-normal glycemic control (HbA1c $< 6.0\%$), the latter could not be correlated with a reduction in the risk of severe hypoglycemia in T2DM patients.

Attempts at the exploration of emergent TDM2 forensic biomarkers are actually performed in the metabolomic field: Nariai and colleagues had performed a case–control study on postmortem diabetic and nondiabetic cases, employing HPLC system analysis of hydrophilic and lipophilic substances extracted from femoral blood. In terms of results, significant variations between DM and non-DM groups were observed for lipid levels, and a reduction was noted in sphingomyelin and plasmalogen levels, as well as an elevation of lyso-phosphatidylcholine and lyso-phosphatidylethanolamine in the DM patient group (Nariai et al. 2022).

4 Conclusions

The blood biomarker disease field is rapidly progressing, mainly driven by the needs of clinical medicine for diagnosis and treatment monitoring. Traditionally, the forensic field remains secondary, but blood biomarkers are also gaining increasing importance. Biomarkers in forensic medicine are used to establish the time since death, determine the cause of death, identify biological traces, and predict the victim's condition. In addition to that, the accuracy of forensic examinations has improved with the transition from traditional methods to molecular technologies, the implementation of which is greatly dependent on financial possibilities.

References

Ahern TP, Rosner BA, Tamimi RM (2014) Reproductive and early life exposures in relation to gene expression-defined breast cancer subtypes in the Nurses' Health Study. Breast Cancer Res Treat 147(2):437–444

Al Kuwari H, Al Khinji A, Al-Hamed D, Al Hor A, Malouche D, Iqbal J (2025) NOTCH1 mutation status as a prognostic biomarker in T-cell acute lymphoblastic leukemia: a systematic review and meta-analysis. BMC Cancer 25(1):1820. https://doi.org/10.1186/s12885-025-15233-2

Al Sharhan NA, Messaoudi SA, Babu SR, Chaudhary AB, Alsharm AA, Alrefaei AF, Kadasah S, Abu-Elmagd M, Assidi M, Buhmeida A, Carracedo Á, Almawi WY (2022) Utility of circulating cell-free DNA in assessing microsatellite instability and loss of heterozygosity in breast cancer using human identification approach. Genes (Basel) 13(4):590

American College of Cardiology (2025) URL: https://www.jacc.org/doi/pdf/10.1016/j.jacc.2021.12.012?_gl=11vouy7a_gcl_auOTY4MzUyOTM5LjE3NjcyODI0NzA._gaNDExMDUyOTEuMTc2NzI4MjQ3MA.._ga_2V8VW4Y237czE3NjcyODI0NzAkbzEkZzEkdDE3NjcyODI1NjckajYwJGwwJGgw. Accessed on 15/11/2025

An JL, Ishida Y, Kimura A, Tsokos M, Kondo T (2009) Immunohistochemical detection of CCR2 and CX3CR1 in sepsis-induced lung injury. Forensic Sci Int 192(1–3):e21–e25. https://doi.org/10.1016/j.forsciint.2009.08.007

Armstrong AJ, Halabi S, Luo J, Nanus DM, Giannakakou P, Szmulewitz RZ, Danila DC, Healy P, Anand M, Rothwell CJ, Rasmussen J, Thornburg B, Berry WR, Wilder RS, Lu C, Chen Y, Silberstein JL, Kemeny G, Galletti G, Somarelli JA et al (2019) Prospective multicenter validation of androgen receptor splice variant 7 and hormone therapy resistance in high-risk castration-resistant prostate cancer: the PROPHECY study. J Clin Oncol Off J Am Soc Clin Oncol 37(13):1120–1129. https://doi.org/10.1200/JCO.18.01731

Astrup BS, Thomsen JL (2007) The routine use of C-reactive protein in forensic investigations. Forensic Sci Int 172(1):49–55. https://doi.org/10.1016/j.forsciint.2006.10.021

Aznar J, Santos MT, Valles J, Sala J (1983) Serum malondialdehyde-like material (MDA-LM) in acute myocardial infarction. J Clin Pathol 36(6):712–715. https://doi.org/10.1136/jcp.36.6.712

Bagacean C, Iuga CA, Bordron A, Tempescul A, Pralea IE, Bernard D, Cornen M, Bergot T, Le Dantec C, Brooks W, Saad H, Ianotto JC, Pers JO, Zdrenghea M, Berthou C, Renaudineau Y (2022) Identification of altered cell signaling pathways using proteomic profiling in stable and progressive chronic lymphocytic leukemia. J Leukoc Biol 111(2):313–325. https://doi.org/10.1002/JLB.4HI0620-392R

Bagheri S, Fard GB, Talkhi N, Rashidi Zadeh D, Mobarra N, Mousavinezhad S, Khamse FM, Hosseini Bafghi M (2024) Laboratory biochemical and hematological parameters: early predictive biomarkers for diagnosing hepatitis C virus infection. J Clin Lab Anal 38(24):e25127. https://doi.org/10.1002/jcla.25127

Bast RC Jr, Knapp RC (1985) Use of the CA 125 antigen in diagnosis and monitoring of ovarian carcinoma. Eur J Obstet Gynecol Reprod Biol 19(6):354–356. https://doi.org/10.1016/0028-2243(85)90130-3

Bettegowda C, Sausen M, Leary RJ, Kinde I, Wang Y, Agrawal N, Bartlett BR, Wang H, Luber B, Alani RM, Antonarakis ES, Azad NS, Bardelli A, Brem H, Cameron JL, Lee CC, Fecher LA, Gallia GL, Gibbs P, Le D et al (2014) Detection of circulating tumor DNA in early- and late-stage human malignancies. Sci Transl Med 6(224):224ra24. https://doi.org/10.1126/scitranslmed.3007094

Boulagnon C, Garnotel R, Fornes P, Gillery P (2011) Post-mortem biochemistry of vitreous humor and glucose metabolism: an update. Clin Chem Lab Med 49(8):1265–1270. https://doi.org/10.1515/CCLM.2011.638

Campo E, Cymbalista F, Ghia P, Jäger U, Pospisilova S, Rosenquist R, Schuh A, Stilgenbauer S (2018) *TP53* aberrations in chronic lymphocytic leukemia: an overview of the clinical implications of improved diagnostics. Haematologica 103(12):1956–1968. https://doi.org/10.3324/haematol.2018.187583

Castro et al (2025). URL: https://www.urotoday.com/conference-highlights/apccc-diagnostics-2025/158645-apccc-diagnostics-2025-new-fashion-blood-barkers-for-metastatic-prostate-cancer.html

Cenin DR, Naber SK, Lansdorp-Vogelaar I, Jenkins MA, Buchanan DD, Preen DB, Ee HC, O'Leary P (2018) Costs and outcomes of lynch syndrome screening in the Australian colorectal cancer population. J Gastroenterol Hepatol 33:1737–1744

Colby B (2023) Decoding the genetic secrets of Mycobacterium Tuberculosis: protection and severity. URL: https://sequencing.com/education-center/medical/mycobacterium-tuberculosis-protection-against?srsltid=AfmBOooLoGfkKZyuvxQ9832wVuPNRWQPGVDeB-9nIsW7hdqaTL_gs3Nv. Accessed on 08/12/2025

Conway JM, Perelson AS, Li JZ (2019) Predictions of time to HIV viral rebound following ART suspension that incorporate personal biomarkers. PLoS Comput Biol 15(7):e1007229

Coskun ME, Alidris A, Temel MT, Akbayram S, Hizli S (2019) Plateletcrit: a possible biomarker of inflammation in hepatitis A infection. Niger J Clin Pract 22(5):727–730. https://doi.org/10.4103/njcp.njcp_331_18

Cox M, Vitello D, Chawla A (2025) Translating the multifaceted use of liquid biopsy to management of early disease in pancreatic adenocarcinoma. Front Oncol 15:1520717. https://doi.org/10.3389/fonc.2025.1520717

Dai X, Li T, Bai Z, Yang Y, Liu X, Zhan J, Shi B (2015) Breast cancer intrinsic subtype classification, clinical use and future trends. Am J Cancer Res 5(10):2929–2943

Diamandis EP, Okui A, Mitsui S, Luo LY, Soosaipillai A, Grass L, Nakamura T, Howarth DJ, Yamaguchi N (2002) Human kallikrein 11: a new biomarker of prostate and ovarian carcinoma. Cancer Res 62(1):295–300

Duffield AS, Mullighan CG, Borowitz MJ (2023) International consensus classification of acute lymphoblastic leukemia/lymphoma. Virchows Arch 482(1):11–26. https://doi.org/10.1007/s00428-022-03448-8

Duffy MJ (2006) Serum tumor markers in breast cancer: are they of clinical value? Clin Chem 52(3):345–351

Epps KC, Wilensky RL (2011) Lp-PLA$_2$- a novel risk factor for high-risk coronary and carotid artery disease. J Intern Med 269(1):94–106. https://doi.org/10.1111/j.1365-2796.2010.02297.x

Eriksen MB, Jakobsen MA, Kringsholm B, Banner J, Thomsen JL, Georgsen J, Pedersen C, Christensen PB (2009) Postmortem detection of hepatitis B, C, and human immunodeficiency virus genomes in blood samples from drug-related deaths in Denmark. J Forensic Sci 54(5):1085–1088. https://doi.org/10.1111/j.1556-4029.2009.01111.x

European Society of Cardiology (2025) URL: https://www.escardio.org/Guidelines/Clinical-Practice-Guidelines/Valvular-Heart-Disease. Accessed on 15/11/2025

Faraj SS, Jalal PJ (2023) IL1β, IL-6, and TNF-α cytokines cooperate to modulate a complicated medical condition among COVID-19 patients: case-control study. Ann Med Surg (Lond) 85(6):2291–2297. https://doi.org/10.1097/MS9.0000000000000679

FDA (2025) URL: https://www.fda.gov/drugs/biomarker-qualification-program/about-biomarkers-and-qualification. Accessed on 10/09/2025

Fujita MQ, Zhu BL, Ishida K, Quan L, Oritani S, Maeda H (2002) Serum C-reactive protein levels in postmortem blood--an analysis with special reference to the cause of death and survival time. Forensic Sci Int 130(2–3):160–166. https://doi.org/10.1016/s0379-0738(02)00381-x

Gao J, Shi W, Wang J, Guan C, Dong Q, Sheng J, Zou X, Xu Z, Ge Y, Yang C, Li J, Bao H, Zhong X, Cui Y (2024) Research progress and applications of epigenetic biomarkers in cancer. Front Pharmacol 15:1308309. https://doi.org/10.3389/fphar.2024.1308309

Gedikli O, Kiris A, Hosoglu Y, Karahan C, Kaplan S (2015) Serum myeloperoxidase level is associated with heart-type fatty acid-binding protein but not troponin T in patients with chronic heart failure. Med Princ Pract 24(1):42–46. https://doi.org/10.1159/000368717

Gharbaran R, Goy A, Tanaka T, Park J, Kim C, Hasan N, Vemulapalli S, Sarojini S, Tuluc M, Nalley K, Bhattacharyya P, Pecora A, Suh KS (2013) Fibroblast growth factor-2 (FGF2) and syndecan-1 (SDC1) are potential biomarkers for putative circulating CD15+/CD30+ cells in poor outcome Hodgkin lymphoma patients. J Hematol Oncol 6:62. https://doi.org/10.1186/1756-8722-6-62

Gisslén M, Price RW, Andreasson U, Norgren N, Nilsson S, Hagberg L, Fuchs D, Spudich S, Blennow K, Zetterberg H (2015) Plasma concentration of the Neurofilament light protein (NFL) is a biomarker of CNS injury in HIV infection: a cross-sectional study. EBioMedicine 3:135–140. https://doi.org/10.1016/j.ebiom.2015.11.036

Gitto L, Mehta R, Mirchia K, Ramos C, Tawil M, Serinelli S (2021) Sudden death due to leukostasis in a subject with undiagnosed chronic lymphocytic leukemia. Forensic Sci Med Pathol 17(4):693–699. https://doi.org/10.1007/s12024-021-00406-9

Gonçalves LG, Bento M, Anjo SI et al (2022) Mass spectrometry-based proteomic and metabolomic profiling of serum samples for discovery and validation of tuberculosis diagnostic biomarker signature. Int J Mol Sci 23(22):13733. https://doi.org/10.3390/ijms232213733

Greenwald MA, Kerby S, Francis K, Noller AC, Gormley WT, Biswas R, Forshee RA (2018) Detection of human immunodeficiency virus, hepatitis C virus, and hepatitis B virus in postmortem blood specimens using infectious disease assays licensed for cadaveric donor screening. Transpl Infect Dis 20(1). https://doi.org/10.1111/tid.12825

Hagberg L, Edén A, Zetterberg H, Price RW, Gisslén M (2022) Blood biomarkers for HIV infection with focus on neurologic complications—a review. Acta Neurol Scand 146:56–60

Hara J, Tanaka Y, Kaneko H, Itoh Y, Ikegaya H (2018) Detection of hepatitis B virus DNA and HBsAg from postmortem blood and bloodstains. Arch Virol 163(3):633–637. https://doi.org/10.1007/s00705-017-3665-x

Hayashi S, Nagaoka K, Tanaka Y (2021) Blood-based biomarkers in hepatitis B virus-related hepatocellular carcinoma, including the viral genome and glycosylated proteins. Int J Mol Sci 22(20):11051. https://doi.org/10.3390/ijms222011051

Ho KM, Lipman J (2009) An update on C-reactive protein for intensivists. Anaesth Intensive Care 37(2):234–241. https://doi.org/10.1177/0310057X0903700217

Hood L, Flores M (2012) A personal view on systems medicine and the emergence of proactive P4 medicine: predictive, preventive, personalized and participatory. New Biotechnol 29(6):613–624. https://doi.org/10.1016/j.nbt.2012.03.004

Huang C, Fang M, Xiao X, Wang H, Gao Z, Ji J et al (2022) Validation of the galad model for early diagnosis and monitoring of hepatocellular carcinoma in Chinese multicenter study. Liver Int 42(1):210–223. https://doi.org/10.1111/liv.15082

Huang C, Xiao X, Zhou L, Chen F, Wang J, Hu X et al (2023) Chinese expert consensus statement on the clinical application of afp/afp-l3%/dcp using galad and galad-like algorithm in hcc. J Clin Lab Anal 37(23–24):e24990. https://doi.org/10.1002/jcla.24990

Inaba H, Mullighan CG (2020) Pediatric acute lymphoblastic leukemia. Haematologica 105(11):2524–2539. https://doi.org/10.3324/haematol.2020.247031

Jones RF, Sunheimer R, Friedman H, Miller D, Ginsburg R, Jumbelic M, Threatte G, Haas GP (2005) Comparison of ante- and post-mortem PSA levels for epidemiological studies. Anticancer Res 25(2B):1263–1267

Kratzer B, Stieger RB, Durmus S, Trapin D, Gattinger P, Ettel P, Sehgal ANA, Borochova K, Dorofeeva Y, Tulaeva I, Grabmeier-Pfistershammer K, Tauber PA, Gerdov M, Perkmann T, Fae I, Wenda S, Kundi M, Wrighton S, Fischer GF, Valenta R et al (2025) Severe COVID-19 induces prolonged elevation of the acute-phase protein pentraxin 3. Front Immunol 16:1672485. https://doi.org/10.3389/fimmu.2025.1672485

Krebs-Drouot L, Karpathiou G, Scolan V, Bidat-Callet C, Boyer B, Péoc'h M (2023) Death due to unsuspected acute myeloid leukaemia: an unusual forensic diagnosis. Forensic Sci Med Pathol 19(1):60–66. https://doi.org/10.1007/s12024-022-00516-y

Labaj W, Papiez A, Polanski A, Polanska J (2017) Comprehensive analysis of MILE gene expression data set advances discovery of leukaemia type and subtype biomarkers. Interdiscip Sci: Comput Life Sci 9(1):24–35. https://doi.org/10.1007/s12539-017-0216-9

Ladikou EE, Sivaloganathan H, Milne KM, Arter WE, Ramasamy R, Saad R, Stoneham SM, Philips B, Eziefula AC, Chevassut T (2020) Von Willebrand factor (vWF): marker of endothelial damage and thrombotic risk in COVID-19? Clin Med (Lond) 20(5):e178–e182. https://doi.org/10.7861/clinmed.2020-0346

Laurier E, Gosset D, Hennache B, Nuttens MC, Debuire B, Lenoir L, Muller PH (1991) La protéine C-réactive péricardique. Un marqueur de la souffrance cardiaque agonique? [Pericardial C-reactive protein. A marker of agonal cardiac disease ?]. Presse medicale (Paris, France: 1983) 20(9):405–408. https://doi.org/10.1038/s41568-019-0199-4

Le DE, García-Jaramillo M, Bobe G, Alcazar Magana A, Vaswani A, Minnier J, Jump DB, Rinkevich D, Alkayed NJ, Maier CS, Kaul S (2021) Plasma oxylipins: a potential risk assessment tool in atherosclerotic coronary artery disease. Front Cardiovasc Med 8:645786. https://doi.org/10.3389/fcvm.2021.645786

Lechuga GC, Morel CM, De-Simone SG (2023) Hematological alterations associated with long COVID-19. Front Physiol 14:1203472. https://doi.org/10.3389/fphys.2023.1203472

Li J, Wang T, Jin B, Li W, Wang Z, Zhang H, Song Y, Li N (2018) Diagnosis accuracy of serum glypican-3 level in patients with hepatocellular carcinoma: a systematic review with meta-analysis. Int J Biol Markers 33:353–363. https://doi.org/10.1177/1724600818784409

Lin X, Lin Z, Zhao X, Liu Z, Xu C, Yu B, Gao P, Wang Z, Ge J, Shen Y, Li L (2022) Serum SELENBP1 and VCL are effective biomarkers for clinical and forensic diagnosis of coronary artery spasm. Int J Mol Sci 23(21):13266. https://doi.org/10.3390/ijms232113266

Lisman D, Zielińska G, Drath J, Łaszczewska A, Savochka I, Parafiniuk M, Ossowski A (2023) Molecular diagnosis of COVID-19 sudden and unexplained deaths: the insidious face of the pandemic. Diagnostics 13(18):2980. https://doi.org/10.3390/diagnostics13182980

Loh KP, Tooze JA, Nicklas BJ, Kritchevsky SB, Williamson JD, Ellis LR, Powell BL, Pardee TS, Goyal NG, Klepin HD (2020) Inflammatory biomarkers, geriatric assessment, and treatment outcomes in acute myeloid leukemia. J Geriatr Oncol 11(3):410–416. https://doi.org/10.1016/j.jgo.2019.03.014

Maeda H, Zhu BL, Bessho Y, Ishikawa T, Quan L, Michiue T, Zhao D, Li DR, Komatsu A (2008) Postmortem serum nitrogen compounds and C-reactive protein levels with special regard to investigation of fatal hyperthermia. Forensic Sci Med Pathol 4(3):175–180. https://doi.org/10.1007/s12024-008-9029-9

MacLean E, Broger T, Yerlikaya S, Fernandez-Carballo BL, Pai M, Denkinger CM (2019) A systematic review of biomarkers to detect active tuberculosis. Nat Microbiol 4(5):748–758

Maiese A, Del Nonno F, Dell'Aquila M, Moauro M, Baiocchini A, Mastracchio A, Bolino G (2017) Postmortem diagnosis of sepsis: a preliminary immunohistochemical study with an anti-procalcitonin antibody. Leg Med (Tokyo) 28:1–5. https://doi.org/10.1016/j.legalmed.2017.07.002

Medscape (2025) URL: https://www.medscape.com/viewarticle/how-help-me-cfs-and-long-covid-disability-claims-2025a1000x2z. Accessed on 07/11/2025

Mitchell GA, Kassovska-Bratinova S, Boukaftane Y, Robert MF, Wang SP, Ashmarina L, Lambert M, Lapierre P, Potier E (1995) Medical aspects of ketone body metabolism. Clin Invest Med 18(3):193–216

Miyashita T, Kakimoto N, Ishida Y et al (2006) Immunohistochemical expression of tumor necrosis factor-α in sepsis-induced lung injury. Forensic Sci Med Pathol 2:103–108. https://doi.org/10.1385/FSMP:2:2:103

Müller C, Pascolo Tièche E, Zieger M (2025) PSA is not suitable as a reliable marker for seminal fluid in rectal samples collected postmortem. Forensic Sci Int Rep 11:100409. https://doi.org/10.1016/j.fsir.2025.100409

Nan Y, Zhao S, Zhang X, Xiao Z, Guo R (2022) Machine learning-based predictive and risk analysis using real-world data with blood biomarkers for hepatitis B patients in the malignant progression of hepatocellular carcinoma. Front Immunol 13:1031400. https://doi.org/10.3389/fimmu.2022.1031400

Nariai M, Abe H, Hoshioka Y, Makino Y, Iwase H (2022) Biomarker profiling of postmortem blood for diabetes mellitus and discussion of possible applications of metabolomics for forensic casework. Int J Legal Med 136(4):1075–1090. https://doi.org/10.1007/s00414-021-02767-w

Netala VR, Hou T, Wang Y, Zhang Z, Teertam SK (2025) Cardiovascular biomarkers: tools for precision diagnosis and prognosis. Int J Mol Sci 26(7):3218. https://doi.org/10.3390/ijms26073218

Nogueira BMF et al (2023) Integrating pathogen- and host-derived blood biomarkers for enhanced tuberculosis diagnosis: a comprehensive review. Front Immunol 14:11341448. https://doi.org/10.3389/fimmu.2024.1438989

O'Neill RS, Stoita A (2021) Biomarkers in the diagnosis of pancreatic cancer: are we closer to finding the golden ticket? World J Gastroenterol 27(26):4045–4087. https://doi.org/10.3748/wjg.v27.i26.4045

Palmiere C, Augsburger M (2015) Pancreatic stone protein as a postmortem biochemical marker for the diagnosis of sepsis. Legal Med 17(1):9–13

Palmiere C, Mangin P (2012) Postmortem chemistry update part I. Int J Legal Med 126(2):187–198. https://doi.org/10.1007/s00414-011-0625-y

Pasternak AO, de Bruin M, Bakker M, Berkhout B, Prins JM (2015) High current CD4+ T cell count predicts suboptimal adherence to antiretroviral therapy. PLoS One 10(10):e0140791. https://doi.org/10.1371/journal.pone.0140791

Patel MA, Knauer MJ, Nicholson M, Daley M, Van Nynatten LR, Martin C, Patterson EK, Cepinskas G, Seney SL, Dobretzberger V, Miholits M, Webb B, Fraser DD (2022) Elevated vascular transformation blood biomarkers in Long-COVID indicate angiogenesis as a key pathophysiological mechanism. Mol Med (Cambridge, Mass.) 28(1):122. https://doi.org/10.1186/s10020-022-00548-8

Pepys MB, Hirschfield GM (2003) C-reactive protein: a critical update. J Clin Invest 111(12):1805–1812. https://doi.org/10.1172/JCI18921

Perreau M, Suffiotti M, Marques-Vidal P, Wiedemann A, Levy Y, Laouénan C, Ghosn J, Fenwick C, Comte D, Roger T, Regina J, Vollenweider P, Waeber G, Oddo M, Calandra T, Pantaleo G (2021) The cytokines HGF and CXCL13 predict the severity and the mortality in COVID-19 patients. Nat Commun 12(1):4888. https://doi.org/10.1038/s41467-021-25191-5

Petricoin EF, Ardekani AM, Hitt BA, Levine PJ, Fusaro VA, Steinberg SM, Mills GB, Simone C, Fishman DA, Kohn EC, Liotta LA (2002) Use of proteomic patterns in serum to identify ovarian cancer. Lancet (London, England) 359(9306):572–577. https://doi.org/10.1016/S0140-6736(02)07746-2

Plantone D, Stufano A, Righi D et al (2024) Neurofilament light chain and glial fibrillary acid protein levels are elevated in post-mild COVID-19 or asymptomatic SARS-CoV-2 cases. Sci Rep 14:6429. https://doi.org/10.1038/s41598-024-57093-z

Rouzioux C, Avettand-Fenoël V (2018) Total HIV DNA: a global marker of HIV persistence. Retrovirology 15(1):30. https://doi.org/10.1186/s12977-018-0412-7

Sabucedo AJ, Furton KG (2003) Estimation of postmortem interval using the protein marker cardiac Troponin I. Forensic Sci Int 134(1):11–16. https://doi.org/10.1016/s0379-0738(03)00080-x

Samraj RS, Zingarelli B, Wong HR (2013) Role of biomarkers in sepsis care. Shock (Augusta, Ga) 40(5):358–365. https://doi.org/10.1097/SHK.0b013e3182a66bd6

Sandmann L, Ohlendorf V, Ehrenbauer A, Bremer B, Kraft ARM, Cornberg M, Deterding K, Wedemeyer H, Maasoumy B (2025) Value and kinetics of virological markers in the natural course of chronic hepatitis D virus infection. Liver Int 45(2):e70003. https://doi.org/10.1111/liv.70003

Siegel RL, Miller KD, Jemal A (2020) Cancer statistics, 2020. CA Cancer J Clin 70(1):7–30. https://doi.org/10.3322/caac.21590

Singer M, Deutschman CS, Seymour CW, Shankar-Hari M, Annane D, Bauer M, Bellomo R, Bernard GR, Chiche JD, Coopersmith CM, Hotchkiss RS, Levy MM, Marshall JC, Martin GS, Opal SM, Rubenfeld GD, van der Poll T, Vincent JL, Angus DC (2016) The third international consensus definitions for sepsis and septic shock (Sepsis-3). JAMA 315(8):801–810. https://doi.org/10.1001/jama.2016.0287

Skates SJ, Horick N, Yu Y, Xu FJ, Berchuck A, Havrilesky LJ, de Bruijn HW, van der Zee AG, Woolas RP, Jacobs IJ, Zhang Z, Bast RC Jr (2004) Preoperative sensitivity and specificity for early-stage ovarian cancer when combining cancer antigen CA-125II, CA 15-3, CA 72-4, and macrophage colony-stimulating factor using mixtures of multivariate normal distributions. J Clin Oncol Off J Am Soc Clin Oncol 22(20):4059–4066. https://doi.org/10.1200/JCO.2004.03.091

Sung H, Ferlay J, Siegel RL, Laversanne M, Soerjomataram I, Jemal A, Bray F (2021) Global cancer statistics 2020: GLOBOCAN estimates of incidence and mortality worldwide for 36 cancers in 185 countries. CA Cancer J Clin 71(3):209–249. doi.org

Takaso M, Hitosugi M, Nakagawa R, Mukaisho K, Moriguchi S, Koh M, Sugihara H (2020) Forensic autopsies can determine latent prostate cancer prevalence. J Forensic Sci 65(5):1557–1562. https://doi.org/10.1111/1556-4029.14472

Taneja S, Sen S, Gupta VK, Aggarwal R, Jameel S (2009) Plasma and urine biomarkers in acute viral hepatitis E. Proteome Sci 7:39. https://doi.org/10.1186/1477-5956-7-39

Tang N, Kido T, Shi J, McCafferty E, Ford JM, Dal Bon K, Pulliam L (2024) Blood markers show neural consequences of LongCOVID-19. Cells 13(6):478. https://doi.org/10.3390/cells13060478
Teachey DT, Pui CH (2019) Comparative features and outcomes between paediatric T-cell and B-cell acute lymphoblastic leukaemia. Lancet Oncol 20(3):e142–e154. https://doi.org/10.1016/S1470-2045(19)30031-2
Tojo M, Shintani-Ishida K, Tsuboi H et al (2019) Postmortem plasma pentraxin 3 is a useful marker of fatal acute coronary syndrome. Sci Rep 9:8090. https://doi.org/10.1038/s41598-019-44472-0
Tsokos M, Reichelt U, Jung R, Nierhaus A, Püschel K (2001) Interleukin-6 and C-reactive protein serum levels in sepsis-related fatalities during the early postmortem period. Forensic Sci Int 119(1):47–56. https://doi.org/10.1016/s0379-0738(00)00391-1
Udeh R, Utrero-Rico A, Dolja-Gore X, Rahmati M, McEVoy M, Kenna T (2023) Lactate dehydrogenase contribution to symptom persistence in long COVID: a pooled analysis. Rev Med Virol 33(6):e2477. https://doi.org/10.1002/rmv.2477
Uhlin-Hansen L (2001) C-reactive protein (CRP), a comparison of pre- and post-mortem blood levels. Forensic Sci Int 124(1):32–35. https://doi.org/10.1016/s0379-0738(01)00558-8
Uno K, Hasan A, Nakayama EE, Rahim R, Harada H, Kaneko M, Hashimoto S, Tanaka T, Matsumoto H, Fujimiya H, Shioda T, Rahman M, Yoshizaki K (2024) Predictive biomarkers of COVID-19 prognosis identified in Bangladesh patients and validated in Japanese cohorts. Sci Rep 14(1):12713. https://doi.org/10.1038/s41598-024-63184-8
Väänänen HK, Syrjälä H, Rahkila P, Vuori J, Melamies LM, Myllylä V, Takala TE (1990) Serum carbonic anhydrase III and myoglobin concentrations in acute myocardial infarction. Clin Chem 36(4):635–638
Van Reenen M et al (2025) Potential biomarkers for tuberculous meningitis diagnosis using metabolomics and proteomics: a systematic review. BMC Infect Dis 25:1298. https://doi.org/10.1186/s12879-025-11740-6
Vassiliou AG, Vrettou CS, Keskinidou C, Dimopoulou I, Kotanidou A, Orfanos SE (2023) Endotheliopathy in acute COVID-19 and long COVID. Int J Mol Sci 24(9):8237. https://doi.org/10.3390/ijms24098237
Vojjala N, Gibatova V, Shah RN, Singal S, Prabhu R, Krishnamoorthy G, Riggins K, Moka N (2025) Integrating circulating tumor DNA into clinical management of colorectal cancer: practical implications and therapeutic challenges. Cancer 17(15):2520. https://doi.org/10.3390/cancers17152520
Voropaeva EN, Pospelova TI, Orlov YL, Churkina MI, Berezina OV, Gurazheva AA, Ageeva TA, Seregina OB, Maksimov VN (2022) The methylation of the p53 targets the genes *MIR-203*, *MIR-129-2*, *MIR-34A* and *MIR-34B/C* in the tumor tissue of diffuse large B-cell lymphoma. Genes 13(8):1401. https://doi.org/10.3390/genes13081401
Wang T, Jiang R, Yao Y, Wang Y, Liu W, Qian L et al (2023) Endometrial cytology in diagnosis of endometrial cancer: a systematic review and meta-analysis of diagnostic accuracy. J Clin Med 12(6):2358
WHO (2025a). URL: https://www.who.int/data/gho/data/themes/mortality-and-global-health estimates. Accessed on 15/11/2025
WHO (2025b). URL: https://www.who.int/news-room/fact-sheets/detail/breast-cancer#:~:text=Breast%20cancer%20caused%20an%20estimated,breast%20cancers%20occur%20in%20men. Accessed on 12/10/2025
Winecker RE, Hammett-Stabler CA, Chapman JF, Ropero-Miller JD (2002) HbA1c as a postmortem tool to identify glycemic control. J Forensic Sci 47(6):1373–1379
Yang JD, Hainaut P, Gores GJ, Amadou A, Plymoth A, Roberts LR (2019) A global view of hepatocellular carcinoma: trends, risk, prevention and management. Nat Rev Gastroenterol Hepatol 16(10):589–604. https://doi.org/10.1038/s41575-019-0186-y
Yang Z, Dong Y, Shang Y et al (2025) Host circulating immunometabolism-associated biomarkers for early diagnosis of active tuberculosis: multi-omics screening with experimental validation. J Inflamm Res 18:10723–10740. https://doi.org/10.2147/JIR.S533116

Part II

Analytical and Investigative Techniques

Bloodstain Pattern Analysis (BPA): Principles and Interpretation

8

Abstract

Understanding the blood spatter pattern holds significance in crime scene investigation and its reconstruction. Two properties of blood, i.e., surface tension and viscosity, help the generation of unique patterns of blood spatters in different circumstances. Various types of bloodstains are found in a crime scene which can be attributed to a particular scenario. Passive bloodstains are created by the force of gravity alone when bloodstains are released from a wound area. The shape and size of passive bloodstains vary with their nature such as drops, drip stain, drip pattern, pools, and clots. A typical blood spatter includes three components, i.e., parent drop, satellite stains, and spines. Careful examination and analysis of bloodstain pattern reveal many useful investigative leads such as angle of impact, height, and origin of the blood source. When a bloodstain hits a surface at 90°, it appears perfectly circular shape. Any deviation from 90° makes the bloodstain elliptical, and the angle of impact can be calculated using the Sin formula. Similarly, in a general notion, the height of fall is directly proportional to the size of the bloodstain. By constructing the path of each bloodstain, a point of convergence can be estimated. From this point of convergence, the height of the blood source can also be found out. Thus, forensic examination of the bloodstain pattern helps in establishing the following: (a) where did the blood originate from? (b) what might be the nature of the wound? (c) what was the direction of blood origin? (d) what was the position of victim and perpetrator? (e) number of potential perpetrators in the crime scene.

Keywords

Bloodstain · Spatter · Angle of impact · Point of convergence · Direction of bloodstain

H. R. Dash et al., *Forensic Investigation of Blood Evidence*,
https://doi.org/10.1007/978-981-95-9183-1_8

1 Introduction

Blood is one of the most frequently found pieces of evidence in the crime scene. The shape, size, orientation, color, and other attributes of the bloodstains vary greatly with the causal mechanism of their origin. Thus, bloodstain pattern analysis (BPA) involves the analysis and interpretation of the nature of bloodstains to ascertain their causal mechanisms (Hicklin et al. 2021). The crime scene examiner examines the shape, size, distribution, and location of the bloodstains to form opinions on what might have occurred during the commitment of the crime. An amalgam of biology, physics, and mathematical principles is applied to obtain conclusions based on blood spatter analysis. A careful observation of the bloodstain patterns and their scientific evaluation might generate conclusions on various aspects such as (a) where did the blood originate from? (b) what might be the nature of the wound? (c) what was the direction of blood origin? (d) what was the position of the victim and perpetrator? (e) number of potential perpetrators in the crime scene.

BPA not only provides information on what might have happened, but it also provides information on what could not have happened. Understanding the impossible events helps in reconstructing the crime, confirming statements from the witnesses, and excluding or including potential perpetrators from the investigation. Though the analysis of blood spatter pattern allows accurate crime scene reconstruction, several factors including the surface texture, viscosity, deposition mechanism, and other environmental conditions determine the reliability of this methodology (Sacco et al. 2025). Thus, the use of standardized protocols, advanced terminological frameworks, and proficiency of the examiners is highly important in scene interpretation using BPA.

Despite the presence of a standardized mechanism for the interpretation of blood spatter, the complex crime scene scenarios such as the involvement of multiple rooms, object displacement, absence of victim, and noncorroboration between injury and blood volume might complicate the BPA. In such complicated cases, digital forensics and the three-dimensional crime scene reconstruction tools can increase the capability of BPA (Rotter et al. 2024). Besides, the BPA can be integrated with other forensic domains such as DNA profiling, clinical assessment of victim, forensic psychology, and forensic toxicology to ensure a legally acceptable conclusion from the BPA.

2 Properties of Liquid Blood

Blood accounts for ~8% of total body weight. Around 40% of blood loss either due to internal or external injury might lead to the death of the individual. However, a loss of around 1.5 L of blood is required to cause incapacitation of an individual (Potente et al. 2020). Liquid blood behaves in a significantly different manner than other liquids. Two important properties of blood make it different from other liquids, i.e., surface tension and viscosity. Surface tension phenomenon of blood is similar to that of water or any other liquid; however, due to its greater contrast, its behavior is more visible. Different molecules of blood are held together by Van der Waals force, resulting in the surface tension of blood. This cohesive force tends to resist penetration and separation. Further, surface tension also decreases the surface area of blood. Thus, when a small volume of blood is detached from its source, it assumes a spherical shape. Viscosity of blood is defined as its internal friction, showing its resistance to flow. The normal blood viscosity is reported to be 3.5–5.5 cP (centipoise) (Nader et al. 2019). Blood is a non-Newtonian fluid. Viscosity of blood plays a crucial role with its motion pattern in a crime scene, where the viscosity changes when force is applied (Baieth 2008). Thus, these two inherent properties of blood, i.e., surface tension and viscosity, lead to the generation of different patterns of bloodstains in different circumstances. The various types and patterns of bloodstains are discussed below.

3 Types of Bloodstains

The routine bloodstains found in a crime scene can be categorized into three types, i.e., passive bloodstain, projected bloodstain, and transfer stains. Passive bloodstains typically result from the result of gravity acting upon an injured body, whereas projected bloodstains result when a force is greater than the gravitational force. Similarly, transfer stains result from the objects coming in contact with the existing bloodstains.

3.1 Passive Bloodstains

Passive bloodstains are drops created by the force of gravity alone. When blood falls without the effect of any external force other than gravity, it maintains its spherical shape and creates a circular pattern when it hits on a surface. Passive bloodstains can be further subdivided into drops, drip stain, drip pattern, pools, and clots (Fig. 8.1). A drip pattern is different than that of a drip stain. A drip pattern is formed when a blood drips into another blood to form a pattern.

The surface nature and texture also affect the shape of a passive bloodstain. Bloodstain can be found on various surfaces such as carpet, wood, tiles, or clothing.

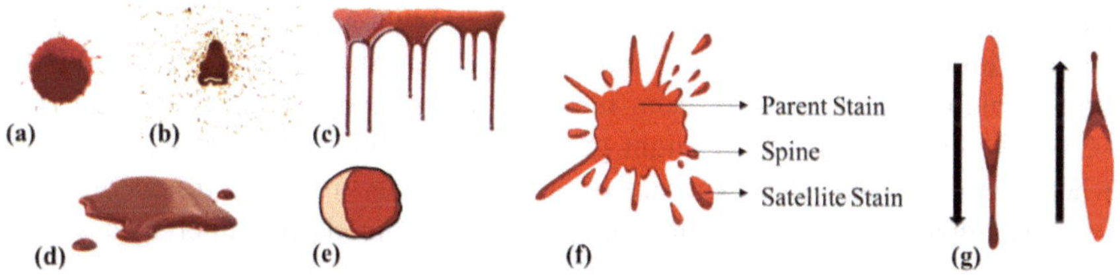

Fig. 8.1 Different types of passive bloodstains: (**a**) drops, (**b**) drip stains, (**c**) drip patterns, (**d**) pools, (**e**) clots, (**f**) typical structure of a blood spatter, and (**g**) directionality of the bloodstain

When a blood drop strikes a surface, the nature of the surface affects the amount of spatter, such as the size and appearance of the stain. When a blood droplet strikes a hard and smooth surface such as glass, little or no distortion around the edge is observed. However, when a blood droplet strikes on surfaces such as linoleum flooring, wood, or concrete, the presence of scalloping (distortion) around the edge of the blood droplets is observed (Dubey et al. 2019). The bloodstains are more distorted on wooden or concrete surfaces producing spines and secondary spatters. On the contrary, the absorbent surfaces generate smaller stains compared to the nonabsorbent surfaces. A typical blood spatter consists of three components, parent drop, satellite spatter, and spines (Fig. 8.1). The droplet from which the satellite spatter originates is called a parent drop. Small blood drops that break from the parent drop when it hits a surface are called satellite spatters. The pointed edges of a bloodstain which radiate out from the spatter are called as spines (Hook et al. 2024). The size and shape of a blood spatter differ with the height of the origin or source and the angle of impact. In a general notion, the size of a bloodstain is directly proportional to the height of fall. Similarly, when a blood drop strikes a surface at a perpendicular angle (90°), a circular bloostain is created. Any deviation of the angle other than 90° generates an elongated bloodstain.

3.2 Projected Bloodstains

Projected bloodstains are created when an exposed blood source is subjected to an action of force, greater than the force of gravity. Blood flows at a high pressure through various arteries, i.e., carotid artery, radial artery in the wrist, the femoral artery in the inner thigh, the brachial artery in the arm, temporal regions of the head, and the aorta (Sanchis-Gimeno et al. 2025). When an amount of force is utilized to strike a blood source, blood is ejected at a greater force. The shape, size, and the number of stains are determined by the amount of force used to hit the source of blood. Blood spatter can be either an impact spatter or a projection spatter. Projection spatter can be either arterial spurts, or expirated sprays, or spatter cast off.

When a source of blood receives a blow or hit, random dispersion of smaller blood droplets occurs. This results in the formation of impact spatters. Based on the nature of impact, impact spatters can be divided into low-velocity impact, medium-velocity impact, and high-velocity impact. High-velocity impact bloodstains move at a speed of ~100 ft./s. Such stains are created when much force is used such as gunshots, high-speed collisions, and explosions. Impact spatters created by high-velocity impact generally create a fine mist of <1 mm size (Stotesbury et al. 2016). Impact blood spatters arising from medium-velocity impact travel at a speed of ~25 ft./s. Such impact stains might generate in stabbings or blunt force injuries when a baseball bat is used. The size of the impact stains resulting from a medium-velocity impact ranges from 1 to 4 mm. Similarly, bloodstain patterns caused by low-velocity impact such as brick or hammer on a blood source are called as low-velocity impact stains. They move at a speed of ~5 ft./s, and their size is mostly >4 mm in diameter. Gunshot injuries can be categorized as high-velocity impact spatter when a blood spatter is generated as a result of a wound generated by thepenetration of a projectile. In such cases, two wounds are formed, the entry wound and exit wound. A forward spatter is generated from the exit wound in the same direction as that of the projectile. However, a back spatter is formed from the entry wound which travels in the opposite direction of the projectile.

Projected spatter is of three types: (i) arterial spurt, (ii) expirated sprays, and (iii) cast off stains. Arterial spurt or gush bloodstain pattern results when blood exits the body under high pressure from a breached artery. As a person is alive and heart is pumping, the blood flows through the arteries under high pressure. When such an artery is breached, blood spurts or gushes from the wound, producing a zig-zag pattern or up-and-down pattern due to the varying pumping action of the heart, resulting in varying blood pressure. Similarly, expirated sprays result from an internal injury, where blood is mixed with air from the lungs and leaves the body through the nose or mouth. Internal injuries in lungs and the airways result in such expirated sprays. Such blood spatters form very fine mist due to a high pressure exerted by the lungs, and the air moves out of the body. In expirated sprays, small air bubbles of <0.5 mm size are mixed with small drops of blood and typically found in such type of blood spatter. In a similar fashion, blood spatters can originate from a bloody object in motion which stops suddenly. For example, when a perpetrator bludgeons a victim with a piece of bat, the criminal's arm comes back to swing the bat again. The blood present on the end of the bat is projected out by centrifugal force due to arc formation. This generates a linear pattern of blood spatter mostly in the opposite direction of the object, and it can be found on the perpetrator's shirt.

3.3 Transfer Bloodstains

Transfer stains or contact stains originate when two surfaces come in contact with each other, where at least one of the surfaces is bloodied. In such cases, blood is transferred to a secondary target generating a unique pattern of bloodstain. Proper examination of transfer bloodstains has the potential to establish the sequence of events at the crime scene, and the movement pattern of objects and/or individuals in the crime scene can also be established. A particular pattern of transfer bloodstain can be attributed to a blood-bearing object. The most common objects such as hand or shoe can generate transfer bloodstains on the surface. Transfer bloodstains may include wipe, swipes, pattern transfers, and general contact of the object.

A contact pattern is formed when a bloody object comes in contact with a clean surface without any movement. In such contact bloodstains, the edges usually remain clean and look like a bloody impression. When a perpetrator steps on a blood pool and puts his/ her step on the clean floor, a bloody shoe/foot print is generated. A variety of investigative leads such as direction, speed of the individual, shoe size, height, and gait pattern can also be established from such contact bloodstains. Alternatively, a swipe pattern is formed when a blood object in motion comes in contact with a clean surface. It leaves behind a smudged impression. Such impressions provide useful information such as shape, size, and pattern of the blood-bearing object. One side of a swipe pattern is usually feathered. It indicates the direction of travel. A wipe pattern is generated when an unstained object moves on a wet bloodstain and alters its appearance. Wipe pattern examination provides clue on the movement and direction of movement of the object. Two unconventional transfer bloodstains can also appear, i.e., perimeter stain and skeletonized stain. A perimeter stain is generated when a stain is disturbed before it is completely dried; hence, it maintains the peripheral characteristics of the original stain. Such stains are generated as the periphery of a bloodstain dries faster compared to its center. Similarly, a skeletonized stain also retains the shape and size of the original stain after wiping.

3.4 Other Bloodstain Types

Few other types of bloodstain pattern can also appear in a crime scene. When an object which was previously present in the crime scene is moved, a void is generated. This leaves an outline or shadow on the final target. The presence of void pattern is the indicator of the movement of discarding of an object after the crime which was present during the criminal process. Similarly, blood stains resulting from the activity of the flies can mimic other patterns of blood stains. Fly activity results in small circular stains called as flyspeck. Such spots should not be confused with the expirated bloodstains where air bubbles are also found. Besides, insects walking through a bloodstain can also generate small additional stains by spreading the blood. Other useful terms associated with a blood spatter analysis is described below.

Bubble ring: An outline within a bloodstain resulting from air in the blood. *Cessation cast-off pattern:* A bloodstain pattern resulting from blood drops released from an object due to its rapid deceleration. *Perimeter stain:* An altered stain that consists of the peripheral characteristics of the original stain. *Saturation stain:* A bloodstain resulting from the accumulation of liquid blood in an absorbent material.

4 Estimation of Directionality of Blood Spatter

Directionality of a bloodstain can be speculated from the shape of the bloodstain. As described earlier, if the shape of the bloodstain is perfectly circular, it resulted from a strike at an angle of 90° to the surface. When the blood strikes the surface other than 90° to the surface, it appears to be more elliptical, and its direction can be estimated. Such bloodstains appear to be elongated or tear drop-shaped. The directionality of the bloodstain can be estimated by observing the pointed end or tail of the bloodstain (Fig. 8.1g). The pointed end is present in the direction of travel.

5 Estimation of Height

The size of a bloodstain is directly proportional to the height of fall. The higher the height, larger is the diameter of the blood spots. Though both the height of fall and blood spot diameter are directly proportional, a direct formula does not exist to calculate the exact height of a fall only from the diameter of the bloodstain. Besides the height of the fall, the size of a bloodstain is influenced by several other factors such as the volume of the blood droplet, surface of landing, and several other factors. If the blood drop volume is known, the height of the suspended blood-stained object can be determined from the size of the blood spots (Leonova et al. 2019).

6 Estimation of the Angle of Impact

The angle of impact provides information on the angle at which a bloodstain hits a surface. For calculating the angle of impact, the width and length measurements of the bloodstains are taken. While measuring such width and length, the tail length of a blood spatter is not included, because the tail of a blood spatter is mostly created by the gravitational force and the weapon force. Thus, the angle of impact (α) is calculated by measuring the minor axis (width, w) and the major axis (length, l) of the parent stain. The following equation is used to estimate the angle of impact of a bloodstain:

$$\sin\alpha = \mathrm{w} / \mathrm{l}, \text{Thus}, \alpha = \sin^{-1}\left(\mathrm{w} / \mathrm{l}\right)$$

For example, when a bloodstain is perfectly circular having its equal width and length, w/l becomes 1. As Sin 90° is 1, hence, the bloodstain moved perpendicular to the surface. When a person having a bloody nose stands still, passive blood drops drip from it and appear to be a perfect circle. However, when such bloodstain drips from the bloody nose when the person is moving, it appears a more elliptical shape. The principle of generating an elliptical bloodstain is given in Fig. 8.2. As blood is fluidic in nature, it may not produce a perfect ellipse when it impacts a surface at an angle other than 90°. In such cases, the pointed end is rounded off to its nearest value. In order to avoid the chance of human error, different bloodstain pattern analysis software can be used to generate the angle of impact in a precise and reproducible manner. The list of the commonly available software is given in Table 8.1.

7 Estimation of Point of Convergence

Point of convergence is the imaginary point in a crime scene where the path of each originating bloodstain meets. However, the actual origin of the blood spatter is at some point above this imaginary intersection. The estimation of the point of convergence might provide the knowledge on where an injured person might have positioned when the assault was taking place. The point of convergence is usually

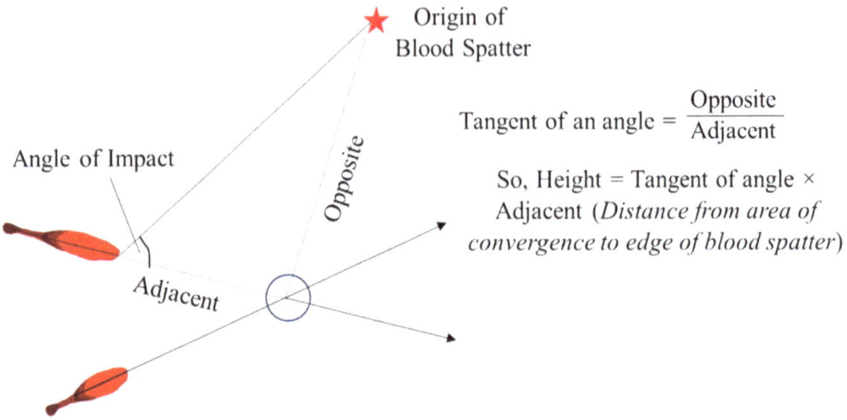

Fig. 8.2 Determination of the height of the origin of blood spatter by estimating the angle of impact and the point of convergence

Table 8.1 List of software available for blood spatter analysis

Sl. No.	Software	AO method	Measurement	Analysis planes	Release year
1	Wilson & Schuessler	String	Physical	Perpendicular surfaces	1987
2	BackTrack™	String	Physical	Orthogonal room	2005
3	No More Strings™	N/A	N/A	N/A	1996
4	ESCrime	N/A	N/A	N/A	2005
5	HemoSpat	String	Digital/Manual	Flat surfaces	2006
6	Microsoft Excel 2003 Autoshapes	Tangent	Digital/Manual	Two perpendicular walls	2003
7	Crime Scene Command	N/A	N/A	N/A	N/A
8	AnTraGoS	Tangent	Digital/Automatic	Single flat surface	2008
9	HemoVision	String	Digital/Automatic	Flat surfaces	2023
10	FARO Scene (Forensic Wizard)	String	Digital/Manual	Flat surfaces	2018
11	CrimeView3D	N/A	N/A	N/A	N/A
12	GigaPan®, Bonaccordo	String	Digital/Manual	Single flat surface	2018
13	FARO Zone 3D	String	Digital/Manual/ Automatic	Any surface	2017
14	Sherlock	String	Physical	Two perpendicular surfaces	2019
15	Leica Map360	String	Digital/ Manual	Any surface	2020

calculated by calculating the impact angle of each bloodstain, and the stains are projected back along their trajectories to intersect at a particular point.

The point of origin is the location from where the bloodstains originate. It can be estimated graphically by plotting the distance from the point of convergence. A point of convergence is established first in the Z-axis by extending the trajectories based on the impact angle of individual bloodstains. The law of tangents is used to calculate the height or the origin of the bloodstains. Going back to the right angle and adding the angle of impact, the height of origin of bloodstains can be determined. The height of the blood source is the side opposite to the angle of impact. Law of tangents is applied to estimate the height of blood origin (Fig. 8.2). The following formula can be used to determine the point of origin of the bloodstain:

$$\text{Height}\left(Z\right) = \text{tan of angle of impact} \times Y\left(\text{distance from the stain to the point of convergence}\right)$$

8 Recovery of Bloodstain Pattern Evidence

Prior to the analysis of the bloodstain pattern, photographic documentation should be carried out. Close-up photographs of all bloodstains are taken normal (90°) to the surface. Each photograph should include a ruler or any other scale of reference. All bloodstains or/and any bloodstain pattern of particular interest should be circled and numbered. Besides, an overall photography should also be carried out to understand the geometric relationship between the bloodstains and the general crime scene. Besides photography, recovery of bloodstains is also practiced to remove the bloodstains. Though lifting the bloodstain pattern is not routinely practiced by the investigators, it can be lifted in a similar fashion to that of fingerprint lifting using the same material, that is, transparent fingerprint tape. One-and-one-half inch fingerprint tape can be pressed over the bloodstain with moderate pressure, followed by careful peeling of the tape. The lifted bloodstain patterns are placed on white plastics. If required, the tape can be lifted again in the laboratory to expose the blood.

9 Conclusions

Understanding the blood spatter pattern holds significance in the crime scene investigation and its reconstruction. Two properties of blood, i.e., surface tension and viscosity, help the generation of unique patterns of blood spatters in different circumstances. Various types of bloodstains are found in a crime scene, which can be attributed to a particular scenario. Passive bloodstains are created by the force of gravity alone when bloodstains are released from a wound area. The shape and size of passive bloodstains vary with their nature such as drops, drip stain, drip pattern, pools, and clots. A typical blood spatter includes three components, i.e., parent drop, satellite stains, and spines. Careful examination and analysis of the bloodstain pattern reveal many useful investigative leads such as angle of impact, height, and origin of the blood source. When a bloodstain hits a surface at 90°, it appears perfectly circular shape. Any deviation from 90° makes the bloodstain elliptical, and the angle of impact can be calculated using the Sin formula. Similarly, in a general notion, the height of fall is directly proportional to the size of the bloodstain. By constructing the path of each bloodstain, the point of convergence can be estimated. From this point of convergence, the height of the blood source can also be found out. Thus, forensic examination of the bloodstain pattern helps in establishing (a) where the blood originated? (b) what might be the nature of the wound? (c) what was the direction of blood origin? (d) what was the position of victim and perpetrator? and (e) the number of potential perpetrators in the crime scene.

References

Baieth HE (2008) Physical parameters of blood as a non - newtonian fluid. Int J Biomed Sci 4:323–329

Dubey MK, Sharma R, Patel KK (2019) A comparative study on variation in bloodstain patterns due to change in surface-using blood from four common animal species. Int J Clin Diagn Pathol 2:167–174

Hicklin RA, Winer KR, Kish PE, Parks CL, Chapman W, Dunagan K, Richetelli N, Epstein EG, Ausdemore MA, Busey TA (2021) Accuracy and reproducibility of conclusions by forensic bloodstain pattern analysts. Forensic Sci Int 325:110856. https://doi.org/10.1016/j.forsciint.2021.110856

Hook E, Fieldhouse S, Flatman-Fairs D, Williams G (2024) Bloodstain classification methods: a critical review and a look to the future. Sci Justice 64:408–420. https://doi.org/10.1016/j.scijus.2024.06.004

Leonova EN, Nagornov MN, Kucha AS (2019) The possibility for determining the height of the fall of the blood drops based on the size of the bloodspots. Sud Med Ekspert 62:21–23. Russian. https://doi.org/10.17116/sudmed20196201121

Nader E, Skinner S, Romana M, Fort R, Lemonne N, Guillot N, Gauthier A, Antoine-Jonville S, Renoux C, Hardy-Dessources MD, Stauffer E, Joly P, Bertrand Y, Connes P (2019) Blood rheology: key parameters, impact on blood flow, role in sickle cell sisease and effects of exercise. Front Physiol 10:1329. https://doi.org/10.3389/fphys.2019.01329

Potente S, Ramsthaler F, Kettner M, Sauer P, Schmidt P (2020) Relative blood loss in forensic medicine-do we need a change in doctrine? Int J Legal Med 134:1123–1131. https://doi.org/10.1007/s00414-020-02260-w

Rotter G, Raffino C, Burrascano G, Ventura Spagnolo E, Baldino G, Asmundo A (2024) The role of Bloodstain Pattern Analysis (BPA) in reconstructing the dynamics of forensic cases. Clin Ter 175:125–129. https://doi.org/10.7417/CT.2024.5099

Sacco MA, Gualtieri S, Princi A, Raffaele R, Aquila I (2025) Application of bloodstain pattern analysis (BPA) to reconstruct a staged crime scene in a complex forensic case. Forensic Sci Int Synergy 11:100624. https://doi.org/10.1016/j.fsisyn.2025.100624

Sanchis-Gimeno J, Loaiza-Giraldo JP, Alruiz Y, Vergara M, Navia MF, Roman C, Suazo-Santibañez A, Nova-Baeza P, Orellana-Donoso M, Oyanedel-Amaro G, Rodriguez-Luengo M, Bruna-Mejias A, Valenzuela-Fuenzalida JJ, León-Rojas JE, Granite G (2025) Prevalence of radial artery variants and their relationship with clinical considerations of the antebrachial region: systematic review and meta-analysis. Diagnostics (Basel) 15:2984. https://doi.org/10.3390/diagnostics15232984

Stotesbury T, Illes M, Vreugdenhil AJ (2016) An impact velocity device design for blood spatter pattern generation with considerations for high-speed video analysis. J Forensic Sci 61:501–508. https://doi.org/10.1111/1556-4029.12975

Collection and Preservation of Blood Evidence

9

Abstract Blood is the most commonly found biological evidence in majority of the crime scenes. Besides blood as a questioned sample, peripheral blood is often collected as a reference sample to be analyzed parallelly. The reference blood samples are collected in a controlled environment from the blood donors. Such blood samples are preserved either in an EDTA vial or an FTA card. Similarly, the blood stains as questioned samples are either collected from the crime scene or from the body of the victim/suspect/deceased. The questioned blood stain from the crime scene is collected either by the forensic expert or the investigating officer, whereas, the sample from the victim/suspect/deceased is collected by the medico-legal expert or the autopsy surgeon. Besides appropriate collection and preservation of blood samples, a proper set of documents also needs to be submitted along with the blood samples. Identification form or the blood collection form should be prepared while collecting the reference blood sample from the donor. Similarly, seizure memo, medico-legal certificate and/or autopsy report should be prepared while collecting blood sample from the victim/suspect/deceased. In both the cases, chain of custody form should be prepared and maintained in each step of forensic investigation.

Keywords Collection · Preservation · Questioned sample · Reference sample · Preservatives

1 Introduction

In contrast to any other forensic evidences, biological evidence needs special attention due to their inherent nature. Biological samples are the nutrient jackpots for the microorganisms as they are rich in proteins, lipids, carbohydrates, and nucleic acids. Hence, the forensic biological samples become more prone to degradation by microorganisms. Blood is the most commonly found biological evidence at the crime scene. 400–1500 μg protein is present in platelets and 100–600 μg protein is

H. R. Dash et al., *Forensic Investigation of Blood Evidence*,
https://doi.org/10.1007/978-981-95-9183-1_9

found in peripheral blood mononuclear cells (PBMC) (Crosley et al. 2009). Besides, plasma and immunoglobins also contribute to the protein richness of blood samples. Thus, blood samples of forensic interest should be collected and preserved with a specified manner to prevent the microbial degradation of the samples and to increase the success rate of obtaining a DNA profile.

Blood as forensic evidence can be found in various forms. Reference samples are mostly collected intravenously and is obtained in liquid form. When blood is deposited on a surface, it dries out and forms dried blood stains. Such dried blood stains can be collected from soil, stone, walls, clothing, metal surfaces or any other surface. The collection strategy of these blood stains varies depending on the nature of the material on which the blood stain is deposited. Similarly, in some cases menstrual blood is also found and such samples are collected with utmost care. When blood stains are found in a crime scene, they may appear in spatter. The blood spatter analysis should be performed before collection of such evidence. The blood spatter analysis provides crucial information for crime scene reconstruction. The types of blood spatters and the procedural framework for the investigation of blood spatter have been discussed in detail in the previous chapter.

2 Collection and Preservation of Blood as a Reference Sample

Reference sample is the biological sample which is collected from a known source, whereas the biological sample whose source is unknown is called as a questioned sample. The differences between questioned sample and reference sample are highlighted in Table 9.1. Blood is the most commonly collected reference sample used in forensic DNA analysis. The reference blood sample is collected either intravenously or by finger-pricking method. By using a vacutainer or a syringe needle, ~1 to 2 ml blood is drawn from the median cubital vein in the inner elbow. After

Table 9.1 Difference between questioned blood sample and reference blood sample

Features	Reference blood sample	Questioned blood sample
Sample source	Known	Unknown
	Single source	May be single source or mixed source
DNA quantity	High quantity	May be high or less quantity
DNA quality	Mostly good quality	May be good or bad quality
Collection	Intravenously or by finger pricking of the sample donor	From the crime scene, environment, body of the victim/suspect/deceased
	Always collected in a controlled environment	Sample is collected as recovered
Chance of contamination	Absence of exogenous contamination	Exogenous contamination might be present
Example	Blood sample from the putative mother and/or father of the deceased	Blood sample collected from the punctured heart postmortem

collection, the blood is transferred into an EDTA vial. Ethylenediamine tetraacetic acid (EDTA) is a known anticoagulant which is routinely used for the preservation of liquid blood samples. EDTA is a polyprotic acid which contains four -COOH groups and two $-NH_2$ groups having lone-pair electrons. Thus, it can chelate calcium (Ca) and several other metal ions (Banfi et al. 2007). During coagulation of blood, Ca acts as a necessary metal for the enzymatic reactions of the coagulation cascade. Thus, the removal of Ca irreversibly by EDTA prevents blood clotting and helps in the stabilization of blood cells. The role of EDTA as an anticoagulant is described in Fig. 9.1. The EDTA vial containing blood sample should be marked and sealed properly. It can be transported to the laboratory by keeping in an ice box, and it should reach the laboratory at the earliest. Once reached in the laboratory, the blood sample in EDTA vial is stored at 4°C till further use.

Alternatively, FTA® card can also be used to collect blood as a reference sample. An alcohol wipe is used to clean the fingertip of the blood donor. The end of the lancet is placed on the cleaned finger and the trigger is pressed to prick the finger. The finger is pressed to deposit a drop of blood on the printed circle in FTA® card without touching its surface. The FTA® card with blood is allowed to air dry completely for 1–2 h at room temperature. Once dried, the FTA® card is closed and placed in a multi barrier pouch. It can be placed in an enveloped, sealed and marked properly and can be transported to the laboratory at room temperature. FTA® card, the Flinders Technology Associates card is a chemically treated filter paper which is designed for the collection, preservation and transport of the biological samples for

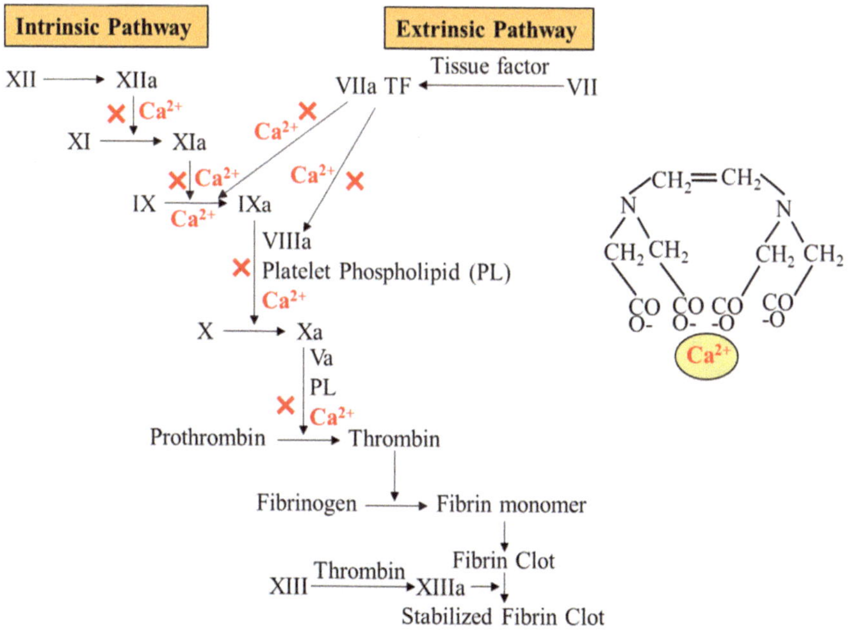

Fig. 9.1 Mode of action of EDTA as a chelating agent to prevent coagulation of blood

forensic DNA examination. In this patented technology, the reagents present in the filter paper help in stabilizing DNA at room temperature which allows long term storage and easy transport. Collection of blood sample on FTA® card has several advantages over collection of blood samples in EDTA vial. The patented chemical formula present in the card lyses the cell membrane, denatures protein and help the DNA come out of the cell (Sahajpal et al. 2019). Thus, the lengthy step of DNA extraction process can be replaced with the washing of FTA® card with wash buffers. This washing step removes all the cell debris except DNA which is embedded on the FTA® card. A small punch of this card can be used for the subsequent steps of forensic DNA analysis.

2.1 Suitable Selection of Reference Blood Sample

A suitable selection of reference sample makes the DNA profile analysis easier for the examiner. In a case of sexual assault and suspected murder, the source of reference blood sample(s) is straight forward. Reference blood samples are collected from the victim and the suspect(s). Similarly, for solving paternity dispute, three blood samples are collected, i.e., blood sample from the questioned child, known mother, and the suspected father. However, selection of reference sample in a case of identification is always challenging and complicated. Depending on the deceased individual and the sample limiting conditions, the blood donors are carefully selected before collection of reference samples from them. Ideally, the reference blood samples are collected from both the putative parents of the deceased individuals. However, in practical scenario, many sample limiting conditions arise. In certain instances, one of the parents is not available or in certain cases both the parents are not available. In such instances, the reverse paternity analysis can be performed to establish the identity of the deceased by autosomal short tandem repeats (STR) analysis. For example, if the deceased is a female, reference blood sample can be collected from its child and the husband. Similarly, if the deceased is a male, reference blood samples can be collected from the child and the wife.

If any of the abovementioned samples are not available, the blood samples can be collected from the siblings of the deceased. When the reference blood samples are collected from the siblings, more than one sibling samples is recommended for autosomal STR-based identification. This is because of the fact that, theoretically, match percentage of autosomal STR profile between two siblings ranges from 0% to 100%, 100% being the two monozygotic twins (Bandah-Rozenfeld et al. 2025). In many cases, neither the parents, nor the child, nor the siblings' samples are available to collect as reference samples for identification purpose. In such instances, reference blood samples can be collected from the putative maternal relatives of the deceased individual. Here, the mitochondrial DNA analysis can be carried out to establish the maternal lineage of the individual which can aid to the identification (Amorim et al. 2019). Alternatively, if the deceased in male, reference blood samples can be collected from the paternal relatives. Here, Y-STR analysis can be performed, and the paternal lineage of the deceased can be established for identification

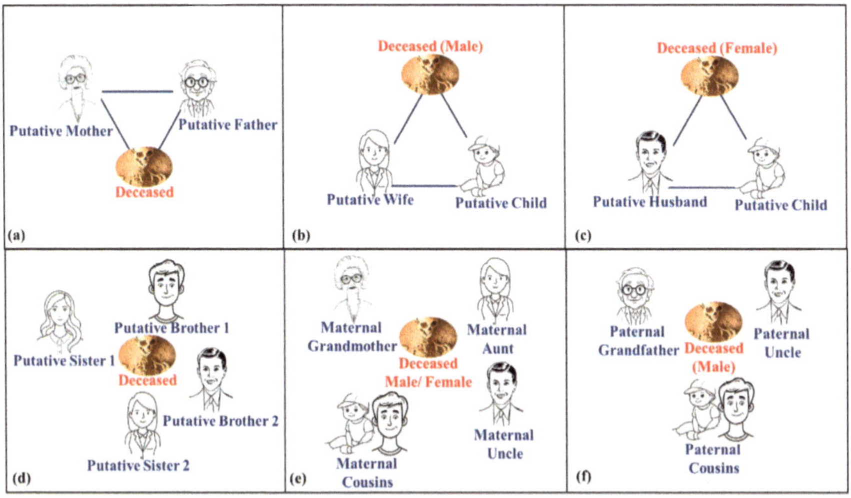

Fig. 9.2 Strategies for selection of blood donors as reference sample in forensic DNA-based identification. (**a**) When putative parents' sample available, (**b**) When deceased is male, wife and child available, (**c**) when deceased is female, husband and child available, (**d**) Siblings available, (**e**) blood samples from maternal relatives, (**f**) deceased male, blood samples from paternal relatives

(Syndercombe Court 2021). Various strategies for the selection of reference samples are given in Fig. 9.2.

2.2 Documentation During Collection of Reference Blood Sample

Two documents are commonly prepared and submitted to the laboratory along with the reference blood samples, i.e., blood collection form, and the chain of custody form. Though the format of the blood collection form varies from laboratory to laboratory, the contents of the form remain the same. A blood collection form can be divided into four sections. In the first section, the personal details of the blood donor such as name, age, sex, address, and ethnicity are mentioned. In the second section, the medical history of the blood donor such as blood transfusion, organ transplant and other known genetic disorders is mentioned. In the third section of the blood collection form the statement of written informed consent of the blood donor is obtained. The fourth section deals with the witnesses. In this section, the signature of the witnesses is obtained, in presence of whom the blood sample was collected. The most common witnesses found in a blood collection form are the sample collector (mostly a medical officer), the investigating officer, and a respectable person of the society. Additionally, a duly attested photograph of the blood donor is also pasted in the blood collection form.

The chain of custody form holds significance in any forensic evidence including the reference blood sample. The chain of custody form ensures integrity of the sample by recording the details of the sample collector, sample information, log of sample transfer along with the signature of the sample custodian. Mostly, a chain of custody form along with the reference sample includes donor information, collector information, details of the sample, transport and storage information. The original chain of custody form travels with the blood sample, whereas a copy of it is retained by the individual at every level.

2.3 Factors Considered in a Reference Blood Sample

The most common intrinsic factor that affects the DNA profile of a reference sample is blood transfusion in the blood donor. The presence of exogenous blood in the blood donor due to transfusion may yield a mixed DNA profile. However, the exogenous blood is replaced with the host blood after few days. The maximum lifespan of RBC is 120 days, whereas the lifespan of WBC is 12–20 days (Bosman 2013). Thus, the status of blood transfusion at the time of blood collection should be informed by the medical officer. Most of the laboratory guidelines suggest to collect reference blood sample after 3 months of last transfusion for DNA tests. However, the present-day blood transfusion practice is different. Once a blood sample is collected for transfusion, it is separated into packed red blood cell (PRBC), platelet concentrate, fresh frozen plasma (FFP), and cryoprecipitate (Basu and Kulkarni 2014). Patients with chronic anemia (Hb level ≤6 g/dl), impaired cardiopulmonary and vascular function are often transfused with PRBCs (Müller et al. 2015). Patients with thrombocytopenia (low platelets) or platelet dysfunction are often transfused with platelet concentrate. Similarly, patients with coagulation deficiencies, abnormal coagulation test results, and active bleeding are often transfused with FFP. Cryoprecipitate is transfused to individuals with low fibrinogen levels during massive trauma, surgery and rare blood clotting disorders (Tinegate et al. 2012). Transfusion of WBC is rare and often practiced when patients with severe infection is unresponsive to antibiotics. Besides, in severe traumatic hemorrhage conditions, transfusion of whole blood is preferred (Cap et al. 2018). In forensic context, a reference blood sample is collected from a sample donor transfused with either whole blood, or WBC, or platelet concentrate, or cryoprecipitate, or FFP, or PRBCs. Cell types such as mature RBCs and platelets do not have nucleus and hence have no genomic DNA. Thus, exogenous DNA from the transfused blood can only be found in patients with whole blood transfusion or WBC transfusion. Thus, except these two types of transfusion, blood sample can be collected from the transfused individual immediately after transfusion.

In addition to blood transfusion, prior knowledge of organ transplantation in the blood donor helps interpreting a DNA profile result by the examiner. Bone marrow transplantation is the most commonly used treatment for blood dyscrasias such as leukemia. As the bone marrow of the recipient is replaced with the bone marrow of the donor's, it starts manufacturing red and white blood cells in the recipient's body. Though mature red cells do not have nucleus and DNA, other cell types such as WBC can have a DNA of the bone marrow donor. Thus, the blood sample of the individual with transfused bone marrow is expected to yield a mixed DNA profile. Due to the presence of lymphatic white cells in the mouth, buccal swab can also yield a mixed profile in bone marrow transplanted individuals (Pope et al. 2006). In such scenario, rooted hair samples (as long as no blood is present in the root) can be collected as a reference sample to yield a single-source DNA profile.

Besides blood transfusion and organ transplantation, the known genetic disorder of the blood donor can also affect the outcome of DNA profiling result. For example, if the person is suffering from Down's syndrome, three peaks (alleles) are expected at D21S11 genetic marker. Individuals with trisomy 18 might generate three peaks (alleles) at D18S51 marker. Similarly, individuals suffering from Patau syndrome generates three peaks (alleles) at D13S317 (Bukyya et al. 2021). Besides, the sex determining marker Amelogenin shows two balanced peaks (X and Y) in single source male samples. However, if the male individual is suffering from Klinefelter Syndrome (XXY), an imbalanced X and Y peaks are observed which should not be confused with a mixed profile. Thus, the prior knowledge of genetic disorders of the individuals donating reference blood samples helps in interpreting a DNA profiling result and developing suitable strategies in solving a case.

3 Collection and Preservation of Blood as a Questioned Sample

Blood samples as a questioned sample is collected either from the crime scene or from the victim/deceased/suspect's body or belongings. A blood sample from the crime scene is collected by the investigating officer or the forensic expert, whereas the blood sample can be collected and preserved by the medicolegal expert or autopsy surgeon during medicolegal examination or autopsy. Such blood samples are recovered either in the liquid form or as a blood stain. When a large amount of fresh blood is found in the crime scene, they can be collected and preserved in a similar fashion like reference sample in an EDTA vial or FTA card. Similarly, liquid blood samples are also collected from the postmortem subjects. In such cases, blood is collected from non-body cavity areas such as heart and major blood vessels. Same preservation, transportation, and storage strategy should be employed like the peripheral reference blood sample as described above.

Blood constitutes 7–8% of an individual's body weight. Most of the injuries lead to the release of blood from the body, only the amount and type of blood vary from case to case depending on the nature of injury and type of case. In routine cases of murder where weapons are used lead to the generation of blood and it is found in the crime scene. Similarly, cases of aggressive sexual assault may lead to injury in the private parts of the victim and blood stains may be deposited in the crime scene. Similarly, menstrual blood can also be found in the crime scene depending on the physiological state of the female victim. Distinguishing peripheral blood from menstrual blood helps in solving infanticide cases.

3.1 Collection Strategies for Blood Stains

When blood stain is present on a small movable object such as coin, wallet, currency notes, pen, knife, the whole object is collected and after air drying and packaged properly. When blood is found on immovable objects in the form of a blood stain, special collection strategies are employed depending on the nature of the surface and the crime scene environment. This can be called as the transfer of stains. After the transfer of stains, the evidence may not be available on the object present in the crime scene. Hence, the whole process of stain transfer should be documented by photography and videography. Broadly, three approaches are used to collect a blood stain from the crime scene, i.e., swabbing, scrapping, and cutting. Swabbing is the most widely used collection strategy of blood stains. It can be used in two ways, i.e., wet swabbing and dry swabbing. A sterile cotton swab moistened in sterile saline solution is applied on the blood stain. As per the general guidelines, the swab is rotated two to four times depending on the nature of the surface followed by proper air drying of the swab and store it in an appropriate container. Conversely, if the blood stain is still found in wet condition, a sterile dry swab can be used and the same procedure can be used to collect the blood stain. Double-swabbing technique, i.e., swabbing with a wet swab followed by a second dry swab has been reported to be more effective than either the dry swabbing or wet swabbing technique alone (Hedman et al. 2021). Swabbing technique of transferring blood stain is more effective on non-absorptive surfaces and can be applied to both indoor and outdoor crime scenes.

Dried blood stains found on various surfaces can be collected by scrapping technique. This technique is mostly used to collect blood stains from solid, non-absorbing and large surfaces where blood stains are found. In this technique, the blood stains are scrapped from the surface using a clean, sterile surgical blade. The powder of blood stains can be collected on a piece of butter paper in the form of a druggist fold. The steps followed for the preparation of a druggist fold are given in Fig. 9.3. The collected sample can be marked properly, sealed and put in a paper envelope and submitted to the laboratory. In outdoor conditions and environment with a high air flow, the chance of sample loss is increased when scrapping technique is used. Alternatively, when blood stains are found on soft, absorptive surfaces, the samples can be collected by cutting technique. The surface can be cut at the periphery of the blood stain and air-dried properly followed by collection in a paper envelope. Cutting technique allows maximum transfer of the stains and the chance loss of evidence is minimum.

Other collection strategies include vacuuming, tape lifting and many others. Such techniques are particularly useful for trace evidence collection when a visible blood stain is not found. The M-Vac® wet-vacuum system has been reported to be more useful than double swabbing technique for collection of trace DNA evidence from brick surfaces (Vickar et al. 2018). Thus, such collection strategies can be adopted in such conditions when the traditional collection strategies fail to gather a usable amount of blood evidence. After collection of blood stains using any of the above discussed techniques, it is highly important to dry those objects/samples

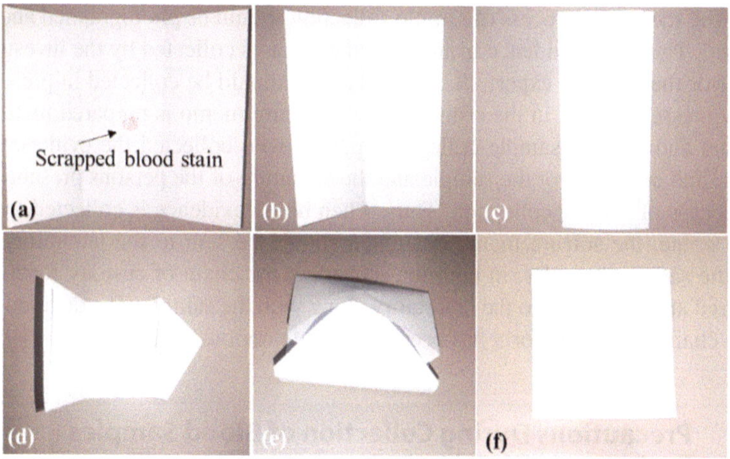

Fig. 9.3 Preparation of druggist fold for collection of blood scrapping. (**a**) Fold a clean, unused paper sheet into thirds by keeping the scrapped material in the middle section, (**b**) Fold one third over the middle section, (**c**) Fold the other third over the middle section, (**d**) Fold in half in the same direction as the thirds were folded, fold the ends up, making one into a point for easier insertion into the other, (**e** and **f**) Insert pointed end into the outermost opening of the straight end, seal it using a cellophane tape

under shade. If the biological samples are dried under direct sunlight, intense heat and UV light from the sun might damage the DNA and enzymes which may hamper the downstream forensic examinations. Drying reduces the water content in the blood samples and reduces the chance of metal growth on them. Reduction of the available water (a_w) for the microorganisms limits their growth and the chance of sample degradation is reduced (Beuchat et al. 2013). However, drying the forensic biological evidence under shade is weather dependent. High moisture content in the environment may increase the drying period significantly generating the chain of custody issues. Hence, for rapid drying of biological evidence in a controlled environment, evidence drying cabinets can be used (Vidua et al. 2024).

3.2 Documentation During Collection of Questioned Blood Sample

Questioned blood samples can be collected and preserved from two places, from the crime scene and from the body of the victim/suspect/deceased. When blood evidence is collected from the body of victim/suspect/deceased, the details of the sample should be mentioned in the medical-legal examination form or autopsy report. Such details are mentioned by the medico-legal expert or autopsy surgeon who conducts medical examination of the individuals. Thus, in such scenario, the medico-legal examination form and/or the autopsy report should be sent to the laboratory along with the sample. Alternatively, when the blood evidence is collected from the crime scene, the process of sample collection should be photographed and video graphed. The blood evidence from the crime scene is collected by the investigating officer or the forensic expert. Such blood stains should be collected in presence of few witnesses present in the crime scene. A seizure memo is prepared mentioning the date and time of sample collection, person who collected the evidence, type, nature, and condition of the sample and the signature of the persons present during the process of sample collection. Thus, when blood evidence is collected from the crime scene, the seizure memo document should be sent to the laboratory along with the sample. Besides, in both the scenarios, the chain of custody form is also prepared and submitted to the laboratory along with the sample. The detailed nature of the chain of custody form has been described previously.

4 Precautions During Collection of Blood Samples for Forensic DNA Analysis

Biological samples including blood are prone to contamination and degradation. Exogenous contamination might lead to the generation of a mixed profile leading to an inconclusive report. Similarly, microbial degradation of blood samples can hamper the enzymatic, antigen–antibody-based experiments routinely carried out for blood examination. Similarly, degradation of DNA will also lead to the generation of either partial or no DNA profile. Hence, the following precautions should be followed to avoid contamination and degradation of the blood evidence.

- Do not touch the crime scene with bare hands.
- Avoid sneezing or coughing over the evidence.
- Use clean gloves while collecting the blood evidence.
- Change the gloves and/or the tweezers between different items having blood evidence.
- Package each item containing blood stains separately.
- Air dry blood stains prior to their sealing and packaging.
- Package samples in paper envelopes or paper bags after drying.
- Mark each package properly with case number, item number, collection date, and initial of authorized person.
- Seal the package properly to maintain a proper chain of custody.
- Transfer blood stains on an unmovable surface (Table/Floor/Stone).

5 Conclusion

Blood is the most commonly found biological evidence in majority of the crime scenes. Besides blood as a questioned sample, peripheral blood is often collected as a reference sample to be analyzed parallelly. The reference blood samples are collected in a controlled environment from the blood donors. Such blood samples are preserved either in an EDTA vial or FTA card. Similarly, the blood stains as questioned sample are either collected from the crime scene or from the body of the victim/suspect/deceased. The questioned blood stain from the crime scene is collected either by the forensic expert or the investigating officer, whereas, the sample from the victim/suspect/deceased is collected by the medico-legal expert or the autopsy surgeon. Besides appropriate collection and preservation of blood samples, a proper set of documents also needs to be submitted along with the blood samples. Identification form or the blood collection form should be prepared while collecting the reference blood sample from the donor. Similarly, seizure memo, medico-legal certificate and/or autopsy report should be prepared while collecting blood sample from the victim/suspect/deceased. In both the cases, chain of custody form should be prepared and maintained in each step of forensic investigation.

References

Amorim A, Fernandes T, Taveira N (2019) Mitochondrial DNA in human identification: a review. PeerJ 7:e7314. https://doi.org/10.7717/peerj.7314

Bandah-Rozenfeld D, Voskoboinik L, Rashkovski K, Avrahami K, Schayek H, Avlas O (2025) The power of STR discrimination: two brothers with high-profile similarity. J Forensic Sci 70:1622–1628. https://doi.org/10.1111/1556-4029.70060

Banfi G, Salvagno GL, Lippi G (2007) The role of ethylenediamine tetraacetic acid (EDTA) as in vitro anticoagulant for diagnostic purposes. Clin Chem Lab Med 45:565–576. https://doi.org/10.1515/CCLM.2007.110

Basu D, Kulkarni R (2014) Overview of blood components and their preparation. Indian J Anaesth 58:529–537. https://doi.org/10.4103/0019-5049.144647

Beuchat LR, Komitopoulou E, Beckers H, Betts RP, Bourdichon F, Fanning S, Joosten HM, Ter Kuile BH (2013) Low--water activity foods: increased concern as vehicles of foodborne pathogens. J Food Prot 76:150–172. https://doi.org/10.4315/0362-028X.JFP-12-211

Bosman GJ (2013) Survival of red blood cells after transfusion: processes and consequences. Front Physiol 4:376. https://doi.org/10.3389/fphys.2013.00376

Bukyya JL, Tejasvi MLA, Avinash A, Chanchala HP, Talwade P, Afroz MM, Pokala A, Neela PK, Shyamilee TK, Srisha V (2021) DNA profiling in forensic science: a review. Glob Med Genet 8:135–143. https://doi.org/10.1055/s-0041-1728689

Cap AP, Beckett A, Benov A, Borgman M, Chen J, Corley JB, Doughty H, Fisher A, Glassberg E, Gonzales R, Kane SF, Malloy WW, Nessen S, Perkins JG, Prat N, Quesada J, Reade M, Sailliol A, Spinella PC, Stockinger Z, Strandenes G, Taylor A, Yazer M, Bryant B, Gurney J (2018) Whole blood transfusion. Mil Med 183:44–51. https://doi.org/10.1093/milmed/usy120

Crosley LK, Duthie SJ, Polley AC, Bouwman FG, Heim C, Mulholland F, Horgan G, Johnson IT, Mariman EC, Elliott RM, Daniel H, de Roos B (2009) Variation in protein levels obtained from human blood cells and biofluids for platelet, peripheral blood mononuclear cell, plasma, urine and saliva proteomics. Genes Nutr 4:95–102. https://doi.org/10.1007/s12263-009-0121-x

Hedman J, Akel Y, Jansson L, Hedell R, Wallmark N, Forsberg C, Ansell R (2021) Enhanced forensic DNA recovery with appropriate swabs and optimized swabbing technique. Forensic Sci Int: Genet 53:102491. https://doi.org/10.1016/j.fsigen.2021.102491

Müller MM, Geisen C, Zacharowski K, Tonn T, Seifried E (2015) Transfusion of packed red cells: indications, triggers and adverse events. Dtsch Arztebl Int 112:507–517. https://doi.org/10.3238/arztebl.2015.0507

Pope S, Chapman H, Lambert J (2006) The effect of bone marrow transplants on DNA profiles; a case example. Sci Justice 46:231–237. https://doi.org/10.1016/S1355-0306(06)71603-3

Sahajpal V, Rajput S, Sharma T, Sharma A, Thakar MK (2019) Development and evaluation of a novel DNA purification buffer and protocol for blood samples on FTA cards. Forensic Sci Int Rep 1:100014. https://doi.org/10.1016/j.fsir.2019.100014

Syndercombe Court D (2021) The Y chromosome and its use in forensic DNA analysis. Emerg Top Life Sci 5:427–441. https://doi.org/10.1042/ETLS20200339

Tinegate H, Allard S, Grant-Casey J, Hennem S, Kilner M, Rowley M, Seeney F, Stanworth S (2012) Cryoprecipitate for transfusion: which patients receive it and why? A study of patterns of use across three regions in England. Transfus Med 22:356–361. https://doi.org/10.1111/j.1365-3148.2012.01158.x

Vickar T, Bache K, Daniel B, Frascione N (2018) The use of the M-Vac® wet-vacuum system as a method for DNA recovery. Sci Justice 58:282–286. https://doi.org/10.1016/j.scijus.2018.01.003

Vidua RK, Arora A, Yadav J, Abraham S (2024) Use of innovative sample dryer for effective sample drying in forensic investigations: an experimental study. Forensic Sci Int 363:112158. https://doi.org/10.1016/j.forsciint.2024.112158

Presumptive Tests for Blood Detection 10

Abstract

Presumptive detection of blood and blood stains is one of the most prerequisite of forensic biological investigation. Most of the presumptive tests rely on the principle of redox reaction which explores the peroxidative nature of Heme found in the hemoglobin of the blood cells. Though most of such tests are highly sensitive, however, they are not specific to either blood or human blood. Many of such tests show cross-reactivity with other body-fluids that might be present in the crime scene. Similarly, plant peroxidases, metal contaminants can also lead to a false positive result. The error rate of such tests can be minimized by using suitable positive and negative controls while using these tests. Hence, though presumptive tests give us an idea of the location of blood stains or latent blood stains, their presence should be confirmed using any of the confirmatory tests.

Keywords

Presumptive test · Blood detection · Blood stain · Redox reaction · Luminol · Peroxidase

1 Introduction

Our ancestors have equated red color as symbolical association with life, vitality, danger as well as murder. The porphyrin-bound iron ions in hemoglobin molecules present in the red blood cells (RBCs) of blood are responsible for blood's crimson color chemistry. In a crime scene, blood is often identified from its red colored appearance. However, all-red colored substances are not blood. Many other substances including the fruits and vegetables such as Beets, strawberries, tomatoes, food coloring agents (FD&C Red No. 3), natural dyes (carmine), red ochre,

H. R. Dash et al., *Forensic Investigation of Blood Evidence*,
https://doi.org/10.1007/978-981-95-9183-1_10

Table 10.1 Generalized differences between a Presumptive and Confirmatory test

Presumptive test	Confirmatory test
Narrows down the possibility and guides further to perform specific confirmatory tests.	Conclusively identifies a body-fluid.
Widely performed on larger surfaces that might contain body fluids.	Not widely performed on larger surfaces.
Can locate a body-fluid not visible to the naked eye.	Confirmatory test is usually performed after locating a possible body-fluid.
High Sensitivity and Low specificity.	High Specificity.
High risk of false positive.	Low risk of false positive.
Generally conducted as on-field Point of Care (PoC) tests.	Generally, requires additional instrumentation(s) to perform such tests.
Less cost involvement.	High-cost involvement.
Less turn-around time.	High turn-around time.

vermilion, red lead (minium) and many other substances can also appear red if found in a crime scene. On contrary, non-finding of red color substances in a crime scene also does not confirm the complete absence of blood or blood stains. Latent or invisible blood stains may also be present in a crime scene. Blood stains become faint either due to their intentional or unintentional removal. In such circumstances, forensic biology comes in rescue to attribute a spot to be of blood origin.

Two types of tests are conducted to identify any body fluid in a forensic set up i.e., (i) Presumptive tests, and (ii) Confirmatory tests. The generalized differences between a presumptive and confirmatory test are given in Table 10.1. Presumptive tests do not guarantee the presence of a particular body fluid; it only indicates the presence of a body fluid. Generally, such presumptive body-fluid examinations are carried out at the crime scene for locating biological evidence. Similarly, latent or invisible body-fluid stains can also be identified using the presumptive tests. Once located, they are collected using standard operating procedures (SOPs) and forwarded to the laboratory for their authentication by confirmatory tests.

2 Requirement of Forensic Identification of Body-Fluids

Body-fluids are one of the most widely encountered biological evidence found in the crime scene. Body-fluid detection and its subsequent analysis can aid the investigation process to establish the crucial link among the donor, body-fluid type, and the nature of activities. In a sexual act, body-fluids such as seminal fluid, vaginal fluid, menstrual blood can be expected. Saliva can also be recovered from the crime scene as well as from the body of the victim. Similarly, violent injury leading to death can generated blood as a body-fluid in the crime scene. In suicidal hanging cases, urine samples may be recovered from the crime scene. Saliva and sweat can be recovered from the bedsheet and pillow to identify the user. Thus, body-fluid identification holds a huge significance in establishing the nature of an act as well as identifying the personal(s).

3 Principle of Body-Fluids Identifications

Composition of each body-fluid is unique. However, there are several components that are commonly found in different body fluids. Though such components are commonly found in different body-fluids, their relative concentration varies. The presence of specific components in one body-fluid vs. another is the basis of forensic body-fluid identification. As the relative concentration of several components vary among the body-fluids, such components can be targeted to make the test(s) more effective. However, due to presence of such common components in different body-fluids, they also show cross-reactivity. This might lead to false positive results and it is one of the shortcomings of the presumptive tests.

For example, saliva, seminal fluid, and vaginal fluid, all contain an enzyme Amylase. However, the relative concentration of amylase in saliva is much higher compared to seminal and vaginal fluids. Hence, Amylase is used for the presumptive detection of saliva. Similarly, both seminal and vaginal fluids contain lactic acid and citric acid. However, vaginal secretions possess a higher amount of lactic acid, whereas, seminal fluids contain a high amount of citric acid. Additionally, urea is found in urine, sweat, and seminal fluids. However, the relative concentration of urea is much higher in urine compared to other body-fluids. Hence, urea is used as an indicator for presumptive identification of urine. Hemoglobin is one of the major constituents of blood. Hence, most of the presumptive tests of blood target hemoglobin for its presumptive identification.

4 Chemical Structure and Nature of Hemoglobin Molecule

Hemoglobin is a type of globular protein present in RBC of the blood. It is a 2-way respiratory carrier (Marengo-Rowe 2006). It has the property to carry O_2 from the lungs to tissues as well as facilitating the return of CO_2 from the tissues to the lungs. A typical adult hemoglobin consists of four subunits. Each subunit has one polypeptide chain and a heme group. In total, a hemoglobin molecule is made up of two α-subunits (made up of α polypeptide chain of 141 amino acids) and two β-subunits (made up of β polypeptide chain of 146 amino acids). Therefore, the adult hemoglobin is also called as $\alpha_2\beta_2$ type. Under normal physiological condition, each subunit contains a heme moiety which binds with oxygen. The heme molecule is also called as Ferroprotoporphyrin (or iron protoporphyrin IX). 'IX' in a protoporphyrin represents a specific isomer of protoporphyrin. The Ferrous (Fe^{2+}) can form six coordinate bonds. Fe^{2+} forms four bonds with the Nitrogen of Protoporphyrin IX (Fig. 10.1). It forms a fifth bond with the globin protein by forming a bond with Nitrogen of a Histidine. The sixth coordinate is free to form a bond with Oxygen. As a tetrameric protein, four oxygen molecules can bind to a hemoglobin molecule.

Hemoglobin exists in 2 states. The T-state is called as deoxygenated-tense state, whereas, the R-state is known as the oxygenated-relaxed state. The unbound deoxygenated-tense state has low affinity towards oxygen. A high partial pressure of O_2 facilitates the binding of O_2 to the T-state hemoglobin. Binding of a single O_2

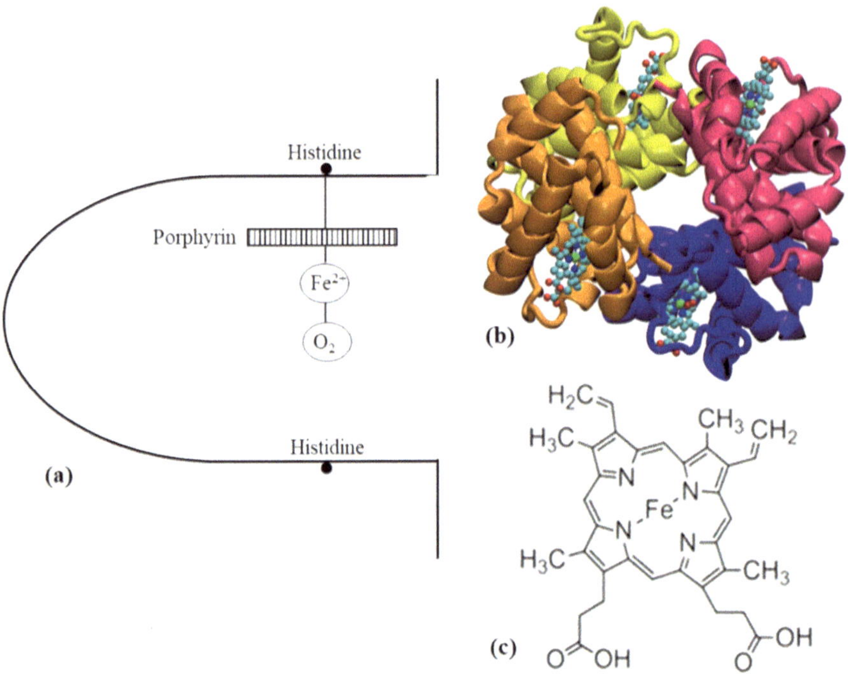

Fig. 10.1 Structure of a hemoglobin molecule (**a**) Heme pocket formed by amino acids, (**b**) A typical $\alpha_2\beta_2$ structure, (**c**) Fe^{2+} forming four bonds with Nitrogen of a Protoporphyrin IX ring

molecule changes the confirmation of T-state hemoglobin to R-state in other subunits and binds to second, third and fourth O_2 at a faster rate (Brauner and Randall 1998). Blood is a buffer system having pH of 7.35–7.45. When the pH decreases oxygen unloading occurs. Thus, both partial pressure of O_2 and decrease in pH allows dissociation of O_2 from hemoglobin and O_2 is delivered to the tissues. 20–25% of CO_2 is carried by hemoglobin from the tissues back to the lungs. The binding of CO_2 to a hemoglobin molecule is facilitated by the increased partial pressure of CO_2 in the tissues compared to the partial pressure of O_2. Similarly, the reversal of partial pressures of CO_2 and O_2 in the lung alveoli facilitated dissociation of CO_2 from carbamino-hemoglobin and uptake of O_2.

4.1 (Pseudo)Peroxidase Activity of Hemoglobin

Peroxidases are a group of enzymes which catalyze the oxidation of a substrate by H_2O_2. Hemoglobin has peroxidative property in presence of H_2O_2. The peroxidase activity of hemoglobin is equivalent to the 5–10% of the activity of horseradish peroxidase. In-vivo conditions, the peroxidase activity of hemoglobin is regulated by the reducing environment. However, once hemoglobin comes out of the cell, the peroxidase activity is greatly expressed. When the red blood cells are damaged or

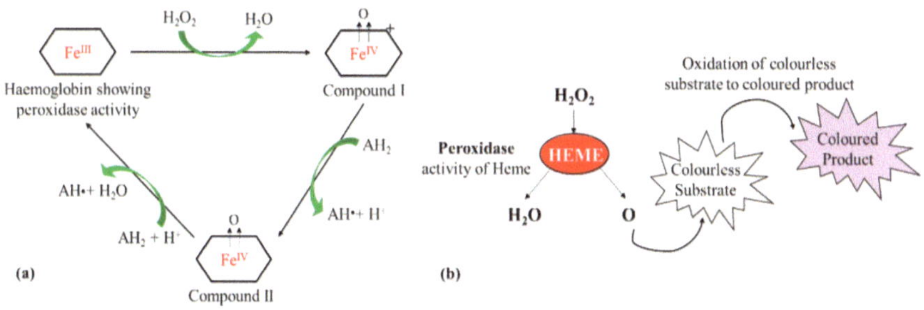

Fig. 10.2 (**a**) Mode of action of Peroxidases. Two arrows represent two unpaired electrons in the bidirectional bond between iron and oxygen. '+' mark indicates positive charge on porphyrin ring. AH_2 and AH. represent the substrate and the free radical product. (**b**) Principle of redox-based presumptive tests for presumptive detection of blood

lysed, hemoglobin exhibits the peroxidase property and can oxidize various substrates in presence of H_2O_2 which acts as a catalyst. The mechanism of action of heme peroxidase is illustrated in Fig. 10.2a. In the first step, oxidation of hemoglobin takes place by H_2O_2 leading to the formation of a ferryl (Fe^{IV}) heme moiety. In this step, the electron is removed either from the porphyrin ring or from the amino acid residue of the protein moiety of hemoglobin. This step leads to the formation of Compound I. Compound I can further react with a substrate to promote its one-electron oxidation and forms compound II. In compound II, heme remains in the ferry (Fe^{IV}) form, but the original structure of porphyrin and amino acid residues is restored. Additionally, a second electron transfer from compound II can restore the native ferric form of hemoglobin.

5 Conventional Presumptive Tests

Most of the conventional presumptive tests for blood explore the peroxidase activity of the hemoglobin found in the red blood cells. A redox reaction is the fundamental behind the outcome of a presumptive peroxidative reaction of hemoglobin. An oxidation-reduction reaction or redox reaction is one of the types of chemical reaction where transfer of electron takes place between molecules. Typically, addition of oxygen or losing one or more electrons is called as oxidation. Conversely, addition of hydrogen or removal or oxygen and gaining one or more electrons is called as reduction. The compounds that can be reduced or can gain electrons from other molecules are called as oxidants. In most of the presumptive tests, heme is used as a catalyst (peroxidative property) and H_2O_2 acts as an oxidant and releases nascent oxygen. This liberated nascent oxygen oxidizes a colorless substrate to generate color, chemiluminescence, or fluorescence (Fig. 10.2b). Most of the presumptive tests as described below, for the detection of blood explore this redox reaction.

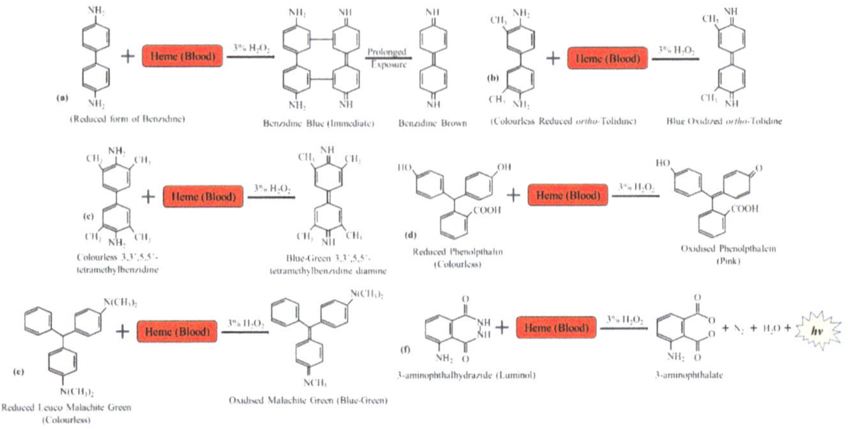

Fig. 10.3 Catalytic chemical reactions of various presumptive tests. (**a**) Benzidine test, (**b**) ortho-tolidine assay, (**c**) Tetra-methyl benzidine assay, (**d**) Phenolphthalin test, (**e**) LMG test, (**f**) Luminol assay

5.1 Reaction with Benzidine and its Derivatives

Benzidine (1,1′-biphenyl-4,4′-diamine) was earlier used as an intermediate in the manufacturing process of dyes. Benzidine was first explored by Adler in 1904 for the detection of blood. Hence, this test is also called as Adler's test. The peroxidase activity of Heme reduces H_2O_2 to liberate O, which ultimately oxidizes benzidine to form oxidized benzidine and it appears blue/dark blue immediately. Thus, the change of colorless benzidine to oxidized-blue benzidine indicates the presence of blood in the tested substance. Prolonged exposure can change the blue color to benzidine brown (Fig. 10.3a); hence, the reading should be taken immediately. However, due to its reported carcinogenic nature, benzidine is no longer being used in forensic casework examination. Similarly, ortho-tolidine, a dimethyl derivative of benzidine can also be catalyzed by peroxidase activity of heme under acidic conditions in presence of H_2O_2 to give blue color (Fig. 10.3b). Ortho-tolidine has also shown potential carcinogenic effect. Hence, nowadays, both benzidine and o-tolidine has been replaced with tetramethylbenzidine (TMB) in most of the forensic biology laboratories. Under acidic conditions, tetramethyl derivative of benzidine can be oxidized in a similar fashion to generate a green to blue-green color product (Fig. 10.3c). The sensitivity of benzidine and its derivatives have been reported to be 1,200,000 times dilution to blood (Sharma 2017).

5.2 Kastle–Meyer Test

Joseph H. Kastle invented this technique of blood detection in 1901 and Erich Meyer modified this method in 1903. Phenolphthalin is a modified form of indicator phenolphthalein, which is prepared by treating phenolphthalein with a reducing agent such as zinc. Phenolphthalin (reduced form) is colorless. When Phenolphthalin is oxidized by the peroxidase activity of Heme in presence of H_2O_2, the oxidized form of Phenolphthalin is produced which is phenolphthalein. Phenolphthalein is pink in color and the generation of pink color product in presence of H_2O_2 indicates the presence of Heme and ultimately identifies the presence of blood presumptively (Fig. 10.3d). Kastle-Meyer test is highly sensitive for blood and can give a positive result for blood diluted up to 1:10000 (Fonseca et al. 2019). Hence, Kastle-Meyer test is one of the most preferred presumptive tests for the detection of blood in forensic science laboratories.

5.3 Leucomalachite Green (LMG) Assay

The leuco base of malachite green (a triphenylmethane dye) is colorless. When leucomalachite green (LMG) is oxidized by the peroxidase activity of Heme in presence of H_2O_2 it gives rise to green or blue-green color (Fig. 10.3e). This reaction takes place in acidic conditions. LMG is also called as MacPhail's reagent. The sensitivity of LMG assay has been reported to be 1: 5000 dilutions of blood in solution or directly on stain (Jaremko et al. 2024).

5.4 Luminol Assay

Many compounds have been found to produce chemiluminescence or fluorescence when oxidized by Heme. One such compound is Luminol. The chemical name of luminol is 3-aminopthalhydrazide. Heme catalyzed oxidation of Luminol produces an intense light which is visible in dark (Fig. 10.3f). Thus, luminol assay can be conducted only in the dark environments.

Luminol powder ($C_8H_7O_3N_3$) is mixed with a liquid containing H_2O_2, a hydroxide (OH^-) and other chemicals. When luminol mixture comes in contact with hemoglobin, luminol loses nitrogen and hydrogen atoms and gains oxygen atoms to form the oxidized luminol, i.e., 3-aminophthalate (Fig. 10.3f). 3-aminophthalate is an energized state and the electrons of oxygen are elevated to higher orbitals. Soon the electrons fall back to their ground state and emit extra energy as light photon generating a chemiluminescence. This generating of light can be observed under dark conditions. The result should be noted quickly as the generated light might fade away quickly. Luminol is capable of detecting latent blood stains diluted up to 10,000 times.

5.5 Fluorescein Assay

Fluorescein is one of the most widely used fluorophores in biological experiments. Fluorescein is highly sensitive to heme components. When fluorescein is oxidized by peroxidase activity of heme its fluorescence property is generated. When fluorescein-treated stains are exposed to 425–485 nm wavelength of light using an alternate light source (ALS), oxidized fluorescein molecules emit intense yellowish-green fluorescence light. This indicates the presence of blood or blood stains. The sensitivity of fluorescein assay has been reported to be 1: 100,000 dilutions of blood on non-absorbent surfaces, whereas, it was found to be 1: 100 on absorbent surfaces (Budowle et al. 2000). A comparative account of the various presumptive tests which rely on the redox reaction principle is given in Table 10.2.

5.6 Point of Care (PoC) Tests

Point of Care (PoC) tests are the tests conducted outside the laboratory in a limited setup mostly by using commercially available kits. With a minimal human intervention and less chance of error such PoC tests are widely recommended for their use in forensic investigation. Some of such commercially available tests have been described below.

5.6.1 Hemastix® Assay

Hemastix® ((Siemens Healthcare Diagnostics Inc., USA) are reagent containing strips designed to detect blood in urine samples. Due to its rapid result and ease to use, it is being used to test for the detection of blood in a forensic set up. The test

Table 10.2 Comparative account of the presumptive tests based on the principle of redox reaction

Test name	Colorless substrate	Product	End product color/ fluorescence	Sensitivity
Benzidine Test/ Adler's test	Reduced Benzidine	Benzidine Blue	Blue	1: 1,200,000
Ortho-tolidine Assay	Ortho-Tolidine Reduced	Ortho-Tolidine Oxidized	Blue	
Tetramethyl benzidine Test	Reduced tetramethyl benzidine	Oxidized tetramethyl benzidine	Blue	
Phenolphthalin Assay/Kastle-Meyer Test	Phenolphthalin Reduced	Phenolphthalein Oxidized	Pink	1: 10,000
Leucomalachite Green (LMG) Assay	Leuco Malachite Green Reduced	Malachite Green Oxidized	Blue-green	1: 5000
Luminol Assay	3-aminopthalhydrazide	3-aminophthalate	Generation of Light	1: 10,000
Fluorescein Assay	Reduced Fluorescein	Oxidized Fluorescein	Yellowish-green fluorescence under 425–485 nm ALS	1: 100,000

strip usually contains diisopropylbenzene dihydroperoxide and 3,3′,5,5′-tetramethylbenzidine. Due to the peroxidative activity of Heme, 3,3′,5,5′-tetramethylbenzidine is converted to a colored oxidized form in presence of H_2O_2 (Matheson and Veall 2014). The resulting color ranges from orange to green depending upon the concentration of blood. The color can be compared with the chat to determine the amount of hemoglobin present in the sample. It is a presumptive test for blood detection as this test cannot differentiate between human and animal blood.

5.6.2 Combur Test®E

Combur3 Test® E (Roche Diagnostics Corporation, Switzerland) is a routinely used test for the semiquantitative determination of hematuria. The plastic strip of Combur3 Test® E contains 3,3′,5,5′-tetramethylbenzidine, 2,5-dimethyl-2,5-dimet yldihydroperoxiesane, and buffering materials which turns green when it comes in contact with blood. The intensity of the color depends on the concentration of blood. As per the claim of the manufacturer, Combur3 Test® E has a limit of detection of 0.15–0.3 μg/ml of hemoglobin corresponding to 5–10 erythrocytes/μl (De Vittori et al. 2016). High sensitivity, low cost, ease of use, and portability make Combur3 Test® E an ideal test for the detection of blood in a forensic set up.

5.6.3 Blood Identifier Kit

Blood identifier kit (HiMedia, India) is based on the principle of redox reaction facilitated by H_2O_2 and the peroxidative activity of Heme using TMB as the substrate. The kit comes with three reagents i.e., BIA, BIB, and BIC. BIA and BIB reagents are mixed well and this mixture is placed on the stain material to be tested. Generation of a dark bluish green color indicates a positive result. If the color is not developed after adding BIA and BIB mixture, a drop of BIC can be added to the stain material.

5.6.4 Bluestar®

Bluestar® is a luminol-based presumptive test useful for the detection of latent blood stains either fresh, dried, cleaned both in neat and diluted form. Bluestar® can detect blood dilutions of 1:100,000 on some substrates (Luedeke et al. 2016). To use the test, three oxidizing tablets are added to the BLUESTAR® FORENSIC chemiluminescent solution. The plunger is then mounted on the spray bottle head and it is tightened properly. The solution is mixed by swirling gently in a circular motion to dissolve the oxidizing tablets completely. The solution is then sprayed on the suspected area. The latent blood stains, if present, generates blue chemiluminescence which can be observed in total darkness.

5.7 Other Presumptive Tests

Blood can absorb a wide range of wavelengths and appear as a dark stain compared to the background. Kearse (2020) reported UV365 for the detection of blood plasma and serum. This will be useful for the detection of blood especially during clotting, cast off and when an injured body is moved from one place to another. Thus, the use of ALS is a useful presumptive test for the detection of blood.

6 Factors Affecting Presumptive Tests for Blood

Most of the presumptive tests used for blood detection are catalytic tests. Though these tests are highly sensitive (Table 10.2), they are not specific to blood. This leads to the observance of false positive or false negative results. Though catalytic tests give positive result in presence of Heme component of blood, such results are interfered with the presence of oxidants and plant peroxidases which are commonly encountered in the crime scene. Similarly, most of these tests also show cross-reactivity, i.e., giving a false positive results for body-fluids other than blood. Most of these blood tests give rise to a false positive result in 79% saliva samples, 50% semen samples, and 90% fecal samples (Stroud et al. 2023).

6.1 Presence of Oxidants

Oxidants are the chemical entities which can be reduced by gaining electrons. In catalytic presumptive tests, H_2O_2 is utilized as oxidant. Chemicals other than H_2O_2 can also act as oxidants. Chemicals such as metal salts, bleach, house cleaning agents can also catalyze the oxidation reaction even in the absence of Heme. This gives rise to the generation of a false positive result. In order to overcome this problem, most of the forensic science laboratories perform a two-step assay for the catalytic test. In the first step, the substrate (TMB/phenolphthalin/LMG) is added to the questioned sample. Ideally color should not be developed at this step. If there is a development of color at this step, it indicates false positive reaction due to presence of other oxidants in the sample. However, if the color changes after the addition of H_2O_2, it indicates a true positive reaction.

6.2 Presence of Reductants

Reductants are the chemical entities which can be oxidized by losing electrons. This process can inhibit the oxidation of our desired substrate. Hence, presence of strong reductant species inhibits the oxidation of substrate to generate a false negative result. Though the strong reductants are rare in occurrence, if present, they can strongly influence the outcome of the test. Example of some strong reductants include lithium metal, metal hydrides, metal alkyls, metal aryls, and silanes (Cibulka 2020).

6.3 Peroxidases

Many plants such as horseradish contain peroxidases. Such plant peroxidases can also catalyze the reactions leading to a false positive outcome (Hiraga et al. 2001). Such plant peroxidases are heat sensitive compared to the hemoglobin. Thus, when the samples are retested after heating plant peroxidases are degraded and the generation of a false positive result can be avoided. Similarly, pus in the body generated after microbial infection also contains peroxidases (Russell 2008). This is a defense mechanism of the body to fight against the infections. Thus, when the samples are contaminated with pus material, it can also give rise to a false positive result due to its peroxidative activity.

7 Conclusion

Presumptive detection of blood and blood stains is one of the most prerequisite of forensic biological investigation. Most of the presumptive tests rely on the principle of redox reaction which explores the peroxidative nature of Heme found in the hemoglobin of the blood cells. Though most of such tests are highly sensitive, however, they are not specific to either blood or human blood. Many of such tests show cross-reactivity with other body-fluids that might be present in the crime scene. Similarly, plant peroxidases, metal contaminants can also lead to a false positive result. The error rate of such tests can be minimized by using suitable positive and negative controls while using these tests. Hence, though presumptive tests give us an idea of the location of blood stains or latent blood stains, their presence should be confirmed using any of the confirmatory tests which will be described in the next chapter.

References

Brauner CJ, Randall DJ (1998) The linkage between oxygen and carbon dioxide transport. In: Perry SF, tufts BL (eds) fish physiology, vol 17. ISBN 9780123504418. Academic Press. https://doi.org/10.1016/S1546-5098(08)60264-2

Budowle B, Leggitt JL, Defenbaugh DA, Keys KM, Malkiewicz SF (2000) The presumptive reagent fluorescein for detection of dilute bloodstains and subsequent STR typing of recovered DNA. J Forensic Sci 45:1090–1092

Cibulka R (2020) Strong chemical reducing agents produced by light. Nature 580:31–32. https://doi.org/10.1038/d41586-020-00872-1

De Vittori E, Barni F, Lewis SW, Antonini G, Rapone C, Berti A (2016) Forensic application of a rapid one-step tetramethylbenzidine-based test for the presumptive trace detection of bloodstains at the crime scene and in the laboratory. Forensic Chem 2:63–74. https://doi.org/10.1016/j.forc.2016.10.002

Fonseca RIB, Ricci EL, Spinosa HD, Bernardi MM, de Abreu GR, Waziry PAF, Nicoletti MA, Ambrosio SR, de Araujo IP, Muñoz JWP, Fukushima AR (2019) Actual trends in the use of the kastle-meyer test: applications in different species and verification of the limit of detection of sensitivity and vestigiality. J Dairy Vet Anim Res 8:166–170

Hiraga S, Sasaki K, Ito H, Ohashi Y, Matsui H (2001) A large family of class III plant peroxidases. Plant Cell Physiol 42:462–468. https://doi.org/10.1093/pcp/pce061

Jaremko KL, Pitts A, Hascall A, Briskie A, Redmond B, Kaur D, Curtis G, LaVena H, West J, Du P, Hernandez S, Silva DSBS (2024) Forensic Sci Med Pathol 20:764–767. https://doi.org/10.1007/s12024-023-00718-y

Kearse KP (2020) Ultraviolet 365 as an alternative light source for detection of blood serum. J Forensic Sci 65:1716–1721. https://doi.org/10.1111/1556-4029.14439

Luedeke M, Miller E, Sprague JE (2016) Technical note: the effects of Bluestar® and luminol when used in conjunction with tetramethylbenzidine or phenolphthalein. Forensic Sci Int 262:156–159. https://doi.org/10.1016/j.forsciint.2016.02.052

Marengo-Rowe AJ (2006) Structure-function relations of human hemoglobins. Baylor Univ Med Cent Proc 19:239–245. https://doi.org/10.1080/08998280.2006.11928171

Matheson CD, Veall MA (2014) Presumptive blood test using Hemastix® with EDTA in archaeology. J Archaeol Sci 41:230–241. https://doi.org/10.1016/j.jas.2013.08.018

Russell DG (2008) *Staphylococcus* and the healing power of pus. Cell Host Microbe 3:115–116. https://doi.org/10.1016/j.chom.2008.02.008

Sharma A (2017) Sensitivity of phenolphthalein and benzidine tests in field of forensic sciences. Int J Res Cult Soc 1:1–7

Stroud A, Gamblin A, Birchall P, Harbison SA, Opperman S (2023) A comprehensive study into false positive rates for 'other' biological samples using common presumptive testing methods. Sci Justice 63:414–420. https://doi.org/10.1016/j.scijus.2023.04.006

Confirmatory Techniques for Blood Identification

11

Abstract

Confirmatory detection of blood evidence holds a huge significance in the criminal justice system. Samples can be collected from the crime scene after conducting the presumptive examination. However, either field based or laboratory-based tests need to be performed before using the samples for source attribution by forensic DNA examination. The primitive examinations such as the crystal tests have been replaced with more advanced immunochromatographic techniques targeting the human-specific antigens present on blood cells. Such tests not only provide confirmatory test results for blood, but also human origin of the blood samples. Besides, immunochromatographic techniques are highly robust and can be used for on filed examinations. mRNA and miRNA-based tests are also emerging. However, such tests need more standardization before their application in the case work samples.

Keywords

Confirmatory test · Blood stain · Takayama · Teichmann · Immunochromatography · RNA expression

1 Introduction

Presumptive tests indicate the possible presence of blood or blood stains on an object. Presumptive tests show cross-reactivity resulting in false positive and/or false negative results. Hence, error rate of such techniques is very high. The objective of forensic analytical techniques is to produce the scientific evidences in the court of law and it should be accepted in the court. The admissibility of any scientific evidence in the court of law is assessed by Daubert's standard. Daubert's standard is primarily used in U.S. Federal courts. It assesses the admissibility of scientific evidences on the basis of four points, (i) whether the theory or technique is generally accepted in the scientific community, (ii) whether the theory or technique is peer reviewed and published, (iii) whether the theory or technique is

H. R. Dash et al., *Forensic Investigation of Blood Evidence*,
https://doi.org/10.1007/978-981-95-9183-1_11

actually tested, and (iv) whether the error rate of the technique is acceptable or not (Pergolizzi and LeQuang 2024). Thus, due to the high error rate of these presumptive tests, they are not accepted in the court of law. Hence, the confirmatory tests having minimal error rates are conducted to ascertain the confirmation regarding the presence of a type of body fluid.

Though the final aim of forensic biology is to individualize the source of biological evidence which can be achieved by forensic DNA profiling, body-fluid identification holds special importance in forensic science. Identification of body fluid is essential for crime scene reconstruction (Sijen and Harbison 2021). For violent crimes involving stab or incised injury, blood or blood stains are expected, not the seminal fluid. Similarly, in a case of sexual assault we may expect seminal fluid, vaginal fluid and/or saliva. Thus, different types of body fluids are related to the nature of the crime. Hence, before individualization of body-fluid stains for personnel identification, confirmatory tests of body fluid need to be conducted. After the identification of blood or blood stain in the crime scene, they should be processed for species determination. In a crime involving humans, human blood is expected, not the animal blood. If required, blood group analysis can also be performed. Determination of blood group plays a pivotal role in excluding a suspect. However, for conclusive identification of the suspect(s), DNA profiling examination should be carried out.

Many microscopic tests were developed in early twentieth century for the confirmatory detection of blood. However, nowadays many immunological tests have been developed targeting specific-antigens present on blood cells. Such antigen-antibody reactions are carried out by immunohistochemistry, ELISA or other immunological techniques. Though these conventional confirmatory tests show high specificity and have high sensitivity, a huge amount of sample is consumed during this process. Most of the forensic biological samples are found in trace amount. When majority of the sample will be consumed for the preliminary examination like body-fluid identification and determination of species origin, limited samples will be available for the confirmatory identification of suspect(s) by DNA profiling. Hence, now-a-days, the genetic testing strategies have been gained popularity. The genetic methods of body-fluid identification techniques can make use of the extracted nucleic acid samples (either DNA or mRNA) for the identification of tissue specific markers (Counsil and McKillip 2010). Considering the prevalence of commensal microbes at specific tissues, microbiome analysis has also shown a huge promise for the confirmatory detection of a body fluid. Thus, in this chapter, all the conventional confirmatory tests for blood detection as well as the advanced immunological and molecular techniques have been discussed in details. Further, investigation of blood can also attribute the nature of the blood to either peripheral blood or menstrual blood. Menstrual blood is also commonly found in many sexual assault cases. Hence, confirmatory tests for the identification of menstrual blood will also be discussed in this chapter.

2 Microscopic Examination of Blood Microcrystals

We can recall the structure of a hemoglobin molecule as described in Chap. 10. The Ferrous (Fe^{2+}) can form six coordinate bonds. Fe^{2+} forms four bonds with the Nitrogen of Protoporphyrin IX, a fifth bond with the globin protein (with Nitrogen of a Histidine), and a sixth coordinate with oxygen. When the hemoglobin molecule is treated with reducing sugar or alkali followed by heating, the ferrous (Fe^{2+}) form is converted into Ferric (Fe^{3+}) form. It also facilitates the dissociation of Ferric (Fe^{3+}) ion from the globin component. This makes the two coordinates of ferric (Fe^{3+}) ion free, which can bind with either two molecules of pyridine or one halide molecule to form a microcrystal structure. Such microcrystal structures are observed under microscope for the confirmatory identification of blood. Two famous tests, the Takayama test and Teichmann tests are routinely performed as microcrystal assay for the confirmatory identification of blood.

2.1 Teichmann's Test

In 1853, Ludwig Karl Teichmann first observed the brownish rhombic crystals when the blood stains are treated with trace amount of sodium chloride and glacial acetic acid. In this process, when blood is mixed with Teichmann's reagent and subsequently heated, RBC is ruptured and hemoglobin is released. In this step, the ferrous iron is also converted into Ferric iron. Heating with acetic acid dentures the globin protein and heme is converted into oxidized heme called the hematin. This hematin binds with halogens such as chloride to form insoluble hemin crystals. Such rhombic crystal of chocolate brown color due to the generation of hemin crystals or hydrochloride of heme are the characteristics of blood (Abid et al. 2024). The reaction of blood in presence of Teichmann's reagent occurs as follows:

$$\text{Haemoglobin}\left(\text{Blood}\right) \underset{\text{Glacial acetic acid}}{\overset{\text{Heat}}{\rightarrow}} \text{Haem} + \text{Globin}$$

$$\text{Haem} \underset{\text{Teichmann's Reagent}}{\overset{KCl/KBr/KI}{\rightarrow}} \text{Haemin Crystals / Hydrochloride of heme}$$

For preparation of Teichmann's reagent 100 mg of KCl, 100 mg of KBr, and 100 mg of KI are dissolved in 100 ml of glacial acetic acid. The blood crusts or liquid blood samples are placed on a glass slide and covered with coverslip. 1–2 drops of Teichmann's reagent is added from the side of the coverslip. The material is gently heated for 10–20 s. The slide is kept at room temperature for 4–5 min and observed under microscope for the visualization of Rhomboid, chocolate brown crystals (Fig. 11.1a).

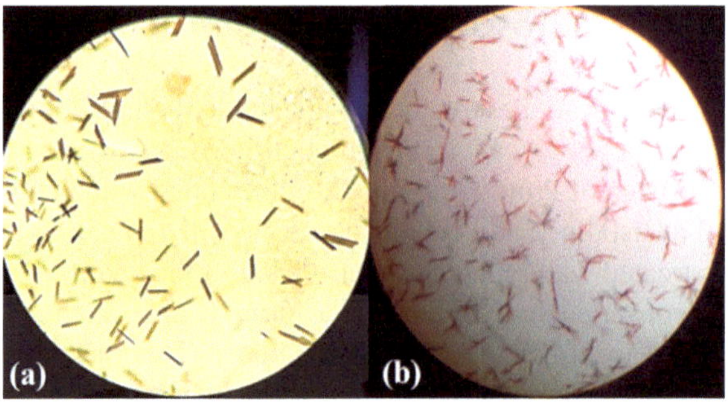

Fig. 11.1 Crystal tests for the detection of blood. (**a**) Rhomboid, chocolate brown hemin crystals, (**b**) pink pyridine hemochromogen crystals

2.2 Takayama Test

Masao Takayama introduced another reagent in Japan in 1912 to develop hemochromogen crystals of blood. Hemochromogen is a compound of heme and globin modified by alkali treatment. In this test, the blood samples are heated in presence of Takayama's reagent. Heating results in the rupturing of RBC and release of hemoglobin. Though the globin protein is also denatured, it remains attached to the heme which is called as ferrohemochrome. Presence of NaOH in the Takayama's reagent converts the iron in ferrous form to ferric form. This oxidized heme is called as alkaline hematin. Subsequently, the hematin binds with pyridine to form insoluble pink colored crystals called pyridine hemochromogen. Saturated glucose solution present in Takayama's reagent acts as a reducing agent which reduces the solubility of the pyridine hemochromogen crystals to develop a large number of visible crystals (Meisari et al. 2024). The reaction of blood in presence of Takayama's reagent occurs as follows:

$$\text{Haemoglobin}\left(\text{Blood}\right) \underset{\text{NaOH}}{\overset{\text{Heat}}{\rightarrow}} \text{Haem} + \text{Globin}$$

$$\text{Haem} \underset{\text{Takayama Reagent}}{\overset{\text{Glucose }+\text{Pyridine}}{\longrightarrow}} \text{Pyridine Haemochromogen}$$

For preparation of Takayama's reagent 3 ml of 10% NaOH, 3 ml of pyridine, and 3 ml of saturated glucose solution is added to 7 ml of distilled water. The blood crusts or liquid blood samples are placed on a glass slide and covered with coverslip. 1–2 drops of Takayama's reagent is added from the side of the coverslip. The slide is gently heated for 10–20 s. The slide is kept at room temperature for 4–5 min and observed under microscope for the visualization of pink pyridine hemochromogen crystals (Fig. 11.1b).

3 Immunological Techniques

Various immunological techniques can be employed for the detection of human hemoglobin utilizing antihuman hemoglobin antibodies. Both enzyme-linked immunosorbent assay (ELISA) and immunochromatographic techniques can be employed for confirmatory detection of human blood in a questioned sample.

3.1 ELISA-Based Methods

In this technique, most of the tests utilize antibodies raised against either the human hemoglobin or glycophorin A for the confirmatory detection of human blood (Harbison and Fleming 2016). The antihuman hemoglobin/ glycophorin A antibody is immobilized on the polystyrene tubes. The questioned blood stain containing antigen (hemoglobin/ glycophorin A) is applied on the immobilized polystyrene tubes. If present, the antigen binds to the immobilized antibody and forms the antigen–antibody complex. It is followed by a washing step, where the unbound antigens are washed away. Further, a secondary antibody targeting a different epitope of the antigen is applied. The secondary antibody binds to the already bound antigen to form an antibody–antigen–antibody sandwich. A second washing step removes the unbound secondary antibodies from the tube. Further, a labeled antiglobulin is applied which mostly targets and binds to the Fc region of the secondary antibody. The bound labeled antiglobulin can be detected by various reporting schemes (Fig. 11.2). In the most commonly used reporting systems, the antiglobulin is conjugated with the enzymes such as peroxidase or phosphatase. When an appropriate substrate (o-phenyldiamine dihydrochloride for peroxidase, p-nitrophenyl phosphate for alkaline phosphatase) is added, a visible color change is observed. Such a color change is recorded by an ELISA plate reader which provides both qualitative and quantitative estimation of blood present in the questioned sample. Most of the ELISA techniques have shown high specificity with no cross-reactivity with pig, sheep, cow, goat, horse and rabbit antigens. Such techniques have also shown a high sensitivity with detection limit of 0.1 μg/ml (Hurley et al. 2009).

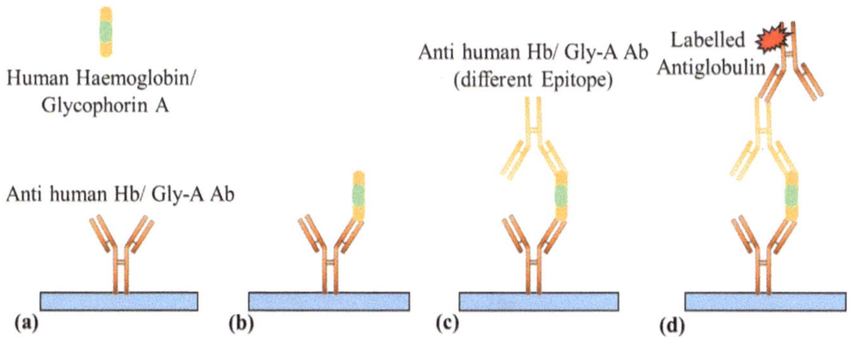

Fig. 11.2 Principle of ELISA-based detection of human blood (**a**) Sample containing hemoglobin or Glycophorin A is added to the tubes immobilized with antihuman Hb/ Gly-A, (**b**) Hb/Gly-A binds with Ab to form Ag-Ab complex, (**c**) Secondary antibody specific for another epitope of Hb/ Gly-A is added and attached to form Ab-Ag-Ab sandwich, (**d**) Labeled antiglobulin binds to the Ab-Ag-Ab sandwich which can be detected by various detection systems

3.2 Immunochromatographic Techniques

In immunochromatographic techniques, the same antigen molecules i.e., hemoglobin/ glycophorin A (found on the surface of RBCs) can also be targeted (Morena et al. 2025). In this technique, the questioned sample containing the antigens is mixed with the running buffer and applied to the sample well. Antigen binds to the labeled antihuman Hb/ Gly-A antibody in the sample well and a labeled antibody–antigen complex is formed. This complex diffuses across the membrane and binds with the immobilized antihuman Hb/ Gly-A antibody (different epitope) in the test zone to form labeled Ab-Ag-Ab sandwich. This produces color/ fluorescence in the test zone. Additionally, the unbound labeled Ab diffuses across the membrane and binds to the antiglobulin in the control zone to generate a positive result. Thus, the presence of antigen in the sample produces positive result in both test zone and the control zone, whereas the absence of antigen in the sample produces positive result in the control zone only (Fig. 11.3).

3.2.1 Commercially Available Immunochromatographic Kits

Most of the commercially available kits rely on the immunochromatographic principles for the confirmatory detection of human blood in forensic samples. Though most of the tests were developed to detect blood in the fecal samples for the screening of colon cancer patients, nowadays, they have been widely adopted in forensic caseworks. Some of the widely used kits include, ABAcard® Hematrace®; 2: Hexagon OBTI; 3: SERATEC® HemDirect; 4: RSID™ blood; 5: Bluestar® OBTI; 6: BLUESTAR® Identi-HEM® (Cuttaia et al. 2024). Various advantages of using immunochromatographic techniques for the detection of human blood include cost-effectiveness (<10 EUR per test), can be performed in the crime scene, and such tests require much lesser time i.e., <1–2 h to perform. Some of the kits utilize a strain extraction buffer to extract the antigens present on the biological samples and

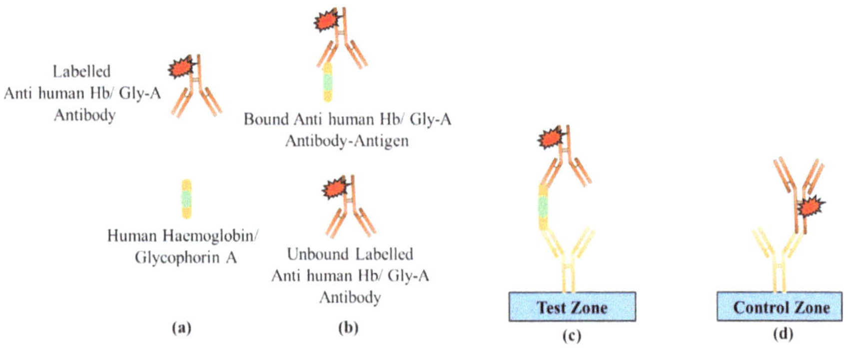

Fig. 11.3 Principle of immunochromatographic based detection of human blood (**a**) Sample containing hemoglobin or Glycophorin A is mixed with labeled antihuman Hb/ Gly-A antibody, (**b**) Hb/Gly-A binds with Ab to form Ag-labeled Ab complex, (**c**) labeled complex binds with immobilized antihuman Hb/ Gly-A antibody for another epitope of Hb/Gly-A in the Test zone, (**d**) labeled antihuman Hb/ Gly-A antibody binds to the antiglobulin at the Control zone

Table 11.1 Comparative account of the commercially available immunochromatographic kits for the forensic detection of human blood

Kit	Antigen	Test	LOD (hHb)	LOD (Blood)
ABAcard® Hematrace®	hHb	mono + poly	50 ng/µl	0.3 nl
Hexagon OBTI	hHb	mono + mono	100 ng/µl	0.6 nl
SERATEC® HemDirect	hHb	mono + mono	40 ng/µl	0.25 nl
RSID™ blood	Gly-A	mono + mono	12.5 ng/µl	75 nl
Bluestar® OBTI	hHb	mono + poly	50 ng/µl	0.3 nl
BLUESTAR® Identi-HEM®	hHb	mono + poly	10 ng/µl	0.06 nl

use it for conducting the test. Many studies have also reported that the residual strain extraction buffer can be used for the successful generation of a DNA profile (Johnston et al. 2008). Similarly, a very high sensitivity of such tests has also been claimed. The kits targeting human hemoglobin enable detection of human blood from as little as 10 to 50 ng/mL, whereas the kits targeting glycophorin A, 75 nL of blood is required (Cuttaia et al. 2024). A comparative analysis of the commercially available kits for forensic detection of human blood is given in Table 11.1.

4 RNA Based Detection of Human Blood

Various types of RNAs are found in humans, i.e., mRNA, miRNA, piRNA, circRNA. Out of which mRNA is the most explored candidate for body fluid detection including blood. The principle of RNA-based detection of body fluids relies on the fact that, RNAs are transcribed exclusively in a particular cell type. Quantitative determination of such RNA type can detect the presence of such body fluid in a forensic biological sample. Over traditional and immunological techniques, RNA-based techniques are highly specific, sensitive, and can be investigated in complex mixtures of trace evidences (Courts et al. 2024). In the routine DNA extraction protocols, RNA is also co-extracted along with DNA. In a typical mRNA analysis, mRNA is first converted into cDNA by reverse transcriptase enzyme. The quantity of a specific cDNA can be measured in quantitative RT-PCR and based on the relative expression of the mRNA; the nature of the body fluid can be determined. Various mRNA markers identified in blood include HBB, ALAS2, PRF1, GlycoA, PF4, SPTB, PBGD, CD93, AMICA1, ANK1, CD3G, SLC4A1 (Bamberg et al. 2021). In a collaborative study on mRNA profiling, HBB was found to be expressed abundantly in blood compared to SPTB and PBGD. All the 16 participating laboratories could successfully isolate and detect RNAs from the mock dried blood samples (Haas et al. 2011). Thus, it can be speculated that, mRNA profiling can be a superior alternative to the conventional serological based detection of blood.

MicroRNAs (miRNAs) are more stable than mRNA. As miRNAs regulate the synthesis of mRNA, hence, miRNAs also show tissue specific expression patterns. Without involving any separate extraction steps, miRNAs can be coextracted with mRNA in a single workflow. Elongation of the poly(A) tails followed by the use of oligo dT primers are alternatively used to reverse transcriptase protocols. Besides qRT-PCR, endpoint RT PCR and capillary electrophoresis can be performed for the analysis of miRNAs for the detection of the nature of the body fluid (Lynch and Flemming 2020). The candidate miRNA species useful for the detection of venous blood include miR-451a, miR-144-3p, miR-16, miR-185, miR-126, miR-150, miR-486, miR-203a, miR-943, miR-200b, and miR-142-3p (Rocchi et al. 2020).

5 Confirmatory Detection of Menstrual Blood

Several techniques have been discovered to distinguish between the peripheral blood and menstrual blood. Distinguishing menstrual blood from peripheral blood can help in solving typical cases of infanticides. Similarly, occurrence of menstrual blood can be detected in many sexual assault cases. The most common constituents of menstrual blood include the vaginal fluid, blood, and the cells of uterine endometrial lining which is shed during menstruation phase. Certain enzymes such as matrix metalloproteinase, lactate dehydrogenase and other proteinases are also detected in menstrual blood. The fundamental differences between the peripheral blood and menstrual blood are given in Table 11.2. Lactate dehydrogenase test and the D-dimer assay are the two most common tests used to distinguish between peripheral blood and the menstrual blood. Other RNA-based detection of menstrual blood is also gaining importance.

5.1 Lactate Dehydrogenase Test

Lactate dehydrogenase is an important enzyme in glycolysis which catalyzes the reversible reduction of pyruvate into lactate under oxygen limiting conditions. The

Table 11.2 Characteristic differences between peripheral blood and menstrual blood

Characteristics	Peripheral Blood	Menstrual Blood
Color	Lighter	Darker
Oxygen	Highly oxygenated	Less oxygenated
Cells	Live cells	Dead cells and no longer functional tissues
Iron/ Hemoglobin	High	Less
RBC count	High	Less
WBC count	More	Less
Bioactive peptides	Not found	Good source of antibacterial peptides
Clotting properties	Cascade mechanism for blood coagulation involving several enzymes	Lacks those enzymes
Clotting properties	More blood coagulation and less fibrinolysis	Less blood coagulation and more fibrinolysis
Fibrinogen degradation product	Less amount	High amount

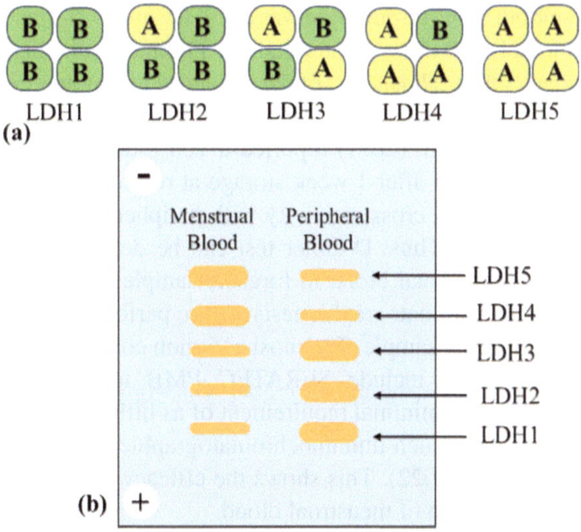

Fig. 11.4 Various isoenzymes of Lactose dehydrogenase. (**a**) LDH1, LDH2, LDH3, LDH4, and LDH5. (**b**) Electrophoretic mobility of various lactose dehydrogenase isoenzymes

tetrameric enzyme of lactate dehydrogenase consists of three subunits. LDHA gene codes for the A subunit which is primarily synthesized in the skeletal muscles, whereas LDHB codes for the B subunit and is synthesized in the cardiac muscle. The C subunit is encoded by the LDHC gene and its expression is confined to the testes. In blood, five isoenzymes of lactate dehydrogenase are detected (Fig. 11.4). LDH1 (B4) has the highest electrophoretic mobility followed by LDH2 (AB3), LDH3 (A2B2), LDH4 (A3B), and LDH5 (A4). The relative abundance of these isoenzymes can be explored to distinguish between the peripheral blood and menstrual blood (Asano et al. 1972). LDH1, LDH2, and LDH3 are the most predominant forms of isoenzymes in peripheral blood, whereas LDH4 and LDH5 are the most predominant forms in menstrual blood. Thus, by electrophoretic separation of lactate dehydrogenase isoenzymes and their detection by colorimetric assay, peripheral blood and menstrual blood can be distinguished.

5.2 D-dimer Assay

Three pairs of polypeptide chains i.e., Aα, Bβ, and γ constitute a circulating fibrinogen. In the site of a vascular injury, fibrinopeptides are cleaved from fibrinogen Aα and Bβ by thrombin, which polymerizes to form monomers of fibrin. Further, factor XIII is activated by thrombin to facilitate binding of γ units to form a fibrin net which stabilizes a blood clot (Gupta et al. 2016). When a clot is not required, plasminogen, a component of the fibrinolytic system is activated by tissue-specific plasminogen activator and urokinase plasminogen activator. Plasminogen activation leads to the formation of plasmin which lyses a clot. Breakdown of clots generates various fibrin degradation products (FDP). One of such FDP is D-dimer. This natural process of fibrinolysis prevents blood clotting inside the uterus and facilitates its smooth flow through the vagina. The mean D-dimer concentration in menstrual blood is 200 times greater than the D-dimer found in the peripheral blood (Miyaishi et al. 1996). Thus, most of the advanced techniques explore D-dimer as an indicator of menstrual blood in a forensic biological sample.

Baker et al. (2011) reported a 100% detection of D-dimer in menstrual blood samples even after 1 week storage at room temperature. Besides, D-dimer test did not show any cross reactivity with peripheral blood or mixture of peripheral blood and semen. Thus, D-dimer test can be deemed to be a suitable stable marker to detect menstrual blood in forensic samples. ELISA based examination as well as immunochromatographic tests can be performed to detect the presence of D-dimer in a forensic sample. The most common commercially available immunochromatographic kits include SERATEC PMB test and OneStep D-dimer RapidCard InstaTest. A minimal requirement of as little as 0.003125 μL of sample is sufficient to carry out such immunochromatographic tests to detect the presence of D-dimer (Tsai et al. 2022). This shows the efficacy of D-dimer test over other tests for the determination of menstrual blood.

5.3 RNA-Based Tests

mRNA of many matrix metallopeptidases (MMPs) has been most commonly used for the forensic detection of menstrual fluid. Out of various MMPs, MMP 7, MMP 10, and MMP 11 are the most explored MMPs for forensic detection of menstrual blood (Albani et al. 2020). Liang et al. (2018) detected the expression of LAPR3 and PAEP exclusively in menstrual blood samples. Thus, in addition to MMPs, LAPR3 and PAEP can also be used for mRNA-based detection of menstrual blood in forensic biological samples. Microarray-based detection of miRNAs has also gained importance in forensic detection of menstrual blood samples. Five miRNAs, i.e., miR-141-3p, miR-373-3p, miR-497-5p, miR-143-5p, and miR-136-5p have shown a higher expression level in menstrual blood samples compared to the peripheral blood (Li et al. 2017). In a recent study, Altmeyer et al. (2024) demonstrated the use of five candidate miRNAs i.e., miR-214-3p, miR-141-3p, miR-200b-3p, miR-412-3p, and miR-3120-3p for accurate detection of menstrual blood. The use of miRNAs has major advantages over the mRNAs as described in the previous section. Hence, in addition to the conventional lactase dehydrogenase and D-dimer tests, RNA based examinations can also be performed for the accurate detection of menstrual blood of forensic interest.

6 Species Determination from Blood

Occasionally, the species determination from the biological samples is performed from forensic interest. Besides the ELISA and immunochromatographic techniques-based detection, various immunodiffusion and immunoelectrophoresis techniques can also be employed for the detection of species origin from the forensic blood samples. The questioned sample contains the antigens, whereas the species-specific antibodies are used to carry out the antigen-antibody reaction.

6.1 Immunodiffusion Assays

Immunodiffusion is a passive process where the antigen and/or antibody (antihuman) are allowed to diffuse in a liquid or semisolid medium to generate a precipitation reaction. Immunodiffusion assays are routinely performed either by single immunodiffusion or double immunodiffusion technique.

Single immunodiffusion in the form of radial immunodiffusion is routinely performed for the determination of species origin from a forensic blood sample. In this technique, the human specific antibody is mixed with the gel matrix and distributed uniformly. Sample wells are prepared and the sample containing antigens are loaded into these wells. Antigens diffuse into the gel containing antibody and a precipitation ring is observed around the well (Fig. 11.5a). The area of the precipitate ring is directly proportional to the concentration of the antigen. A standard curve can be prepared with the known antigen concentrations and the ring area and the concentration of the unknown antigen can be estimated.

Similarly, double immunodiffusion can be performed either by Ring assay or Ouchterlony assay. In a typical double immunodiffusion assay, a concentration gradient is established for both the antigen and the antibody. Both the antigen and antibody diffuse towards each other and a line of precipitation is observed. A capillary or a test tube is used for a ring assay to determine the species origin. Antihuman

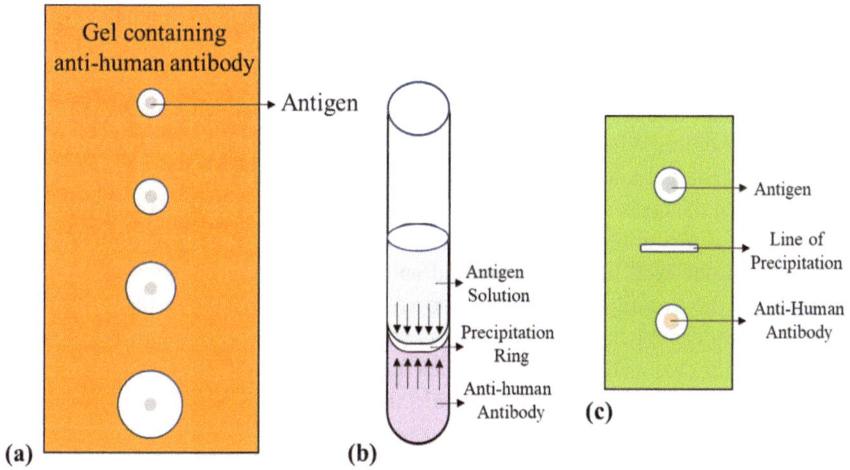

Fig. 11.5 Different methods of immunodiffusion techniques for the determination of species origin of blood. (**a**) Single immunodiffusion assay, (**b**) Ring assay, and (**c**) Ouchterlony assay

antibody is placed at the bottom of the capillary tube as it is denser compared to the antigens. On the top pf the antibody solution, a layer of antigen solution is applied without mixing. Both the antigen and antibody migrate towards each other by diffusion and a ring of precipitate is observed at the interface (Fig. 11.5b). Similarly, in Ouchterlony assay, a layer of agarose is prepared on a glass slide or a polyester film. Two wells are prepared side by side by puncturing holes on the agarose gel. In one well the antigen solution is applied, whereas in the other well, the antihuman antibody is placed. When the agarose slide is incubated, both the antigen and antibody migrate towards each other and a line of precipitation can be observed (Fig. 11.5c). Occasionally, a central well surrounded by six wells are prepared and more than one antigen can be tested in a single reaction.

6.2 Electrophoretic Techniques

Diffusion is a passive process. Thus, electrophoresis process and the diffusion are combined together to enhance the test results. Four types of electrophoretic techniques are commonly used, including immunoelectrophoresis (IEP), crossed immunoelectrophoresis (CRIE), rocket immunoelectrophoresis, and crossed over immunoelectrophoresis (CIE).

In IEP, first agarose gel electrophoresis is carried out to separate the mixture of antigens. In the second step a trough is cut parallel to the line of the separated antigens. Further, the antibody is placed in the trough and the gel is incubated to facilitate a double diffusion process. In a positive reaction, an arc shaped precipitate line is observed (Fig. 11.6a). CRIE is a modified version of IEP and is also called as two-dimensional immunoelectrophoresis. In the first step of CRIE, electrophoresis separates antigen mixtures in the sample. Further, the gel is turned 90° angle and separated by a second round of electrophoresis. During second round of electrophoresis, the antigens react with the antibodies uniformly distributed on the agarose gel and an arc shaped precipitate line is observed (Fig. 11.6b). In rocket immunoelectrophoresis, the antigens are loaded into the wells made in the agarose gel containing the antibodies. When an electric current is applied, the antigens migrate from the wells to the gel and rocket-shaped precipitation line is observed (Fig. 11.6c). The height of the rocket is directly proportional to the concentration of the antigens. In crossed-over/ counter immunoelectrophoresis, two wells are prepared on an agarose gel. In one well antigen is applied and in another well antibody solution is applied. As the overall charge of the normal antigens are negative, they are placed near the cathode. When electrophoresis is conducted, a precipitate line is observed in between the two wells (Fig. 11.6d). During electrophoresis, the movement of the antibody occurs by the influence of the electroendosmosis.

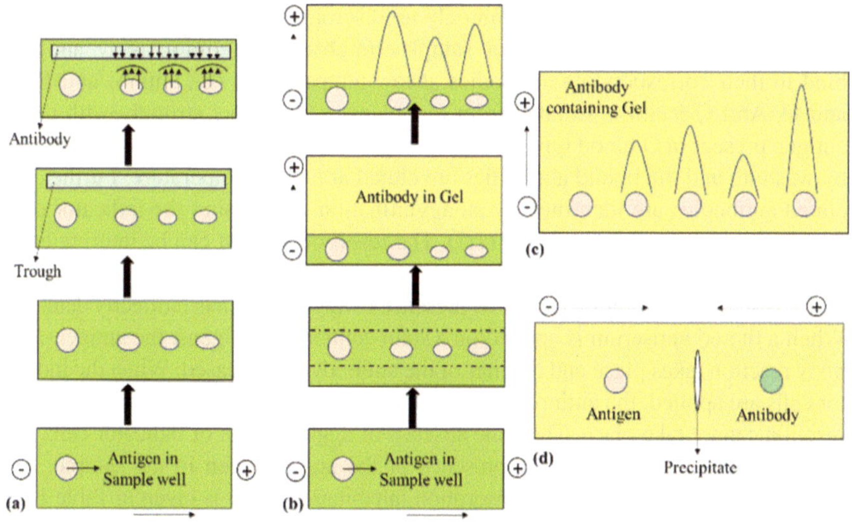

Fig. 11.6 Different methods of electrophoretic techniques for the determination of species origin of blood. (**a**) immunoelectrophoresis (IEP), (**b**) crossed immunoelectrophoresis (CRIE), (**c**) rocket immunoelectrophoresis, and (**d**) crossed over immunoelectrophoresis (CIE)

7 Blood Group Analysis

After the inception of DNA profiling technique, the use of blood group analysis from forensic biological samples has shown limited scope. However, blood group analysis by serological techniques is also being practiced in many laboratories. Out of >30 blood group systems, ABO system is more prevalent in different populations and determination of ABO groups is occasionally performed by forensic serological analysis. The blood group examination relies on the basic principle of antigen-antibody reaction to initiate an agglutination reaction. Type A individuals have anti-B antibodies, type B individuals have anti-A antibodies, type O individuals develop both anti-A and anti-B antibodies, whereas type AB individuals do not develop any antibodies. When the plasma of type B individuals (containing anti-A antibodies) is mixed with type A cells agglutination of type A cells take place. This fundamental aspect of agglutination is explored in the determination of blood group.

In lattes crust assay, fresh blood cells are collected from A, B, and O individuals and such cells are called as the indicator cells. Agglutination reaction is performed with the naturally occurring corresponding antibodies present in the questioned sample. Though lattes crust assay is a simple and rapid method, it is difficult to test the aged blood stains using this approach. Most of the laboratories have replaced the lattes crust assay with the microplate assay (Mudd and Adams 1990). For a successful agglutination reaction, intact cells containing surface antigens are required. However, in dried forensic samples, cells lyse and antigens remain immobilized on the surface of the cells. Absorption-elution assay, absorption-inhibition assay, and mixed agglutination assay are commonly used serological tests to determine blood groups from the forensic biological samples. In absorption-elution assay, antigens bind to their corresponding antibodies at low temperatures i.e., anti-A, anti-B, or anti-O. Anti-O lectin is isolated from few plants which react strongly with the O antigen present in O blood type. After binding, the unbound antibodies are removed by washing and the bound antibodies are eluted at a high temperature. Further, the eluted antibodies are identified by an agglutination assay using the indicator cells (Wanio Sungoh et al. 2023) (Fig. 11.7). The result of this test can be interpreted as given in Table 11.3.

In Absorption Inhibition assay, the presence of agglutinogen is indirectly detected. When a titered antiserum is applied on a stain containing antigens, the antigen-antibody reaction takes place and the titer of the antibody is decreased. When the indicator cells are applied, the antibodies are no longer available to react and agglutination reaction cannot take place. Thus, the absence of agglutination of indicator cells in a reaction indirectly indicates the presence of the agglutinogen in the sample. The result interpretation of a typical absorption inhabitation assay is given in Table 11.3. Additionally, mixed agglutination test is also performed to identify the presence of

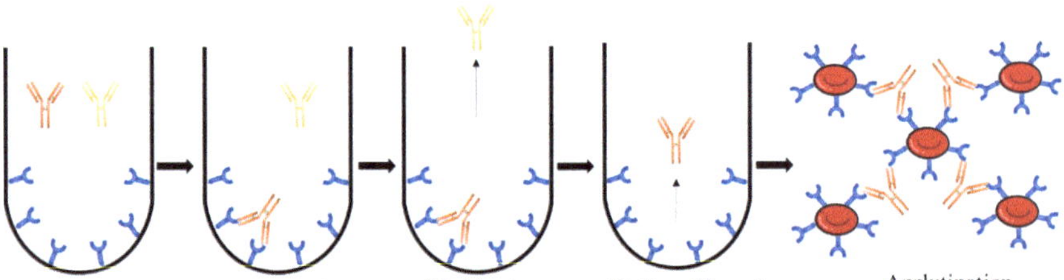

Fig. 11.7 Principle and steps of Absorption Elution assay

Table 11.3 Result interpretation of Absorption Elution and Absorption Inhibition assays

Absorption Elution Assay				
Specimen	Agglutination of Indicator Cells			Inference
	Anti-A	Anti-B	Anti-H	
Specimen 1	+		+	A
Specimen 2	–	+	+	B
Specimen 3	–	–	+	AB
Specimen 4	–	–	–	O
Absorption Inhibition Assay				
Specimen 1	+	+	+	Nonsecretor
Specimen 2	–	+	–	Secretor A
Specimen 3	+	–	–	Secretor B
Specimen 4	–	–	–	Secretor AB
Specimen 5	+	+	–	Secretor O

blood group antigens in a forensic biological sample. Initially, a known antiserum is added to the sample ached to a surface followed by washing with cold saline to wash away the unbound antibodies. When indicator cells are applied, homologous cells bind to the free end of the bound antibody and the presence of cells attached to the stain indicates a positive reaction. However, none of these serological tests are being performed routinely in forensic science laboratories.

8 Conclusion

Confirmatory detection of blood evidence holds a huge significance in the criminal justice system. Samples can be collected from the crime scene after conducting the presumptive examination. However, either field based or laboratory-based tests need to be performed before using the samples for source attribution by forensic DNA examination. The primitive examinations such as the crystal tests have been replaced with more advanced immunochromatographic techniques targeting the human-specific antigens present on blood cells. Such tests not only provide confirmatory test results for blood, but also human origin of the blood samples. Besides, immunochromatographic techniques are highly robust and can be used for on filed examinations. mRNA and miRNA-based tests are also emerging. However, such tests need more standardization before their application in the case work samples.

References

Abid IF, Rezakola E, Hanura IY (2024) Teichmann test: assessment of hemoglobin crystals on blood spots exposed to powder detergent. Jurnal Biosains Pascasarjana 26:23–27. https://doi.org/10.20473/jbp.v26i1.2024.23-27

Albani PP, Patel J, Fleming RI (2020) Detection of five specified menstrual fluid mRNA markers over the uterine cycle. Forensic Sci Int Genet 49:102359. https://doi.org/10.1016/j.fsigen.2020.102359

Altmeyer L, Baumer K, Hall D (2024) Differentiation of five forensically relevant body fluids using a small set of microRNA markers. Electrophoresis 45:1785–1795. https://doi.org/10.1002/elps.202400089

Asano M, Oya M, Hayakawa M (1972) Identification of menstrual blood stains by the electrophoretic pattern of lactate dehydrogenase isozymes. Forensic Sci 1:327–332. https://doi.org/10.1016/0300-9432(72)90030-1

Baker DJ, Grimes EA, Hopwood AJ (2011) D-dimer assays for the identification of menstrual blood. Forensic Sci Int 212:210–214. https://doi.org/10.1016/j.forsciint.2011.06.013

Bamberg M, Dierig L, Kulstein G, Kunz SN, Schmidt M, Hadrys T, Wiegand P (2021) Development and validation of an mRNA-based multiplex body fluid identification workflow and a rectal mucosa marker pilot study. Forensic Sci Int Genet 54:102542. https://doi.org/10.1016/j.fsigen.2021.102542

Counsil TI, McKillip JL (2010) Forensic blood evidence analysis using RNA targets and novel molecular tools. Biologia 65:175–182. https://doi.org/10.2478/s11756-010-0001-2

Courts C, Gosch A, Rothschild M (2024) RNA analysis in forensic molecular biology. Dtsch Arztebl Int 121:363–369. https://doi.org/10.3238/arztebl.m2024.0051

Cuttaia C, Di Stefano B, Ciglieri SS, Vetrini R, Previderè C, Fattorini P (2024) Immunochromatographic detection of human blood: a forensic review. Separations 11:66. https://doi.org/10.3390/separations11030066
Gupta S, Biswas A, Akhter MS, Krettler C, Reinhart C, Dodt J, Reuter A, Philippou H, Ivaskevicius V, Oldenburg J (2016) Revisiting the mechanism of coagulation factor XIII activation and regulation from a structure/functional perspective. Sci Rep 6:30105. https://doi.org/10.1038/srep30105
Haas C, Hanson E, Bär W, Banemann R, Bento AM, Berti A, Borges E, Bouakaze C, Carracedo A, Carvalho M, Choma A, Dötsch M, Duriancikovẚ M, Hoff-Olsen P, Hohoff C, Johansen P, Lindenbergh PA, Loddenkötter B, Ludes B, Maroñas O, Ballantyne J (2011) mRNA profiling for the identification of blood—results of a collaborative EDNAP exercise. Forensic Sci Int: Genet 5:21–26. https://doi.org/10.1016/j.fsigen.2010.01.003
Harbison S, Fleming R (2016) Forensic body fluid identification: state of the art. Res Rep Forensic Med Sci 6:11–23. https://doi.org/10.2147/RRFMS.S57994
Hurley IP, Cook R, Laughton CW, Pickles NA, Ireland HE, Williams JH (2009) Detection of human blood by immunoassay for applications in forensic analysis. Forensic Sci Int 190:91–97. https://doi.org/10.1016/j.forsciint.2009.05.018
Johnston E, Ames CE, Dagnall KE, Foster J, Daniel BE (2008) Comparison of presumptive blood test kits including Hexagon OBTI. J Forensic Sci 53:687–689
Li Z, Bai P, Peng D, Wang H, Guo Y, Jiang Y, He W, Tian H, Yang Y, Huang Y, Long B, Liang W, Zhang L (2017) Screening and confirmation of microRNA markers for distinguishing between menstrual and peripheral blood. Forensic Sci Int Genet 30:24–33. https://doi.org/10.1016/j.fsigen.2017.05.012
Liang Q, Sun H, Wu X, Ou X, Gao G, Jin Y, Tong D (2018) Development of new mRNA markers for the identification of menstrual blood. Ann Clin Lab Sci 48:55–62
Lynch C, Flemming R (2020) RNA-based approaches for body fluid identification in forensic science. WIREs Forensic Sci 3:1–23. https://doi.org/10.1002/wfs2.1407
Meisari S, Lestari TD, Legowo D, Widiyatno TV, Hestianah EP, Wahjuni RS, Damayanti R, Khairullah AR, Moses IB, Raissa R (2024) Identification of different blood concentration of domestic cat (*Felis catus*) with Leucomalachite Green and Takayama reagent. Open Vet J. 14:2609–2617. https://doi.org/10.5455/OVJ.2024.v14.i10.10
Miyaishi S, Kitao T, Yamamoto Y, Ishizu H, Matsumoto T, Mizutani Y, Heinemann A, Puschel K (1996) Identification of menstrual blood by the simultaneous determination of FDP-D dimer and myoglobin contents. Jpn J Legal Med 50:400–403
Morena D, Manta AM, Santurro A, Scopetti M, Turillazzi E, Fineschi V (2025) Bloody evidence: the validity of Glycophorin A in the determination of wound vitality-a systematic review of the literature. Int J Mol Sci 26:5308. https://doi.org/10.3390/ijms26115308
Mudd JL, Adams DE (1990) A modification of the microplate method for reverse ABO typing of bloodstains and additional validation studies. J Forensic Sci 35:447–451
Pergolizzi J, LeQuang JAK (2024) Black robes and white coats: Daubert standard and medical and legal considerations for medical expert witnesses. Cureus 16:e69346. https://doi.org/10.7759/cureus.69346
Rocchi A, Chiti E, Maiese A, Turillazzi E, Spinetti I (2020) MicroRNAs: an update of applications in forensic science. Diagnostics (Basel) 11:32. https://doi.org/10.3390/diagnostics11010032
Sijen T, Harbison SA (2021) On the identification of body fluids and tissues: a crucial link in the investigation and solution of crime. Genes (Basel) 12:1728. https://doi.org/10.3390/genes12111728
Tsai LC, Liu KL, Lin WY, Lin YC, Huang NE, Lee JCI, Linacre A, Hsieh HM (2022) Evaluation of three commercial kits effective identification of menstrual blood based on the D-dimer. Forensic Sci Int 338:111389. https://doi.org/10.1016/j.forsciint.2022.111389
Wanio Sungoh JDO, Singh KP, Devi TM (2023) A comparison of absorption inhibition and absorption elution techniques for detection of ABO blood groups in saliva in Meitei population: a cross sectional study. J Indian Acad Forensic Med 45:64–67. https://doi.org/10.48165/jiafm.2023.45.1.17

Estimating the Age of Blood Stains and Its Application in Forensic Science

12

Abstract

Accurate estimation of the age of bloodstains—also known as time since deposition (TSD)—remains a critical yet unresolved challenge in forensic science. Traditional methods, such as visual inspection and colorimetric assays, offer limited temporal resolution and are highly sensitive to environmental variability. In contrast, recent advancements have introduced a diverse array of analytical techniques, including ultraviolet-visible (UV–Vis) spectroscopy, attenuated total reflectance Fourier-transform infrared (ATR-FTIR) spectroscopy, and fluorescence-based methods. These tools enable the detection of chemical and structural changes in hemoglobin and other biomolecules as bloodstains age. Concurrently, molecular approaches—such as mRNA degradation profiling, DNA methylation analysis, and microRNA (miRNA)-based models—have shown promise in providing stable, time-sensitive biomarkers. The integration of machine learning, particularly through neural networks and random forest models, further enhances the accuracy and field applicability of these methods. Portable devices, including handheld near-infrared (NIR) spectrometers and mobile imaging systems, are now being developed for real-time, on-site analysis. Despite these advancements, challenges persist in the form of environmental sensitivity, substrate dependency, lack of standardization, and legal admissibility. This paper reviews the current landscape of bloodstain aging techniques, highlights their strengths and limitations, and outlines future directions for developing robust, admissible, and field-ready solutions for forensic timelines.

Keywords

Bloodstain age estimation · Time since deposition · Forensic spectroscopy · RNA degradation · DNA methylation · microRNA · Machine learning · Portable forensic tools · NIR spectroscopy · Forensic timelines

H. R. Dash et al., *Forensic Investigation of Blood Evidence*,
https://doi.org/10.1007/978-981-95-9183-1_12

1 Introduction

In the field of forensic science, time plays a crucial role in how evidence is interpreted. While considerable advancements have been made in identifying individuals through DNA and fingerprint analysis, determining the *time since a biological material was deposited*, especially *blood*, remains a significant challenge. Knowing *when* blood was shed can help investigators reconstruct timelines, confirm or challenge alibis, and narrow the window in which a crime could have occurred. Despite its forensic value, estimating the age of bloodstains—also known as *Time Since Deposition (TSD)*—has no standardized method currently accepted for courtroom use.

Historically, investigators have relied on visual cues—such as color changes in blood—to make subjective judgments about its age. However, such visual assessments are inherently unreliable due to the influence of environmental conditions such as temperature, humidity, UV exposure, and the nature of the surface the blood lands on (Bergmann et al. 2025). A bloodstain on metal exposed to direct sunlight will degrade very differently from one on fabric kept in a cool, shaded room. These variables introduce inconsistencies that make visual estimation practically unusable in formal forensic processes.

To overcome these limitations, modern research is turning toward *objective, measurable changes* in blood components as it ages. One promising approach involves the degradation of *messenger RNA (mRNA)*. Since mRNA molecules are relatively unstable, they degrade over time in a predictable manner, offering a biochemical clock that can be monitored under controlled conditions. Fonneløp et al. (2025) developed a capillary electrophoresis-based method that tracks six specific mRNA targets, showing promising accuracy within the first 24–48 hours post-deposition. This is particularly relevant in short-term forensic cases, where precise timelines are essential for criminal investigations.

However, mRNA is not the only molecular marker of interest. *Circular RNA (circRNA)* has gained attention for its enhanced stability compared to linear RNA. Wei et al. (2022) demonstrated that combining circRNA and mRNA markers significantly improved the accuracy of bloodstain age estimation. Because circRNAs are more resistant to enzymatic degradation and environmental stressors, they may be more suitable for longer time windows—potentially bridging the gap between fresh and several-day-old stains.

Spectroscopic methods also play a vital role in current research. These methods are grounded in the idea that chemical compounds within blood undergo predictable changes in their absorption spectra as they oxidize and degrade. *UV–Visible spectroscopy*, for example, is used to monitor the transformation of hemoglobin and its derivatives. Bergmann et al. (2025) highlighted how specific absorbance peaks—particularly those corresponding to oxyhemoglobin and methemoglobin—can serve as indicators of bloodstain age. While effective, this method requires careful calibration and often assumes controlled environmental conditions, limiting its use in uncontrolled crime scene scenarios.

Another nondestructive technique that has shown considerable promise is *Attenuated Total Reflectance Fourier-Transform Infrared (ATR-FTIR) spectroscopy*. This method allows for detailed chemical fingerprinting of biological samples. In a recent study, Nagesh and Nagarajamurthy (2025) applied ATR-FTIR to bloodstains on various surfaces and demonstrated that, when combined with chemometric models, the spectral data could accurately estimate TSD across different substrates. This versatility makes ATR-FTIR particularly valuable for field applications where bloodstains may be found on anything from glass to concrete to fabric.

Beyond traditional analytical methods, the rise of *machine learning* has introduced new possibilities for integrating multiple sources of data into robust predictive models. For example, Giulietti et al. (2023) used *neural networks* in conjunction with *hyperspectral imaging* to estimate the age of bloodstains regardless of the surface on which they were deposited. Hyperspectral data captures hundreds of narrow spectral bands, providing a detailed map of how light interacts with the sample. The neural network then learns to associate specific spectral patterns with known time intervals, allowing for substrate-independent TSD estimation. This represents a significant leap forward, particularly in real-world scenarios where environmental and material variability is the norm.

Colorimetric methods have also seen innovation through data-driven modeling. Seki et al. (2024) introduced a *random forest regression model* trained on the subtle color changes in bloodstains over time. Unlike human visual inspection, which can be easily influenced by lighting and subjective perception, machine learning models can detect patterns invisible to the naked eye. Their model successfully estimated TSD up to 72 hours with relatively high precision, indicating that low-cost imaging systems paired with AI might serve as practical tools for on-site investigation.

Parallel to this, *metabolomics*—the study of small molecules involved in metabolism—offers a biochemical approach to time estimation. Blood contains a wide array of metabolites that degrade or transform over time. Lee et al. (2022) identified a panel of blood metabolites that change in concentration over a seven-day period. By tracking these metabolic changes, researchers were able to estimate the age of bloodstains within a window of accuracy comparable to genetic and spectroscopic methods. Since metabolomics reflects real-time biochemical processes, it provides an added layer of biological context that can be especially useful in cases involving environmental stress.

Finally, *epigenetic markers* such as *DNA methylation patterns* are being explored for their potential use in forensic timing. Unlike RNA or protein markers, DNA methylation changes occur slowly and may be more suited for long-term estimates. Han et al. (2022) validated a panel of methylation-sensitive CpG sites for estimating biological age, which could be adapted for forensic applications by observing post-depositional methylation shifts in bloodstains. Although still in early stages, this avenue suggests a potential for developing long-range time indicators beyond the first few days of deposition.

Despite these promising developments, major challenges remain. Environmental conditions still introduce substantial variability in results, and many models require calibration under tightly controlled settings before they can be generalized.

Furthermore, few techniques have yet reached the stage of forensic admissibility, meaning they cannot currently be used as standalone evidence in court. Nevertheless, the field is rapidly evolving, and the integration of multiple modalities—combining spectroscopic, molecular, and machine learning approaches—may ultimately lead to a robust, standardized method for bloodstain age estimation.

2 Mechanisms of Bloodstain Aging

Understanding how blood degrades over time is fundamental to forensic efforts in estimating the *time since deposition (TSD)*. Bloodstains do not remain chemically or physically static once they leave the body. Instead, they undergo a sequence of well-documented changes—some visible, others biochemical—that unfold in response to intrinsic biological decay and external environmental conditions.

2.1 Physical Changes: Color and Texture

One of the earliest observable indicators of bloodstain aging is a change in *color*. Fresh blood is typically bright red due to oxygenated hemoglobin (oxyHb), but it gradually darkens as hemoglobin oxidizes to methemoglobin (metHb) and then to hemichrome (Cossette et al. 2021). Over time, this results in a shift from a vivid red to a dull brown-black hue. Though visually noticeable, this transformation is heavily influenced by *substrate type, light exposure, and humidity*, making it unreliable for precise age estimation on its own.

Changes in *texture* can also offer clues. As blood dries, it contracts and forms crusty edges, which may flake or crack. On porous surfaces, stains may sink deeper into fibers, whereas on non-porous surfaces they often remain superficial, leading to differential drying rates and morphological cues (Balbudhe et al. 2025). However, while physical cues can suggest relative aging, they remain qualitative and subjective without supporting chemical data.

2.2 Biochemical Degradation: Hemoglobin, RNA, and DNA

Beyond visual cues, *biochemical degradation* offers a more quantifiable path toward TSD estimation. Hemoglobin, the dominant protein in red blood cells, is particularly useful due to its oxidative properties. As bloodstains age, hemoglobin's structure alters chemically, progressing from oxyHb to metHb and ultimately to denatured forms such as hemichrome (Tiessen et al. 2022). These transitions are detectable through spectroscopic and electrochemical methods, providing time-correlated markers that can be tracked in both laboratory and field settings.

The breakdown of *nucleic acids*—specifically RNA and DNA—has emerged as another reliable aging indicator. *RNA*, in particular, degrades rapidly after exposure to ambient conditions, often within hours to days. Gao et al. (2025) demonstrated that certain mRNA transcripts exhibit predictable degradation profiles, making them suitable as short-term molecular clocks. These patterns are further influenced by UV light, enzymatic activity, and surface type, requiring careful calibration across different environments.

DNA, though more stable than RNA, also shows signs of deterioration over time. Strand breaks, oxidative damage, and base modifications accumulate in exposed bloodstains, especially under fluctuating environmental conditions (Lee et al. 2023). While DNA degradation progresses more slowly, its analysis can be combined with RNA data to span both short- and long-term post-depositional periods.

2.3 Environmental Factors: Climate and Substrate Effects

Environmental conditions profoundly impact all physical and biochemical aspects of bloodstain aging. Temperature is a dominant factor—higher temperatures accelerate *oxidation*, *enzymatic activity*, and *nucleic acid degradation*. In contrast, lower temperatures tend to preserve both the structure and biochemistry of stains for longer periods (Cossette et al. 2021).

Humidity interacts with drying rates and protein denaturation. In moist environments, bloodstains remain hydrated longer, which can lead to bacterial growth and rapid RNA degradation. Conversely, dry conditions promote faster crust formation, potentially slowing biochemical activity but accelerating color change.

The *type of substrate* on which the blood lands also plays a significant role. Porous materials such as cotton or carpet can retain moisture longer and promote deeper seepage, complicating spectroscopic or imaging analyses. Nonporous surfaces like glass or tile may yield clearer morphological and chemical signatures but can be more susceptible to environmental contamination or photobleaching (Russell and Amankwaa 2025).

Controlled studies have quantified these impacts. In one experiment, bloodstains aged at 25 °C with 60% humidity showed significantly faster RNA degradation and hemoglobin oxidation than those stored at 4 °C under dry conditions (Heneghan et al. 2021). Such findings underline the importance of contextualizing all TSD estimates with detailed environmental metadata.

Figure 12.1 illustrates the flowchart of the main processes that drive bloodstain aging. Physical changes appear first, such as the gradual darkening of color (from bright red oxyhemoglobin to brownish hemichrome) and the way stains dry, crack, or flake. At the same time, biochemical shifts occur inside the stain, including the oxidation of hemoglobin and the breakdown of RNA and DNA, both of which can act as time markers. These processes don't happen in isolation—factors like temperature, humidity, and the type of surface all play a major role in how fast or slow blood degrades, which directly affects the accuracy of estimating the time since deposition (TSD).

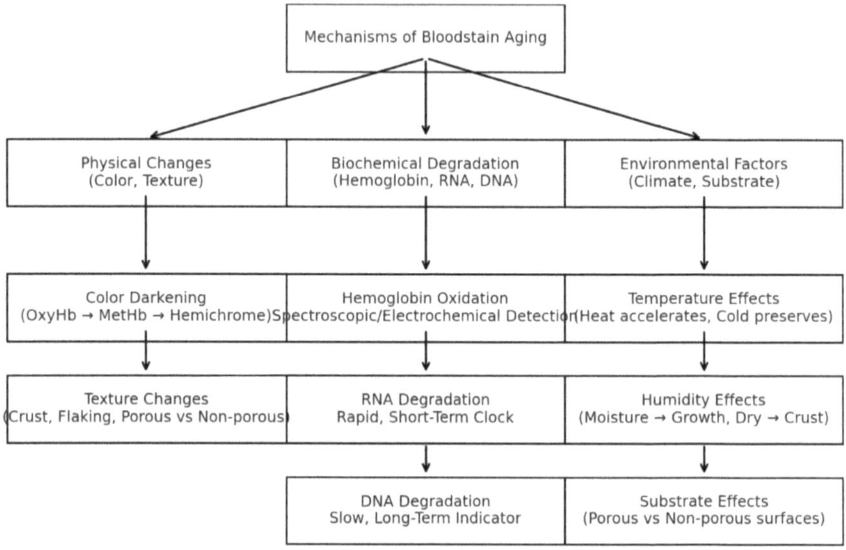

Fig. 12.1 Flowchart of bloodstain aging, showing physical changes, biochemical breakdown, and environmental influences that shape time since deposition (TSD)

3 Traditional Techniques

While forensic science continues to adopt advanced tools like spectroscopy and machine learning, the field of bloodstain aging was built on more traditional, foundational methods. These *classical techniques*, such as *visual analysis* and *colorimetric assays*, remain prevalent in routine crime scene investigations due to their low cost, accessibility, and ease of use. However, their reliability and objectivity have long been questioned, especially under uncontrolled environmental conditions.

3.1 Visual and Morphological Analysis

The earliest—and still most common—approach to estimating the age of a bloodstain relies on *visual examination*. Investigators often assess the *color, shape, and texture* of stains to form a preliminary hypothesis. Fresh bloodstains are generally bright red due to oxyhemoglobin but shift to a dark brown or nearly black as oxidation progresses. Likewise, the drying pattern, edge crusting, and central pooling may offer clues about elapsed time since deposition.

Morphologically, the *shape and size* of a stain are also examined. A relatively circular stain with uniform borders may indicate a recent deposition, while irregular margins or fragmentation may suggest aging or environmental disturbance (Sirisathitkul and Sirisathitkul 2025). However, such patterns are profoundly influenced by the *angle of impact, surface type, and environmental exposure*—factors that introduce considerable variability and reduce the objectivity of the technique.

Moreover, while useful in the immediate moments following deposition, *visual cues degrade rapidly*, losing diagnostic value after just a few hours. As Cossette et al. (2021) noted, even under controlled indoor conditions, the color and texture of bloodstains began to stabilize after 12–24 hours, making it difficult to distinguish between stains that are 1 day versus 3 days old. This inherent limitation has motivated the forensic community to seek more quantitative and consistent methods.

3.2 Colorimetric Assays

Another staple of traditional forensic blood detection is the use of *colorimetric tests*, which rely on chemical reactions between blood components and reagents to produce a visible color change. Among the most widely used are *Leucomalachite Green (LMG)* and *Tetramethylbenzidine (TMB).*

These reagents react with the *peroxidase-like activity of hemoglobin*, causing a color change that confirms the presence of blood. In the case of LMG, the color shift is typically from colorless to blue-green when in contact with hemoglobin and an oxidizer such as hydrogen peroxide. TMB, on the other hand, yields a blue color under similar conditions. These tests are highly sensitive and can detect even trace amounts of blood, including in old or diluted samples.

However, their utility in estimating the *age of bloodstains* is less reliable. Russell and Amankwaa (2025) evaluated the robustness of these tests under different environmental conditions and concluded that while both LMG and TMB remained reactive on buried fabrics after extended periods, the *intensity and speed of the reaction* varied significantly depending on *humidity, temperature, and substrate type*. This inconsistency limits their use as quantitative age indicators, although they are still invaluable for *presumptive blood detection.*

Further complicating matters is the risk of *false positives* and *chemical interference*. Colorimetric reagents may react with substances such as plant peroxidases, rust, or certain cleaning agents, especially under outdoor conditions where contamination is likely. Additionally, extended exposure to UV light or moisture can degrade the reagent's sensitivity or produce ambiguous results. Nevertheless, colorimetric tests remain *foundational tools* in forensic practice due to their speed and simplicity. When interpreted alongside other techniques—such as RNA degradation or spectroscopic profiles—they can provide a useful initial screening mechanism, particularly in fieldwork settings where time and resources are constrained.

Table 12.1 *summaries the traditional methods for bloodstain age estimation. Visual and morphological analysis offers quick, low-cost insights but is highly subjective, while colorimetric assays such as LMG and TMB provide sensitive presumptive tests yet remain unreliable for precise time estimates.*

Table 12.1 Summary of traditional methods for bloodstain age estimation

Technique	Description	Strengths	Limitations
Visual & morphological analysis	Observation of color, texture, drying, and shape to infer stain age. Fresh blood is bright red (oxyhemoglobin), darkens to brown/black with oxidation. Texture changes like crusting or flaking provide additional cues.	Low cost, rapid, requires no equipment; useful for preliminary assessment at crime scenes.	Subjective and qualitative; environmental factors (light, substrate, humidity) reduce reliability; cues stabilize after ~12 to 24 hours.
Colorimetric assays (LMG, TMB)	Chemical reagents (e.g., LMG and TMB) react with hemoglobin to produce visible color change, confirming presence of blood. Often shifts to blue-green (LMG) or blue (TMB).	Highly sensitive, can detect trace blood, effective even on old or diluted stains; widely used in fieldwork.	Limited utility for precise aging; influenced by temperature, humidity, substrate; risk of false positives (e.g., plant peroxidases, rust, cleaning agents).

4 Spectroscopic Approaches

Spectroscopic techniques have become increasingly central in the effort to establish more objective and quantifiable methods for estimating the *age of bloodstains*. Unlike visual inspection or colorimetric assays, which rely on subjective interpretation or binary chemical reactions, spectroscopic approaches leverage precise measurements of how blood components interact with light. These methods provide non-destructive, real-time insights into *biochemical changes* such as protein denaturation, pigment oxidation, and nucleic acid degradation—making them highly suited for forensic timelines.

4.1 UV–Vis Spectroscopy

Ultraviolet-visible (UV–Vis) spectroscopy is one of the most established methods for tracking oxidative changes in hemoglobin, the primary pigment responsible for blood color. As hemoglobin degrades from *oxyhemoglobin* to *methemoglobin* and *hemichrome*, it exhibits distinct absorbance peaks in the visible spectrum. Monitoring the relative intensities and positions of these peaks allows researchers to model the age of a bloodstain based on its spectral signature.

In their comprehensive review, Bergmann et al. (2025) outlined how absorbance peaks at around 415 nm (the Soret band) and 540–580 nm (Q-bands) shift over time in predictable patterns. These shifts correspond to structural transitions in the hemoglobin molecule as it oxidizes, particularly under light and heat exposure. Because UV–Vis measurements can be taken rapidly and without sample destruction, they are particularly useful for developing *in situ forensic tools*. However, the method's sensitivity to environmental noise—particularly humidity and substrate reflectivity—necessitates robust calibration models to ensure reliability.

4.2 ATR-FTIR Spectroscopy

Another spectroscopic method gaining traction in forensic bloodstain analysis is *Attenuated Total Reflectance–Fourier Transform Infrared (ATR-FTIR) spectroscopy*. This technique focuses on measuring how blood absorbs infrared light, with peaks corresponding to the *vibrational modes of chemical bonds* in proteins, lipids, and other biomolecules.

Time-dependent changes in amide I and II bands, reflecting protein denaturation, alongside lipid-associated bands in the 2800–3000 cm^{-1} region, have been widely reported in ATR-FTIR studies of biological stain ageing (Cano-Trujillo et al. 2026; Barney et al. 2025). Their findings showed that as bloodstains aged, the intensities and positions of these bands changed due to *protein denaturation* and *hydrolysis*. Additionally, lipid-associated bands in the 2800–3000 cm^{-1} region exhibited time-dependent behavior, offering a multidimensional approach to age estimation.

One advantage of ATR-FTIR is its *surface insensitivity*, which allows bloodstains on fabric, tile, or glass to be analyzed without extensive sample preparation. Moreover, its combination with *chemometric modeling*—including principal component analysis (PCA) and partial least squares regression (PLSR)—enables high-accuracy predictions across varied environmental conditions. Still, signal interference from water vapor and organic contaminants remains a technical hurdle, particularly in field settings.

4.3 Fluorescence Spectroscopy

While less commonly used than UV–Vis or IR, *fluorescence spectroscopy* presents an emerging frontier in bloodstain age analysis, particularly for detecting *time-dependent changes in nucleic acids and proteins*. This method measures the emission of light by fluorophores—molecules that absorb and re-emit light—after excitation, often revealing subtle structural changes in biological targets.

Elliott et al. (2025) explored the use of fluorescence spectroscopy to assess DNA-bound dyes in aging bloodstains. They found that fluorescence intensity and emission wavelength shifted over time due to *DNA fragmentation* and alterations in chromatin binding. Notably, dyes such as SYBR Green and PicoGreen, which bind preferentially to double-stranded DNA, exhibited decreasing fluorescence as the DNA degraded—a potential marker for short-term time estimation.

Fluorescence spectroscopy's key advantage lies in its *high sensitivity*, often capable of detecting minute changes undetectable by other methods. Additionally, its potential for miniaturization—via handheld fluorometers or smartphone-based systems—could eventually enable *on-site forensic diagnostics*. However, its current limitation lies in *photobleaching*, background autofluorescence from substrates, and dye stability, which may compromise reproducibility.

Table 12.2 summarizes how spectroscopic methods contribute to estimating the age of bloodstains. UV–Vis spectroscopy focuses on hemoglobin oxidation by tracking changes in characteristic absorbance peaks, offering rapid and non-destructive analysis but with sensitivity to environmental noise.

Table 12.2 Spectroscopic approaches for bloodstain age estimation, highlighting their focus, strengths, and limitations

Technique	Focus/description	Strengths	Limitations
UV-Vis spectroscopy	Tracks hemoglobin oxidation; absorbance peaks at 415 & 540–580 nm shift predictably with aging.	Rapid, non-destructive; useful for in situ analysis.	Sensitive to environment (humidity, substrate reflectivity).
ATR-FTIR spectroscopy	Monitors protein & lipid changes; amide I & II bands and lipid bands (2800–3000 cm^{-1}) reflect structural shifts.	Effective on varied surfaces; chemometric modeling enhances accuracy.	Signal interference from water vapor and contaminants.
Fluorescence spectroscopy	Detects DNA/protein changes using DNA-binding dyes; fluorescence intensity decreases with nucleic acid degradation.	High sensitivity; potential for miniaturization and field use.	Prone to photobleaching, autofluorescence, and dye instability.

ATR-FTIR spectroscopy captures protein and lipid degradation patterns and is effective across different surfaces, though it can suffer from interference caused by water vapor and contaminants. Fluorescence spectroscopy detects subtle DNA and protein changes using dyes, providing high sensitivity and potential for portable applications, yet it remains limited by photobleaching and autofluorescence. Together, these approaches demonstrate the shift from subjective visual techniques toward more quantifiable, technology-driven methods.

5 Molecular Techniques

In recent years, the field of forensic science has seen a paradigm shift from purely morphological and spectroscopic methods toward the use of *molecular biology* to estimate the age of bloodstains. Among the most promising approaches are those focusing on *RNA degradation*, *DNA methylation*, and *microRNA (miRNA) expression*—all of which offer the potential for *quantitative, time-dependent biomarkers* that remain detectable under varying conditions.

One of the earliest and most studied molecular strategies involves *messenger RNA (mRNA) profiling*. Since mRNA molecules are inherently unstable, their degradation begins almost immediately after blood is deposited. This makes them highly suitable for *short-term time-since-deposition (TSD)* estimation. Specific transcripts such as *GAPDH*, *β-actin*, and *18S rRNA* have been shown to degrade at reproducible rates under controlled conditions. Fonneløp et al. (2025) developed a capillary electrophoresis-based method that targeted six such transcripts, producing consistent TSD predictions within the first 48 hours of bloodstain aging. While the method is highly sensitive, it requires careful handling, as RNA degrades rapidly in response to environmental stressors like heat, UV radiation, and moisture.

To address these limitations, researchers have turned to more *stable molecular markers*, particularly *DNA methylation* and *microRNAs (miRNAs)*. DNA methylation—chemical modifications on cytosine bases—has been extensively studied for estimating biological age and is now being adapted for forensic timing. Gao et al. (2025) demonstrated that combining methylation profiles with miRNA data using droplet digital PCR (ddPCR) significantly enhances predictive accuracy. Their study revealed that the integration of both markers could track bloodstain aging not just over hours but over multiple days, withstanding moderate environmental fluctuations.

Among these, *miRNAs have emerged as especially robust* due to their small size and resistance to enzymatic degradation. They regulate gene expression and are often more stable than mRNAs under adverse conditions. This stability makes them well-suited for forensic applications where bloodstains are exposed to fluctuating temperature or humidity. In a recent study, Lu et al. (2025) developed a miRNA-based model that accurately estimated the age of bloodstains over a week, outperforming mRNA-based models in degraded samples. The authors suggested that select miRNAs—such as *miR-16*, *miR-21*, and *miR-451*—retain their expression patterns even when traditional biomarkers begin to fail, making them particularly promising for *field-level forensic analysis*.

Taken together, molecular techniques are redefining what is possible in forensic bloodstain analysis. By moving beyond surface-level observation and into the molecular landscape, these approaches offer a *quantitative, reliable, and potentially courtroom-admissible method* for estimating the time since blood was shed. Future developments may involve combining multiple molecular markers with machine learning algorithms to produce hybrid models capable of compensating for environmental variability and sample degradation.

6 AI and Machine Learning Applications

While traditional and molecular methods have brought bloodstain aging closer to scientific rigor, they often rely on single-variable analysis or require controlled lab settings. To address these limitations, researchers are increasingly turning to *artificial intelligence (AI)* and *machine learning (ML)* techniques. These computational approaches offer the unique ability to integrate multiple types of data—visual, spectral, molecular, and environmental—and identify subtle patterns that may elude human analysis.

One of the most promising areas is the use of *supervised learning algorithms* trained on large datasets of aged bloodstains. In particular, *random forest regression models* have shown success in correlating subtle colorimetric shifts with time. Seki et al. (2024) developed a model trained on high-resolution images of bloodstains over a 72-hour period, achieving impressive accuracy in predicting age from color features alone. Unlike visual observation, the algorithm could detect minuscule changes invisible to the human eye, making it ideal for forensic photography applications in both indoor and outdoor scenes.

Beyond color, *spectral imaging combined with deep learning* is making it possible to assess bloodstains across different substrates and environmental conditions. Giulietti et al. (2023) applied *convolutional neural networks (CNNs)* to hyperspectral data, enabling age estimation regardless of whether the blood was deposited on fabric, tile, or plastic. Their model not only distinguished fresh from older stains but also correctly grouped samples by hourly and daily time intervals. The advantage of such models lies in their ability to generalize—learning time-based degradation patterns independent of surface reflectance or ambient light.

Another emerging tool is the integration of *machine learning with molecular data*, particularly RNA and miRNA profiles. Algorithms such as *support vector machines (SVM)* and *gradient boosting* are now being trained to recognize degradation signatures in molecular datasets. Sancı and Kaya (2024) outlined a framework that combines miRNA expression levels with supervised learning models, significantly improving accuracy under variable temperature and humidity. Their study suggested that AI can effectively handle the complexity and variability that often hinder molecular techniques when applied outside of laboratory settings.

AI also holds potential for *portable, field-deployable applications*. Paired with smartphone cameras or handheld spectrometers, AI-driven systems could perform rapid, on-site bloodstain age estimation. As such, forensic teams could receive instant time approximations even in remote or uncontrolled environments, drastically speeding up investigative timelines. Although such systems are still in early stages of development, the convergence of hardware miniaturization and cloud-based computing makes this a highly feasible direction.

Despite these advancements, challenges remain. Training data must be diverse enough to represent the full spectrum of environmental conditions and biological variability. Additionally, forensic admissibility will require algorithmic transparency—an area where black-box AI models still face scrutiny in the legal system. Nevertheless, AI is rapidly moving from proof-of-concept to practical deployment, and its ability to handle high-dimensional, noisy data may make it the linchpin of next-generation forensic bloodstain analysis.

Figure 12.2 illustrates the complementary roles of molecular and AI/ML techniques in advancing bloodstain age estimation. On the molecular side, mRNA profiling provides valuable short-term markers but is limited by the rapid degradation of RNA under environmental stress. DNA methylation extends the timeframe by offering relatively stable chemical modifications, while microRNA (miRNA) models stand out for their robustness, maintaining reliability even in degraded or variable conditions. In contrast, AI and machine learning methods leverage computational power to integrate diverse data sources. Supervised learning excels at detecting subtle color shifts invisible to the human eye, while deep learning applied to spectral imaging delivers substrate-independent predictions across different surfaces. Hybrid models that combine molecular data with machine learning show particular promise, as they can compensate for environmental variability and biological noise more effectively than single-method approaches. Together, these innovations represent a move toward more objective, data-driven forensic tools that can operate reliably both in controlled laboratories and in the field.

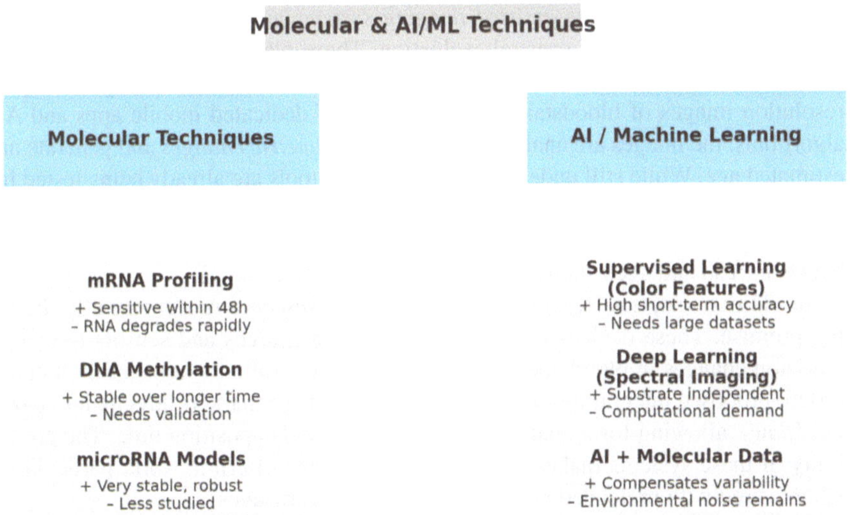

Fig. 12.2 Overview of molecular and AI/ML techniques for bloodstain age estimation. Molecular approaches (mRNA, DNA methylation, and miRNA) provide biological markers with varying stability, while AI/ML methods enhance prediction accuracy through color, spectral, and molecular data integration

7 Portable and On-Site Techniques

As forensic investigations increasingly demand fast, reliable, and field-ready tools, there is growing interest in *portable technologies* capable of estimating the age of bloodstains *on-site*. Traditional lab-based methods, while precise, often require significant delays due to sample transport, preservation, and complex instrumentation. In contrast, portable devices offer the promise of *real-time analysis*, empowering investigators to make time-sensitive decisions directly at crime scenes.

One of the most promising developments in this space is the use of *handheld near-infrared (NIR) spectrometers*. These compact devices analyze bloodstains by detecting *molecular vibrations* associated with chemical bonds, particularly those in hemoglobin and its degradation products. Fonseca et al. (2025) demonstrated that when combined with *chemometric modeling*, handheld NIR spectrometry could accurately estimate bloodstain age across a variety of substrates, including ceramic tile, vinyl, and untreated wood. Their results showed that time-sensitive spectral shifts, particularly in the 1400–2400 nm range, reliably distinguished between stains aged for hours versus several days.

The key advantage of NIR devices lies in their *nondestructive and contactless nature*, which preserves evidence while delivering near-instant feedback. When integrated with statistical tools such as *Partial Least Squares Regression (PLSR)* or *Support Vector Regression (SVR)*, these systems can produce TSD estimates with minimal operator training. This makes them particularly attractive for field personnel, especially in resource-limited or remote settings.

In parallel, *smartphone-assisted imaging systems* are emerging as a low-cost alternative for visual and spectral evaluation. These platforms utilize smartphone cameras, sometimes augmented with external lenses or filters, to capture high-resolution images of bloodstains. With the help of dedicated mobile apps and AI algorithms, the images are analyzed to assess *colorimetric changes* and generate an estimated age. While still under development, such tools are already being tested in controlled field trials. Seki et al. (2024) noted that when paired with cloud-based random forest models, even consumer-grade phone cameras could differentiate between fresh and aged stains with surprising accuracy.

In addition to optics-based systems, *portable fluorescence readers* are also showing promise. These devices use compact excitation sources and sensors to detect emission changes in blood-specific dyes or intrinsic autofluorescence. Elliott et al. (2025) proposed integrating such fluorescence-based age markers directly into *DNA workflows*, allowing for a dual analysis of identity and deposition time. The portability of these systems makes them suitable for scene-of-crime work, particularly when paired with preloaded reagents or disposable cartridges.

Still, despite their potential, field-ready tools come with challenges. Environmental factors such as *ambient light, surface reflectance, dust, or moisture* can interfere with signal accuracy. Substrate variability also remains a hurdle—what works on ceramic may fail on textile or soil. Additionally, many portable systems still rely on models trained in controlled lab conditions, requiring *contextual calibration* to perform reliably in the field.

To mitigate these concerns, some teams are integrating *multisensor platforms*—combining NIR, imaging, and humidity/temperature sensors into one device. This allows the system to self-correct based on environmental inputs and adjust its predictions accordingly. While these systems are still in development, they signal a shift toward *adaptive forensic tools* that blend portability with precision.

In essence, portable and on-site technologies are no longer speculative—they are rapidly becoming essential components of modern forensic practice. As they continue to evolve, these tools promise to *democratize forensic timing analysis*, making bloodstain age estimation accessible, efficient, and real-time for a broader range of practitioners.

8 Limitations

Despite impressive advances in the estimation of bloodstain age through molecular, spectroscopic, and machine learning methods, the field continues to face *significant limitations* that hinder both scientific consensus and courtroom applicability. These challenges span technical, environmental, and practical domains, underscoring the complexity of translating laboratory breakthroughs into reliable forensic tools.

8.1 Environmental Sensitivity and Variability

Perhaps the most persistent challenge in bloodstain aging research is the *high sensitivity of most methods to environmental conditions.* Temperature, humidity, light exposure, and airflow can dramatically influence the rate of chemical and biological degradation in bloodstains. For example, RNA-based methods—such as those described by Fonneløp et al. (2025)—are highly effective under controlled conditions but degrade rapidly under heat or moisture, often limiting their use to fresh stains. Similarly, spectroscopic signatures, especially in UV–Vis and fluorescence methods, are prone to distortion by substrate reflectance, ambient light, and surface porosity (Bergmann et al. 2025; Elliott et al. 2025).

Even portable NIR systems, while promising, require calibration against environmental data to yield accurate results in the field (Fonseca et al. 2025). A bloodstain on ceramic tile in a dry indoor setting may degrade very differently from one on fabric outdoors, making *generalized models difficult to achieve.*

8.2 Lack of Standardization and Forensic Admissibility

Another critical barrier is the *absence of standardized protocols* across different labs, instruments, and forensic agencies. Variability in sample handling, storage, analysis timelines, and even data interpretation makes it difficult to replicate findings across studies or jurisdictions. As a result, many of these techniques—especially those involving AI, miRNA, or spectroscopy—remain in the *experimental stage*, not yet admissible in court under the Daubert or Frye standards.

This lack of standardization also affects *dataset diversity*. Most machine learning models are trained on curated lab samples with known conditions, making them poorly suited to real-world crime scenes where variables are uncontrolled and unknown (Seki et al. 2024; Giulietti et al. 2023).

8.3 Substrate and Surface Dependencies

Surface interaction remains another major challenge. While some methods attempt to be substrate-independent—such as hyperspectral imaging (Giulietti et al. 2023)—many tools still perform inconsistently across different materials. For instance, ATR-FTIR spectroscopy is highly effective on smooth, non-porous surfaces but struggles with fabric or textured substrates due to signal interference (Liu et al. 2025). This substrate dependence requires investigators to make on-site judgments about whether a tool or model can be trusted in a given situation.

8.4 Limited Time Window and Degradation Overlap

Most techniques perform best within a *narrow time window*, typically ranging from a few hours to a few days. Outside this window, the degradation curves tend to *plateau*, making it difficult to distinguish between stains that are, for example, 5 days old versus 8 days. Both colorimetric and molecular methods suffer from this, as the biomarkers they rely on degrade to undetectable levels or stabilize beyond meaningful interpretation (Gao et al. 2025; Wei et al. 2022).

This plateau effect not only reduces temporal resolution but also increases the risk of *false negatives*—where older bloodstains are incorrectly identified as undetectable or misdated entirely.

8.5 Ethical and Legal Considerations in AI Use

The adoption of *machine learning and AI in forensic science* raises its own set of limitations. While these models can analyze high-dimensional data with great accuracy, they often operate as *black boxes*, making their decision-making process opaque. This lack of transparency poses challenges for legal admissibility, where expert witnesses must explain how conclusions were drawn. Without interpretability, even the most accurate AI model may be dismissed in court.

Moreover, many AI systems are trained on *homogeneous datasets* that may not account for variation in blood type, ethnicity, or pathological conditions such as anemia or clotting disorders, raising concerns about *bias and generalizability* (Sancı and Kaya 2024).

9 Future Directions

As forensic science moves steadily toward data-rich, interdisciplinary frameworks, the field of *bloodstain age estimation* stands at the intersection of molecular biology, analytical chemistry, artificial intelligence, and portable technology. While recent developments have laid a strong foundation, the journey toward a standardized, universally applicable method is still underway. Several key areas are emerging as focal points for future innovation and research.

9.1 Multimodal Integration

One of the most promising directions is the *integration of multimodal data*—combining visual, spectroscopic, molecular, and environmental inputs into a unified analytical framework. Many current studies focus on a single data stream, whether it's mRNA decay, hemoglobin oxidation, or colorimetric change. However, bloodstains are complex biological systems affected by numerous simultaneous factors. Creating hybrid models that combine time-sensitive RNA markers, spectral shifts, and machine learning classification could significantly improve both accuracy and resilience across diverse conditions.

Tools like *hyperspectral imaging*, when paired with *molecular profiling*, may allow for *layered interpretations* that consider both surface-level and molecular-level degradation. These systems would not only estimate time since deposition but potentially differentiate between multiple events in scenes involving *mixed bloodstains* or secondary transfers (Giulietti et al. 2023).

9.2 Real-Time, On-Scene Estimation

The development of *portable, real-time devices* for forensic timing is rapidly gaining momentum. While early prototypes exist for handheld NIR and fluorescence spectrometers, future tools are expected to be more *modular, AI-integrated*, and capable of field calibration. These devices may include *multi-sensor platforms* with environmental probes (temperature, humidity, UV index) that adjust TSD estimates based on local conditions.

Additionally, we can expect greater use of *edge computing and mobile AI models*. With increased processing power in smartphones and field tablets, image- and signal-based analysis could be performed directly on-site without requiring an internet connection or lab processing. Such autonomy would significantly reduce the lag between evidence collection and analysis, enabling *faster decision-making* during active investigations.

9.3 Advanced Molecular Targets

On the molecular front, ongoing research is shifting from single biomarkers to *biomarker panels*. Rather than relying solely on traditional mRNA or hemoglobin derivatives, future assays may involve *combinatorial profiling* of microRNAs, circular RNAs, epigenetic markers, and even protein degradation fragments. These molecular signatures could be mapped into *time-curves* that stretch across short (0–24 h), medium (1–7 days), and long (>1 week) timeframes.

Newer technologies like *droplet digital PCR (ddPCR)* and *next-generation sequencing (NGS)* are enabling high-sensitivity quantification of these markers, even from small or degraded samples (Gao et al. 2025). The challenge ahead will be to miniaturize and automate these techniques for field deployment.

9.4 Environmental Modeling and AI Generalization

To improve the reliability of machine learning systems in real-world conditions, future models must learn to *generalize across diverse environments and substrates*. Current algorithms often struggle when applied outside their training conditions, which limits their forensic value.

Future models will likely include *environment-aware architecture*, where inputs like ambient temperature, relative humidity, and surface type are incorporated into the prediction process. In addition, *synthetic data generation* and *transfer learning* may be employed to simulate wide-ranging conditions and reduce the dependency on large labeled datasets (Sancı and Kaya 2024).

There is also a growing emphasis on *explainable AI (XAI)*, especially in forensic contexts where transparency is crucial for legal admissibility. Next-gen models may include embedded justifications for each decision, allowing forensic experts to trace and validate the algorithm's logic—an essential step toward courtroom use.

9.5 Standardization and Cross-Laboratory Validation

Perhaps the most pressing need is for *interlaboratory validation and international standardization*. Many promising tools remain isolated within research settings due to inconsistent protocols, varied datasets, and proprietary algorithms. Coordinated efforts—such as shared open-access bloodstain datasets, cross-institution calibration studies, and proficiency testing—are necessary to bridge the gap between research and forensic practice.

Professional bodies in forensic science may eventually issue *certification standards for bloodstain aging techniques*, much like those already established for DNA profiling. Such frameworks would support the *legal admissibility* and widespread adoption of these tools.

10 Conclusion

The ability to estimate the age of bloodstains holds profound significance in forensic science, offering insights not only into the presence of blood but into the *timeline of events* at a crime scene. Over the past two decades, this field has evolved from subjective visual inspection to sophisticated molecular and spectroscopic methods, powered increasingly by artificial intelligence and field-deployable technologies.

Traditional approaches such as morphological observation and colorimetric assays, while still useful for preliminary analysis, lack the precision and reliability required for formal time estimation. In contrast, *spectroscopic techniques*—including UV–Vis, ATR-FTIR, and fluorescence—offer quantifiable data on the biochemical changes in hemoglobin, proteins, and nucleic acids. These methods have proven effective in lab conditions but are sensitive to environmental variables and substrate differences.

The integration of *molecular techniques*, particularly mRNA decay, DNA methylation, and miRNA profiling, has opened new avenues for time-dependent biomarker analysis. These approaches allow for higher resolution in age estimation, especially when combined with advanced analytical platforms like *droplet digital PCR*. Their potential is further amplified when interpreted through machine learning models capable of handling multi-dimensional data.

Meanwhile, *AI-powered tools* and *portable spectrometers* have begun to bridge the gap between laboratory science and on-scene application. Devices capable of delivering real-time estimations—especially when equipped with environmental sensors and adaptive algorithms—are pushing forensic science toward a future where timely, accurate, and minimally invasive analysis becomes routine.

However, the field still faces challenges. Environmental sensitivity, lack of standardization, substrate variability, and legal admissibility remain significant hurdles. Future research must prioritize *cross-laboratory validation*, *model generalizability*, and *transparent AI systems* to ensure these technologies are not only scientifically sound but also accepted in legal contexts.

In conclusion, bloodstain age estimation is rapidly transforming into a *multidisciplinary science*, where chemistry, biology, data science, and engineering converge. With continued innovation, rigorous validation, and collaborative standardization, it is likely that this once speculative area will become a robust, court-admissible tool—changing the way investigators reconstruct time at crime scenes.

References

Balbudhe MS, Shetty BSK, Mazumder N, Narayan Singh B, Arun Kumar N (2025) Time-dependent morphological changes in bloodstains on tile and vinyl flooring. Egypt J Forensic Sci 15(1):54

Barney A, Trojan V, Hrib R, Newland A, Halámek J, Halámková L (2025) From spectra to signatures: detecting fentanyl in human nails with ATR–FTIR and machine learning. Sensors 25(1):227

Bergmann T, Oberdörster M, Schade F (2025) Chronology of forensic bloodstain age estimation using UV–Vis spectroscopy. Forensic Sci Int 376:111789. https://doi.org/10.1016/j.forsciint.2025.111789

Cano-Trujillo C, Ortega-Ojeda FE, García-Ruiz C, Montalvo G (2026) Forensic estimation of semen stains' time since deposition on porous substrates using ATR FTIR and Raman spectroscopy supported by OPLS-DA models. Forensic Chemistry:100734

Cossette ML, Stotesbury T, Shafer AB (2021) Quantifying visible absorbance changes and DNA degradation in aging bloodstains under extreme temperatures. Forensic Sci Int 318:110627

Elliott CI, Simmons DBD, Stotesbury T (2025) Integrating time since deposition estimation of bloodstains into a DNA profiling workflow: a novel approach using fluorescence spectroscopy. ChemRxiv. https://chemrxiv.org/engage/api-gateway/chemrxiv/assets/orp/resource/item/66e49b24cec5d6c142fa9ce5/original/integrating-time-since-deposition-estimation-of-bloodstains-into-a-dna-profiling-workflow-a-novel-approach-using-fluorescence-spectroscopy.pdf

Fonneløp AE, Hänggi NV, Derevlean CC (2025) A CE-based mRNA profiling method including six targets to estimate the time since deposition of blood stains. Forensic Sci Int Genet 77:103240. https://doi.org/10.1016/j.fsigen.2025.103240

Fonseca ACS, Pereira JFQ, Honorato RS (2025) Age estimation of bloodstains deposited on different floor tiles using handheld NIR spectrometer and chemometrics. Microchem J 183:108447. https://doi.org/10.1016/j.microc.2025.108447

Gao N, Li J, Yang F, Yu D et al (2025) Forensic age estimation from blood samples by combining DNA methylation and microRNA markers using droplet digital PCR. Electrophoresis 46(5):e8133. https://doi.org/10.1002/elps.8133

Giulietti N, Discepolo S, Castellini P, Martarelli M (2023) Neural network based hyperspectral imaging for substrate independent bloodstain age estimation. Forensic Sci Int 349:111742

Han X, Xiao C, Yi S, Li Y, Chen M, Huang D (2022) Accurate age estimation from blood samples of Han Chinese individuals using eight high-performance age-related CpG sites. Int J Legal Med 136(6):1655–1665

Heneghan N, Fu J, Pritchard J, Payton M (2021) The effect of environmental conditions on the rate of RNA degradation in dried blood stains. Forensic Sci Int Genet 53:102544. https://doi.org/10.1016/j.fsigen.2021.102544

Lee YR, Lee S, Kwon S, Lee J, Kang HG (2022) Bloodstain metabolite markers: discovery and validation for estimating age of bloodstain within 7 days. Anal Chem 94(39):13377–13384

Lee YR, Lee S, Kwon S, Lee J, Kang HG (2023) Effect of environmental conditions on bloodstain metabolite analysis. Environ Res 220:114970. https://doi.org/10.1016/j.envres.2023.114970

Lu H, Zhang X, Yu X, Lin F (2025) Development of a microRNA-based age estimation model using whole-blood microRNA expression profiling. Forensic Sci Int Rep 7:100357. https://doi.org/10.1016/j.fsir.2025.100357

Nagesh D, Nagarajamurthy B (2025) Estimation of time-since-deposition of bloodstains on different surfaces using ATR-FTIR spectroscopy and chemometrics. Forensic Sci Med Pathol 21(1):123–137

Russell C, Amankwaa AO (2025) The persistence of bloodstains on buried fabrics and robustness of the leucomalachite green test. Forensic Sci Int 374:111678. https://doi.org/10.1016/j.forsciint.2025.111678

Sancı A, Kaya B (2024) Applications of artificial intelligence in age estimation: a review. J Radiol Med 1(4):77–83

Seki T, Hsiao YY, Ishizawa F, Sugano Y, Takahashi Y (2024) Establishment of a random forest regression model to estimate the age of bloodstains based on temporal colorimetric analysis. Legal Med 69:102343

Sirisathitkul C, Sirisathitkul Y (2025) Bloodstain in forensics: from visual inspections to AI-assisted pattern analysis and age estimation. Hist Sci Technol 15(1):26–46

Tiessen M, Fruehwald HM, Easton EB, Stotesbury T (2022) Insights into bloodstain degradation and time since deposition estimation using electrochemistry. Front Anal Sci 2:900483

Wei Y, Wang J, Wang Q, Cong B, Li S (2022) The estimation of bloodstain age utilizing circRNAs and mRNAs biomarkers. Forensic Sci Int 338:111408

Determining Postmortem Interval from Blood Analysis

13

Abstract

Determination of postmortem time interval holds significance in defining the direction of a forensic investigation. The conventional methods of estimating PMI explore the early postmortem changes appear in the body including rigor mortis, livor mortis, and algor mortis. Though such techniques are widely used in PMI estimation, such methods are relatively inaccurate and offer limited practical relevance. Many biochemical and molecular techniques have been explored using the postmortem blood samples for the accurate, reliable estimation of PMI. Besides the analysis of metabolites and proteins, molecular techniques measuring the level of DNA and RNA in the postmortem blood has shown a huge potential in estimating PMI. The descendent trend of the amount of DNA postmortem has shown its correlation with the PMI. The degradation pattern of DNA can be studied by using flow cytometer, image analysis techniques, single cell gel electrophoresis (SCGE), DNA amplification assay, and other assays such as the terminal deoxynucleotidyl transferase assay. Similarly, RT-PCR-based analysis of expression pattern of mRNA, miRNA, and circRNA can also provide crucial information on the PMI. mRNA of *Actb*, *Gapdh*, *Ppia*, *Srp72*, β-actin, and glyceraldehyde-3-phosphate dehydrogenase (GAPDH) have shown potential use in PMI estimation.

Keywords

PMI estimation · Rigor mortis · Proteomics · mRNA · RT-PCR · GAPDH

1 Introduction

From forensic medicine context, death can be defined as the permanent irreversible cessation of the basic tripod of life, i.e., brain, heart, and lungs. After death of an individual, various changes categorized as immediate, early and late postmortem changes occur in the body. Careful observation of such postmortem changes might lead to the accurate estimation of postmortem time interval in a carcass. Establishment

H. R. Dash et al., *Forensic Investigation of Blood Evidence*,
https://doi.org/10.1007/978-981-95-9183-1_13

of postmortem time interval (PMI) plays a crucial role in the direction of an investigation. In cases of homicide, and recovery of unidentified dead body, determination of PMI provides many investigative leads which leads to the linking of suspect and victim to a crime scene which helps in crime scene reconstruction. Though the field of PMI analysis advancing at a rapid pace, the accuracy and reliability of the available technique pose a huge challenge to the investigation agencies.

When death occurs, certain immediate changes such as stoppage of the function of nervous system, respiratory system, and circulatory system occurs. Various early postmortem changes are also characterized to establish the PMI. Some of the early postmortem changes include livor mortis, algor mortis, rigor mortis, pupil dilation, etc. Livor mortis or hypostasis is defined as the red–blue–purple discoloration on the skin within 30 min to 3 h of death. When a person is alive, heart continuously pumps blood and blood is circulated throughout the body through arteries and veins. However, when a person is dead and heart stops pumping blood, the circulation of blood comes to a halt in the body. Blood moves to the low-lying dependent region of the body due to gravity and settle within the dermal capillaries. This results in the development of red–blue–purple staining pattern in the low-lying portions of the body, called as livor mortis, post mortem staining, post mortem lividity, or post mortem hypostasis. Though postmortem lividity appears within 30 min to 3 h postmortem, in moderate to cool climatic conditions, it is completely developed within 4–8 h postmortem and reach its maximum intensity after ~10 h.

Similarly, another early postmortem change is the algor mortis. Algor mortis is defined as the cooling of the body after death. As humans are the homeothermal animals, they maintain their body temperature at 98.6 °F irrespective of the body temperature. However, after death of an individual, such homeostasis mechanisms do not act and the body temperature begins to cool and equilibrate with the ambient temperature. Measurement of rectal temperature is commonly used as a standard to determine the decedent temperature. After immediate death, the residual metabolic activity of the body slows the cooling rate of the body initially which follows a sigmoid curve. However, after few hours, the body temperature drops steadily in the next 12 h at a rate of 1–1.5 °C/h (Strete et al. 2025). Various environmental factors such as ambient temperature, humidity, air circulation, and closeness to heat sources affect algor mortis, which makes the PMI estimation challenging based on the algor mortis of the body.

Out of all early postmortem changes, rigor mortis is the most widely used technique for the determination of PMI. Rigor mortis is defined as the stiffening of the body muscle after death. In a live body, ATP is produced during normal glycolysis through a series of biochemical changes. When a person dies, various biochemical events of the body cease and ATP could not be generated. This leads to the depletion of ATP and the influx of calcium which ultimately prevents the detachment of actin-myosin crosslink and muscle cannot be relaxed. The march of onset of rigor is governed by the Nysten's Law. Rigor appears first in the smaller muscles over the face followed by the muscles of the upper limbs and the lower limbs. Rule of 12 defines the primary estimation of PMI from rigor mortis. After the onset of rigor within 2 h after death, the complete body undergoes rigor after 12 h. The body remains in the rigor for 12 h (24 h postmortem) and then starts disappearing due to proteolysis (Dalal et al. 2006). Rigor mortis disappears completely after 36 h after which the body enters into a state called as secondary flaccidity. Though such early/immediate postmortem changes provide preliminary information on the PMI, such methods are relatively inaccurate and offer limited practical relevance (Sampaio-Silva et al. 2013). Thus, other biochemical and molecular techniques have been explored for an accurate estimation of PMI.

2 Difference Between Antemortem and Postmortem Blood

Human antemortem blood is comprised of many proteins and metabolites. A total of 7518 blood proteins and 4229 blood metabolites have been identified and annotated in the human proteome database and human metabolome database. The abundant antemortem blood proteins include albumin, haptoglobin, fibrinogen, lipoproteins, coagulation factors etc. Similarly, the most abundant metabolites include glucose, lactate, insulin, growth hormones and many others. During early postmortem, the concentration of some proteins like fibrinolysins and lactic dehydrogenase increase drastically. Similarly, metabolites like glucose, lactate, hypoxanthine, bilirubin, uric acid and ammonia also increase in postmortem blood. Besides their composition, the pH of postmortem blood decreases to 5.0–5.5. Blood hematocrit is the ratio of red blood cells (RBCs) in the total blood volume expressed in percentage. Blood hematocrit (HCT) increases from 40–45 antemortem to 47–78 postmortem. Due to the loss of function of Na^+/K^+ pump in the cell membrane as ATP is not generated, the blood cells begin to lyse postmortem. Lysis of cells result in the leakage of intracellular components into the surrounding tissue spaces. These cellular components are widely used as blood biomarkers for the determination of PMI.

3 Biochemical Markers of Blood for PMI Estimation

3.1 Postmortem Blood Proteins

Though the total amount of blood proteins has been explored as a marker for the estimation of PMI, they do not show a significant difference than that of the antemortem total blood protein level. The increase in the total protein content in postmortem blood remains within the range of antemortem values. Thus, the total protein level alone cannot be considered as a good marker for the estimation of PMI. Two proteins Aspartate Aminotransferase and lactate dehydrogenase have shown potential as biomarkers for the determination of PMI from blood.

Aspartate aminotransferase (AST) found in the heart, skeletal muscles, liver, kidney, erythrocyte, and brain tissue is responsible for the conversion of aspartic acid to glutamate. Postmortem release of AST from tissues can be detected and enumerated for the estimation of PMI. The severity of cell lysis governs the amount of AST in the blood stream. As RBCs contain a high amount of AST, lysis of RBCs postmortem leads to the elevation of AST concentration in blood. It has been estimated that the increase in AST amount is linear during the first 60 h postmortem. Thus, AST can be considered as a good biomarker for the estimation of PMI by estimating its concentration in postmortem blood.

Similarly, determining the amount of lactate dehydrogenase (LDH) in postmortem blood can also be useful in estimating the PMI. The terminal step of anaerobic glycolysis is catalyzed by LDH by converting pyruvate to lactate. Two subunits (A and B) of LDH combine together in five different confirmations to generate LDH1 (B4), LDH2 (A1B3), LDH3 (A2B2), LDH4 (A3B1), and LDH5 (A4). Due to the lysis of cells postmortem, the level of LDH also increases in the blood. A rapid increase in the LDH concentration has been observed within few hours' postmortem followed by a steady increase for 2–3 days. The highest concentration of serum LDH is reached after fourth day postmortem. As RBCs are the rich source of LDH in blood, their lysis provides the main source of LDH in the serum which can be detected to estimate the PMI from the postmortem blood sample.

3.2 Postmortem Blood Metabolites

Qualitative and quantitative analysis of various metabolites in the blood sample can also provide a clue for the estimation of PMI. The useful blood metabolites for PMI estimation include sodium, chloride, potassium, calcium, magnesium, phosphate, lactate, hypoxanthine, urea, creatinine, uric acid, ammonia, catecholamines, and ethanol which has been discussed below. The detailed information of such metabolic biomarkers for the estimation of PMI is given in Fig. 13.1.

The concentration of sodium and chloride in the serum is dependent on various factors including the status of hydration of the individual, chronic kidney problems or other associated illness. The serum sodium and chloride level is mostly used for the detection of possible kidney disease and/ or electrolyte imbalance in an individual. Despite of its variation and clinical use antemortem, a decrease of sodium

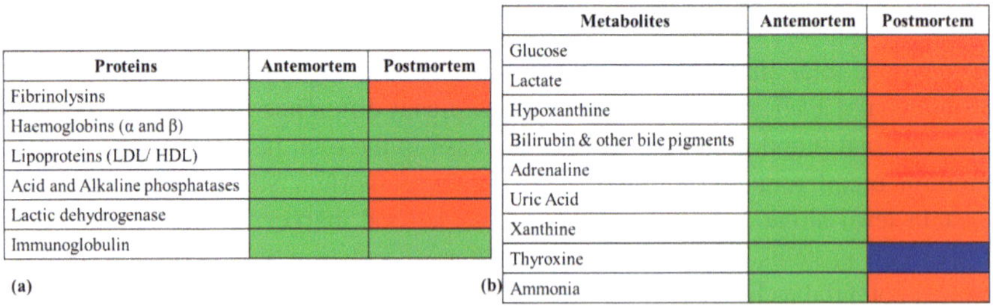

Fig. 13.1 Postmortem variation of (**a**) proteins and (**b**) metabolites in the serum. Green indicates the antemortem level, red is the elevated level, and blue is the decreased level

and chloride at a rate of 1 mM/l per hour has been observed in the serum within 3–52 h postmortem. However, due to the effect of many intrinsic and extrinsic factors on serum the sodium and chloride concentration postmortem, these two metabolites are not considered as reliable markers for the estimation of PMI. Besides, some studies have reported that, postmortem vitreous humor sample can provide more accurate estimation of PMI than serum level of sodium and chloride.

The level of potassium in the serum has also shown a rapid spike after 1–2 h of death. This rapid increase in serum potassium level is observed within 1–2 min of death. Hence, measurement of serum potassium level for the estimation of PMI becomes highly challenging. To overcome this challenge, the potassium level in vitreous humor becomes highly useful as vitreous humor shows a rapid increase in the potassium concentration in the first 6–12 h postmortem followed by a linear decrease in potassium after 24 h. However, various indigenous and exogenous factors such as sampling technique, temperature, intereye difference, age of the person as well as the terminal episode duration. Hence, no reliable technique has been developed till date for an accurate estimation of PMI from the level of potassium either from the serum or from the vitreous humor.

The serum calcium level has been reported to be relatively stable during the early postmortem period, i.e., 12–24 h. After 24 h, the serum calcium concentration rapidly increases at a rate of 4 mM/l per hour. However, variable results reported from different researchers limits the application of using this biomarker for the estimation of PMI. If the serum calcium level can be revisited with the advanced instruments, a more accurate result might be obtained. Similarly, till the integrity of a cell is maintained, the serum magnesium level remains constant postmortem. As hemolysis occurs after death, the serum magnesium level also further increases swiftly to 10–15 mM/l. Though the serum magnesium concentration has been reported to increase postmortem, the concentration varied widely in different sampling sites of the body such as right and left heart, subclavian vein, external iliac vein. Thus, it becomes immensely difficult to estimate PMI from the serum magnesium level. In an interesting study, the serum magnesium concentration has been reported to be very high in serum of victims of fire fatalities, salt water drowning victims, and poisoning by methamphetamine. Hence, besides PMI, serum magnesium level can be useful in determining the cause of death.

The level of both organic and inorganic phosphorous in serum can be used as an indicator of PMI estimation. In human serum, 0.6–0.9 mM/l of inorganic phosphate is found which increases further to 6.6 mM/l after 18 h of death. However, antemortem blood does not contain organic phosphate, hence, estimation of organic phosphate in postmortem blood can be a good indicator of estimating the PMI than estimating the inorganic phosphate in the serum. Similarly, other biochemical metabolites have also been explored to estimate PMI in a human cadaver. The summery of which is given in Fig. 13.1.

4 Molecular Markers in Blood for PMI Estimation

The physiological, biochemical, archaeological and entomological approach of estimation of PMI lack precision, unreliable and demand expertise for authentic estimation of PMI (Sangwan et al. 2021). Thus, with the technological advancements, the degradation pattern of biological markers such as DNA, RNA, and proteins is widely explored for reliable estimation of PMI.

4.1 DNA Markers for PMI Estimation

A total of 25 randomly selected samples with known time of death showed that hepatocyte DNA degradation is better correlated with PMI compared to the spleen and peripheral blood. Similarly, a descendent trend of the amount of DNA has been reported in viscera sample postmortem correlating with the PMI. Cytofluorometric analysis and flow cytometer evaluation of dental pulp tissue have shown that DNA degradation rate is a good indicator of PMI in the late degradation stage. However, the rate of DNA degradation postmortem is tissue specific. The rate of DNA degradation is slower in brain tissues compared to the spleen.

The estimation PMI based on the rate of DNA degradation can be studied by using advanced molecular biology techniques such as flow cytometer, image analysis techniques, single cell gel electrophoresis (SCGE), DNA amplification assay, and other assays such as the terminal deoxynucleotidyl transferase assay (Tozzo et al. 2020). Integration of computer technology, mathematics, optics, and morphological analysis can be employed to determine the relative amount of the substances in a cell. Thus, the measurement of index of distortion (ID), integral optical density (IOD), average optical density (AOD) and average grey (AG) can result in the quantification of DNA. Similarly, SCGE or comet assay is widely used for the determination of DNA integrity. In this technique, various parameters such as tail length (μm) are determined which ultimately determines the size of the DNA fragments (Bhoyar et al. 2025). Real time PCR is considered to be the gold standard technique for the quantification of DNA. Determination of copy number and degradation pattern by RT-PCR allows accurate estimation of PMI by DNA analysis. Additionally, terminal deoxynucleotidyl transferase (TDT) which catalyzes random addition of dNTPs to the 3′-OH of DNA has also been used to detect the DNA for accurate estimation of PMI.

4.2 RNA Markers for PMI Estimation

Compared to DNA, RNA degradation has been widely explored for its application in PMI estimation. RNA being the more liable biological entity to degradation compared to DNA and protein and the rate of degradation increases swiftly after death, has shown its potential in accurate estimation of PMI (Sangwan et al. 2021). mRNA, microRNA, and circRNA have been explored as biomarkers for the estimation of

PMI. Sampaio-Silva et al. (2013) identified the relative quantity of four muscle genes mRNA, i.e., *Actb*, *Gapdh*, *Ppia,* and *Srp72* showing a significant correlation with the PMI. They also developed a mathematical model for accurate estimation of PMI at 95% confidence level. mRNA from β-actin and glyceraldehyde-3-phosphate dehydrogenase (GAPDH) have also shown huge potential in estimating PMI within few hours of death (Cianci et al. 2024). The list of various mRNA genes explored for the estimation of PMI is given in Table 13.1.

Besides mRNA, candidate miRNAs such as miR-9, miR-125b have also shown their potential for accurate estimation of PMI (Cianci et al. 2024). Other miRNA targets including miR-3185, miR-122, miR-133a, and miR-1 have also been explored out of which miR-3185 did not show any correlation with the PMI. Whereas, miR-122, miR-133a, and miR-1 showed stable expression pattern postmortem (Pasaribu et al. 2023). In a recent study, circFat3/mt-co1 and circFat3/28S rRNA models was reported to show accurate prediction for both short-term and long-term PMI. This study showed that the circRNA circFat3 is a novel biomarker which can be used as a supplementary tool for the estimation of PMI (Song et al. 2025). RT-PCR is the most widely used technique for the estimation of PMI using RNAs.

4.3 PMI Estimation by Protein Degradation

Protein degradation approach of estimating PMI has gained importance in recent years. Protein degradation occurs primarily by the hydrolysis of peptide bonds. Inside a living cell, hydrolysis of peptide bonds is mainly catalyzed by the proteolytic enzymes. The non-enzymatic proteolysis governed by calpain system is usually inhibited in living cells by calpastatin and low concentration of calcium in the cytoplasm. Postmortem increase in intracellular calcium level activates calpain, and ATP-independent postmortem proteolysis takes place. Besides, microbial activity and necrophagous insects also facilitate the degradation of proteins postmortem (Zissler et al. 2020). Various key protein components such as tropomyosin, GAPDH, desmin, tubulin, and eEF1A2 have shown huge potential for estimating PMI by studying their degradation patterns. A comprehensive list of protein biomarkers and their degradation pattern useful for estimating PMI is given in Table 13.2.

Table 13.1 List of mRNA genes used for the estimation of PMI

mRNA	Tissue	PMI Significance	Reference
β-actin HIF1α	Brain, heart, liver, skeletal muscle	Between 2 and 24 h	Deng et al. (2013)
GAPDH Hprt Cyp2E1	Heart, muscle, liver	Up to 11 h	Sampaio-Silva et al. (2013)
5S rRNA 18S rRNA	Brain	Up to 24 h	Lü et al. (2016)

Table 13.2 List of protein biomarkers useful for estimating PMI

Protein	Antemortem function	Reference
Troponin T (TnT)	Muscle contraction	Hughes et al. (2024)
Troponin I (TnI)	Regulation of contraction by blocking myosin's attachment to actin	
Desmin	Cytoskeletal protein linking Z bands to membranes	Battistini et al. (2023)
Titin	Structural scaffold in muscle sarcomeres	
Nebulin	Structural scaffold in muscle sarcomeres	
Vinculin	Cytoskeletal protein, regulator of cell adhesion	
α-actinin	Cytoskeletal actin-binding protein	Pittner et al. (2020)
PDZ and LIM domain 7 (PDLIM7)	Scaffold protein, links cytoskeleton to signaling proteins	Battistini et al. (2023)
Tropomyosin alpha-1 chain (TPM1)	Structural constituent of cytoskeleton	
Calcium-transporting ATPase SERCA2 (ATP2A2)	Ca pump utilizing ATP for active transport of Ca^{2+}	
Glyceraldehyde-3-phosphate dehydrogenase (GAPDH)	Crucial glycolytic enzyme	Elghamry et al. (2017)
eEF1A2	Eukaryotic Elongation Factor 1 Alpha 2; essential for protein synthesis	Choi et al. (2019)
K1C13	Keratins	Garcés-Parra et al. (2024)
CO3A1 and PGS1	Collagen-related proteins	

A mass spectrometry approach of analyzing PDLIM7, TPM1, and ATP2A2 proteins was found to be useful in estimating PMI up to 120 h. Besides, the postmortem degradation of desmin and cardiac troponin T has also been reported in humans. Appearance of specific degradation products of calcineurin A, protein phosphatase 2A, alpha-actinin, calsequestrin 1, desmin, nebulin, titin, SERCA 1 and SERCA 2 (sarco-endoplasmic reticulum calcium ATPase), tropomyosin, troponin T cardiac protein (cTnT), laminin and calpain, troponin, tropomyosin, vinculin, AMP-activated protein kinase, caspase 3 and glycogen synthase have also been detected by using SDS-PAGE and western blotting (Battistini et al. 2023).

Enzyme linked immunosorbent assay (ELISA) and western blotting are two widely used techniques for the estimation of candidate protein degradation postmortem. ELISA has been widely used in forensic analysis of suspicious toxins and to assess the postmortem redistribution of toxins. Besides, ELISA has been used in various cases to determine the exact cause and nature of death. Western blotting is usually explored for the quantification of protein expression and can also be employed for the estimation of PMI by protein analysis. Differential expression pattern of autophage related factors such as LC3-II and p62 identified by western blotting have been found useful for PMI estimation during 1–4 days postmortem.

With the advent of omics technology, proteomics approach of abnormal expression pattern has also been explored for the estimation of PMI. Besides, protein postmortem degradation is a complex biochemical phenomenon influenced by a variety of biotic and abiotic factors. The biotic factors may include microbes and parasitic insects, whereas, the abiotic factors include the change in temperature and pH. Thus, the routine protein investigation might not provide an accurate estimation of PMI (Huang et al. 2024). Thus, the multi omics approach involving genomics, proteomics, metabolomics, transcriptomics, and metagenomics can come together to develop a suitable model for accurate estimation of PMI.

5 Conclusion

Determination of postmortem time interval holds significance in defining the direction of a forensic investigation. The conventional methods of estimating PMI explore the early postmortem changes appear in the body including rigor mortis, livor mortis, and algor mortis. Though such techniques are widely used in PMI estimation, such methods are relatively inaccurate and offer limited practical relevance. Many biochemical and molecular techniques have been explored using the postmortem blood samples for the accurate, reliable estimation of PMI. The composition and pH of blood changes postmortem due to the ceasing of physiological activities postmortem. Lysis of blood after death releases various intracellular components such as glucose, lactate, hypoxanthine, bilirubin, uric acid and ammonia. Similarly, many intracellular proteins including Aspartate aminotransferase and Lactate dehydrogenase also increase in the postmortem blood.

Besides the analysis of metabolites and proteins, molecular techniques measuring the level of DNA and RNA in the postmortem blood has shown a huge potential in estimating PMI. The descendent trend of the amount of DNA postmortem has shown its correlation with the PMI. The degradation pattern of DNA can be studied by using flow cytometer, image analysis techniques, single cell gel electrophoresis (SCGE), DNA amplification assay, and other assays such as the terminal deoxynucleotidyl transferase assay. Similarly, RT-PCR based analysis of expression pattern of mRNA, miRNA, and circRNA can also provide crucial information on the PMI. mRNA of *Actb*, *Gapdh*, *Ppia*, *Srp72*, β-actin and glyceraldehyde-3-phosphate dehydrogenase (GAPDH) have shown potential use in PMI estimation. Similarly, various miRNA such as miR-9, miR-125b, miR-3185, miR-122, miR-133a, and miR-1 have been found useful in estimating PMI. Candidate proteins including tropomyosin, GAPDH, desmin, tubulin, and eEF1A2 have also shown huge potential for estimating PMI by studying their degradation patterns. Thus, postmortem blood can be analyzed for estimation of PMI by using proteomics, metabolomics, genomics and transcriptomics approach.

References

Battistini A, Capitanio D, Bailo P, Moriggi M, Tambuzzi S, Gelfi C, Piccinini A (2023) Proteomic analysis by mass spectrometry of postmortem muscle protein degradation for PMI estimation: a pilot study. Forensic Sci Int 349:111774. https://doi.org/10.1016/j.forsciint.2023.111774

Bhoyar L, Mehar P, Chavali K (2025) Assessing the forensic implications of DNA degradation for PMI estimation using comet assay: a systematic review. J Forensic Leg Med 109:102801. https://doi.org/10.1016/j.jflm.2025.102801

Choi KM, Zissler A, Kim E, Ehrenfellner B, Cho E, Lee SI, Steinbacher P, Yun KN, Shin JH, Kim JY, Stoiber W, Chung H, Monticelli FC, Kim JY, Pittner S (2019) Postmortem proteomics to discover biomarkers for forensic PMI estimation. Int J Legal Med 133:899–908. https://doi.org/10.1007/s00414-019-02011-6

Cianci V, Mondello C, Sapienza D, Guerrera MC, Cianci A, Cracò A, Omero F, Gioffrè V, Gualniera P, Asmundo A, Germanà A (2024) Potential role of mRNA in estimating postmortem interval: a systematic review. Int J Mol Sci 25:8185. https://doi.org/10.3390/ijms25158185

Dalal JS, Tejpal HR, Chanana A, Kaur N (2006) Medicolegal study of rigor mortis to estimate postmortem interval. J Indian Acad Forensic Med 28:49–51

Deng W, Lv M, Wang L, Bai P, Liang W, Zhang L (2013) mRNA degradation pattern analysis in post-mortem normalized using the DNA. Forensic Sci Int Genet Suppl Ser 4:e266–e267

Elghamry HA, Mohamed MI, Hassan FM, Abdelfattah DS, Abdelaal AG (2017) Potential use of GAPDH m-RNA in estimating PMI in brain tissue of albino rats at different environmental conditions. Egypt J Forensic Sci 7:24. https://doi.org/10.1186/s41935-017-0024-8

Garcés-Parra C, Saldivia P, Hernández M, Uribe E, Román J, Torrejón M, Gutiérrez JL, Cabrera-Vives G, García-Robles MLÁ, Aguilar W, Soto M, Tarifeño-Saldivia E (2024) Enhancing late postmortem interval prediction: a pilot study integrating proteomics and machine learning to distinguish human bone remains over 15 years. Biol Res 57:75. https://doi.org/10.1186/s40659-024-00552-8

Huang W, Zhao S, Liu H, Pan M, Dong H (2024) The role of protein degradation in estimation postmortem interval and confirmation of cause of death in forensic pathology: a literature review. Int J Mol Sci 25:1659. https://doi.org/10.3390/ijms25031659

Hughes C, Ackland G, Shelley B (2024) Perioperative myocardial injury. BJA Educ 24:352–360. https://doi.org/10.1016/j.bjae.2024.06.001

Lü YH, Li ZH, Tuo Y, Liu L, Li K, Bian J, Ma JL, Chen L (2016) Correlation between RNA degradation patterns of rat's brain and early PMI at different temperatures. Fa Yi Xue Za Zhi 32:165–170. https://doi.org/10.3969/j.issn.1004-5619.2016.03.002

Pasaribu RS, Auerkari EI, Suhartono AW, Auerkari P (2023) A small RNA, microRNA as a potential biomolecular marker to estimate post mortem interval in forensic science: a systematic review. Int J Legal Med 137:1313–1325. https://doi.org/10.1007/s00414-023-03015-z

Pittner S, Gotsmy W, Zissler A, Ehrenfellner B, Baumgartner D, Schrüfer A, Steinbacher P, Monticelli F (2020) Intra- and intermuscular variations of postmortem protein degradation for PMI estimation. Int J Legal Med 134:1775–1782. https://doi.org/10.1007/s00414-020-02355-4

Sampaio-Silva F, Magalhaes T, Carvalho F, Dinis-Oliveira RJ, Silvestre R (2013) Profiling of RNA degradation for estimation of post morterm interval. PLoS One 8:e56507. https://doi.org/10.1371/journal.pone.0056507

Sangwan A, Singh SP, Singh P, Gupta OP, Manas A, Gupta S (2021) Role of molecular techniques in PMI estimation: an update. J Forensic Leg Med 83:102251. https://doi.org/10.1016/j.jflm.2021.102251

Song B, Fu J, Cheng J, Chiampanichayakul S, Anuchapreeda S, Fu J (2025) Circular RNA circFat3 as a biomarker for construction of postmortem interval estimation models in mouse brain tissues at multiple temperatures. Sci Rep 15:21577. https://doi.org/10.1038/s41598-025-07998-0

Strete G, Sălcudean A, Cozma AA, Radu CC (2025) Current understanding and future research direction for estimating the postmortem interval: a systematic review. Diagnostics (Basel) 15:1954. https://doi.org/10.3390/diagnostics15151954

Tozzo P, Scrivano S, Sanavio M, Caenazzo L (2020) The role of DNA degradation in the estimation of post-mortem interval: a systematic review of the current literature. Int J Mol Sci 21:3540. https://doi.org/10.3390/ijms21103540

Zissler A, Stoiber W, Steinbacher P, Geissenberger J, Monticelli FC, Pittner S (2020) Postmortem protein degradation as a tool to estimate the PMI: a systematic review. Diagnostics (Basel) 10:1014. https://doi.org/10.3390/diagnostics10121014

Part III

Advanced Analytical Methods

DNA Profiling from Blood Evidence

14

Abstract

Blood is among the most commonly encountered types of biological evidence in forensic investigations, offering both physical context and a rich source of genetic material. Modern DNA profiling techniques—particularly short tandem repeat (STR) analysis and probabilistic genotyping—have transformed how investigators identify individuals and resolve complex cases. This paper explores the full process of handling blood evidence, from on-scene collection through laboratory analysis, highlighting technical challenges such as contamination, mixture interpretation, and low-template samples. It also reviews key international standards and recent innovations in forensic genetics that are shaping global best practices. Emphasis is placed on the need for consistent protocols, evaluative reporting, and scientifically validated methods to ensure DNA evidence remains robust and reliable in court. By examining both current techniques and emerging tools, this study underscores the importance of integrating scientific rigor with legal accountability in forensic casework.

Keywords

Forensic DNA · STR · Blood evidence · Probabilistic genotyping · Forensic guidelines

1 Introduction

Biological evidence has always played a central role in criminal investigations. Among such materials, blood remains one of the most informative, offering both a genetic profile and contextual clues about how a crime unfolded. Among these, blood is particularly significant because of its prevalence at violent crime scenes and its dual utility as both physical and genetic evidence. While bloodstains may reveal patterns indicative of the manner of assault, they also contain DNA, which provides a unique genetic signature for human identification (Butler 2015).

H. R. Dash et al., *Forensic Investigation of Blood Evidence*,
https://doi.org/10.1007/978-981-95-9183-1_14

Since the introduction of forensic DNA typing in the late 1980s, DNA profiling has become a cornerstone of modern investigations. Its ability to distinguish individuals with a high degree of accuracy has not only aided in criminal prosecutions but also facilitated exonerations through post-conviction reviews. The reliability of these outcomes, however, depends on proper collection, preservation, analysis, and interpretation of blood evidence (Gill et al. 2012).

Before the advent of DNA profiling, forensic serology was the primary tool for analyzing blood evidence. Early techniques included blood group typing (ABO system), enzyme polymorphism testing, and protein-based methods, which provided limited discriminatory power. These approaches could exclude suspects but rarely identify individuals uniquely.

The modern era of DNA profiling began in 1984 when Sir Alec Jeffreys developed the first DNA fingerprinting technique using restriction fragment length polymorphism (RFLP) analysis (Jeffreys et al. 1985). The landmark case of *R v. Pitchfork* (1987) in the United Kingdom demonstrated the power of DNA to both convict a perpetrator and exonerate an innocent suspect. This shifted forensic science from population-level probabilities to individual-level identification, establishing DNA as a gold standard in criminal investigations.

While blood is a rich source of DNA, its spatial patterns at a crime scene can also provide essential clues about the sequence of events. Locating and identifying blood at a crime scene is the first step before DNA analysis. Presumptive tests such as Kastle–Meyer and luminol are widely used for their sensitivity, although they can yield false positives from substances like rust or bleach (Virkler and Lednev 2009). Confirmatory tests, including ABAcard® HemaTrace® immunochromatographic assays, detect human hemoglobin specifically, thereby reducing the risk of misinterpretation.

Bloodstain pattern analysis (BPA) is often performed alongside DNA profiling. The physical distribution of blood—whether spattered, transferred, or pooled—can reconstruct events and guide investigators to relevant samples for DNA recovery (Bevel and Gardner 2008). By combining BPA with DNA profiling, forensic experts can reconstruct both the event timeline and identify the individuals involved.

1.1 Detection of Blood at Crime Scenes

Before blood can be used for DNA profiling, it must first be located. Traditional presumptive tests include colorimetric assays such as Kastle–Meyer (phenolphthalein) and leucomalachite green, both of which react to the peroxidase activity of hemoglobin, producing a rapid color change (Virkler and Lednev 2009). Although highly sensitive, these tests can yield false positives from plant peroxidases or cleaning agents like bleach.

Chemiluminescent methods such as luminol and Bluestar® have been extensively used to detect latent blood traces. Luminol, for example, can reveal cleaned stains by producing a blue glow in the presence of hemoglobin. However, its application may dilute DNA, emphasizing the need for careful collection strategies.

1.2 Confirmatory Testing

Immunochromatographic assays such as ABAcard® HemaTrace® target human hemoglobin specifically, thereby distinguishing human blood from animal sources. The use of confirmatory tests is critical to avoid misinterpretation, especially in cases where environmental contamination may mimic blood reactions.

1.3 Integration with Bloodstain Pattern Analysis

Bloodstain pattern analysis (BPA) complements DNA profiling by providing contextual information about the mechanism of injury. Patterns such as spatter, transfer, and pooling can suggest weapon type, sequence of events, and positions of individuals at the crime scene (Bevel and Gardner 2008). When combined with DNA profiling, BPA allows investigators to answer both the "what happened" and the "who was present" questions, enhancing evidentiary strength.

2 Collection and Preservation of Blood Evidence

The probative value of DNA extracted from blood evidence depends greatly on how the sample is handled from the moment it is discovered. Unlike fingerprints or trace fibers, blood is both a biological fluid and a chemical substance that can degrade rapidly when exposed to unfavorable conditions. Forensic practitioners must therefore follow strict protocols to preserve the genetic integrity of bloodstains. Mishandling at the scene—such as packaging in plastic bags, exposing samples to direct sunlight, or delaying drying—can result in contamination, bacterial growth, or complete loss of usable DNA (Butler 2015).

Figure 14.1 illustrates the workflow of DNA profiling from blood evidence.

The FBI *Handbook of Forensic Services* provides clear guidance: liquid blood samples should be collected on sterile swabs or in tubes containing preservatives such as ethylenediaminetetraacetic acid (EDTA), while dried stains should be carefully cut from substrates or collected with sterile cotton swabs (Suchma et al. 2012). EDTA prevents clotting and inhibits nuclease activity, thereby protecting the DNA until analysis. However, blood collected in anticoagulant-free

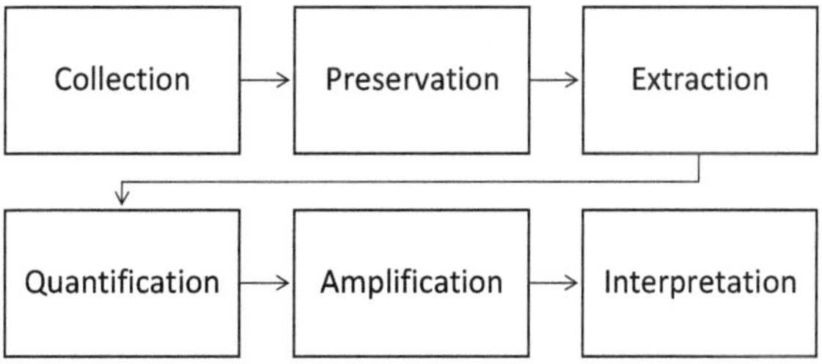

Fig. 14.1 Workflow of DNA profiling from blood evidence

containers may degrade within hours if not properly refrigerated (Linacre and Tobe 2013).

Environmental factors also pose significant risks. High temperatures accelerate enzymatic activity, humidity promotes microbial growth, and exposure to ultraviolet (UV) light causes DNA fragmentation. A study by Alaeddini (2012) demonstrated that blood left at room temperature in humid conditions showed pronounced allelic drop-out within weeks, while stains stored in dry, cool environments remained stable for months. Accordingly, forensic guidelines recommend air-drying all blood samples, packaging them in breathable paper envelopes rather than plastic, and storing them under refrigeration or freezing for long-term preservation (Ballou et al. 2013).

Another challenge is the presence of PCR inhibitors within blood itself. Heme and related compounds can interfere with DNA polymerase activity, reducing amplification efficiency. To counteract this, laboratories employ purification techniques such as silica column extraction or magnetic bead-based methods, which remove inhibitors while concentrating DNA (Al-Soud and Rådström 2001). These steps are critical in cases where only trace amounts of blood are available, such as touch-transfer stains or aged samples.

Standardization of collection and preservation practices is equally important to maintain evidential integrity across jurisdictions. The Organization of Scientific Area Committees (OSAC) stresses strict adherence to contamination prevention, including the use of gloves, masks, and disposable tools, as well as meticulous documentation of collection procedures (Lee et al. 2023). Consistent training and accreditation ensure that evidence collected in one jurisdiction can be reliably analyzed and compared in another.

2.1 Chain of Custody and Contamination Control

Maintaining the chain of custody is essential to uphold the credibility of blood evidence in court. Every transfer—from the crime scene to the laboratory bench—must be logged with time, date, and personnel details. This documentation ensures traceability and prevents defense challenges regarding mishandling. Adherence to ISO/IEC 17025 accreditation requirements provides an additional layer of assurance by mandating quality control systems and routine audits (Willis et al. 2015).

Contamination control is another cornerstone of reliable forensic practice. Secondary transfer of DNA, though less common with blood than with trace epithelial cells, has been documented. For example, bloodstains collected with improperly sterilized instruments may contain DNA from multiple individuals, complicating interpretation (Meakin and Jamieson 2013). Forensic teams now employ single-use swabs, sealed containers, and clean working areas to minimize this risk.

2.2 Sampling Strategies

The substrate on which blood is deposited often dictates the most appropriate sampling method. Porous materials such as fabric or paper allow absorption, making direct cutting or double-swabbing (wet followed by dry swab) effective strategies (Van Oorschot et al. 2019). On nonporous surfaces like glass, metal, or tile, wet swabbing with sterile distilled water increases DNA recovery. Comparative studies have shown that flocked swabs (e.g., FLOQSwabs®) outperform traditional cotton swabs by maximizing cellular capture and DNA yield.

In addition to swabbing, adhesive tape lifts and cutting of small substrate portions are sometimes employed, especially in cases where stains are faint or microscopic. Each method must balance maximizing DNA recovery with minimizing damage to the sample, preserving enough for both forensic analysis and potential retesting.

2.3 Long-Term Storage

For long-term preservation, proper packaging is paramount. Blood samples should never be sealed in plastic bags unless fully dried, as residual moisture can promote bacterial growth and DNA degradation. Breathable paper envelopes or cardboard boxes with desiccants are preferred to maintain dryness (Ballou et al. 2013).

Refrigeration at 4 °C is adequate for short-term storage, while freezing at −20 °C or – 80 °C ensures long-term stability. Research indicates that DNA in dried bloodstains can remain amplifiable for decades under optimal storage conditions, provided samples are shielded from light and moisture (Alaeddini 2012). Thanks to modern techniques, analysts can now recover useful genetic information even from older or poorly stored blood samples.

Table 14.1 shows the comparison of DNA extraction methods from blood Samples.

Table 14.1 Comparison of DNA Extraction Methods from Blood Samples

Method	Principle	DNA yield	Inhibitor removal	Suitability for degraded samples
Silica column extraction	Binds DNA to silica in high salt	High	Moderate	Moderate
Magnetic beads	DNA binds to coated beads	Very high	High	High
Phenol-chloroform	Organic extraction	High	Low	Poor
Chelex 100	Heat and resin-based	Moderate	Low	Poor

3 Analytical Methods in DNA Profiling

Forensic DNA profiling has evolved considerably over the past three decades, moving from single-locus markers to highly multiplexed systems capable of producing profiles from challenging samples. Blood, being a common biological material at crime scenes, is particularly valuable for profiling due to its abundance of nucleated cells. However, the success of analysis depends on the selection of appropriate methods suited to the condition and quantity of DNA.

3.1 Short Tandem Repeats (STRs)

Short tandem repeats (STRs) are the most widely used genetic markers in forensic identification. These are short sequences of DNA, typically 2–6 base pairs in length, repeated in tandem arrays. The high variability in the number of repeats among individuals provides a powerful tool for human identification (Butler 2015).

The Combined DNA Index System (CODIS), maintained by the FBI, initially required 13 core loci but was expanded to 20 loci in 2017 to enhance discrimination and improve international compatibility (Hares 2015). This expansion reduced the likelihood of coincidental matches in large databases and increased robustness in kinship testing. STR kits such as GlobalFiler™ and PowerPlex® Fusion 6C now enable the simultaneous amplification of these loci in a single multiplex PCR reaction, increasing efficiency while conserving sample material.

Table 14.2 illustrates different types of STR kits used in forensic blood profiling.

3.2 Low-Template and Degraded DNA

Blood evidence is often recovered from compromised environments, resulting in low-template or degraded DNA. Modified PCR methods and miniSTR assays target shorter fragments to improve success rates in such samples (Butler 2015).

The introduction of probabilistic genotyping software has sparked debate in forensic and legal communities. While these systems provide a statistically rigorous framework for complex mixtures, their proprietary algorithms have been criticized for lack of transparency. Nonetheless, courts in the United States, the United Kingdom, and New Zealand have increasingly admitted probabilistic DNA

Table 14.2 Types of STR kits used in forensic blood profiling

Kit name	Core loci covered	Multiplex capacity	Use case
GlobalFiler™	21 autosomal STRs	High	National database compatibility
PowerPlex® Fusion	24 loci	Very high	Complex mixture deconvolution
NGM Select™	16 loci	Moderate	European standard profiles

evidence, provided that validation studies and expert testimony accompany its use (Bright et al. 2017; Gill et al. 2012).

A landmark example includes the use of TrueAllele® in the Pennsylvania cold case, where degraded blood evidence produced a match probability exceeding one in a billion, leading to conviction decades after the crime (Perlin et al. 2014).

3.3 Probabilistic Genotyping

Probabilistic genotyping has transformed the way complex DNA profiles are interpreted. Traditional binary approaches—"include" or "exclude"—were often insufficient when interpreting mixtures or low-template samples. Probabilistic software such as STRmix™, EuroForMix, and TrueAllele® use statistical models to evaluate the probability of observed profiles under competing hypotheses, generating likelihood ratios (Bright et al. 2017).

For instance, in the Pennsylvania cold case, TrueAllele® was applied to degraded blood evidence, producing a match probability of over one in a billion and leading to conviction decades later (Perlin et al. 2014). These systems have been increasingly admitted in courts across the U.S., U.K., and New Zealand, though concerns about proprietary algorithms and transparency continue to prompt debate (Buckleton et al. 2019).

3.4 Y-STR and X-STR Analysis

Autosomal STRs remain the standard, but in specific scenarios, sex-chromosome STRs provide critical information. Y-STR profiling is highly useful in sexual assault cases involving mixed male-female DNA, as it isolates paternal lineages without interference from the female background (Purps et al. 2014). This has also been applied in cases of mass disasters and missing persons where paternal kinship is the only link available.

X-STRs, although less commonly used, are valuable in kinship testing, particularly when female inheritance patterns are relevant. They play a role in deficient pedigree cases such as disputed sibling relationships or missing father cases where only maternal relatives are available (Szibor et al. 2003). Figure 14.2 demonstrates how single-source and mixed DNA profiles appear in electropherogram form.

3.5 Mitochondrial DNA (mtDNA)

When nuclear DNA is highly degraded or absent, mitochondrial DNA (mtDNA) offers an alternative. Each cell contains hundreds to thousands of mtDNA copies, making it more likely to survive harsh environmental conditions. Sequencing the hypervariable regions (HVR-I and HVR-II) of the mtDNA control region is common in forensic casework, particularly for old bloodstains, skeletal remains, or hair shafts lacking roots (Parson and Bandelt 2007).

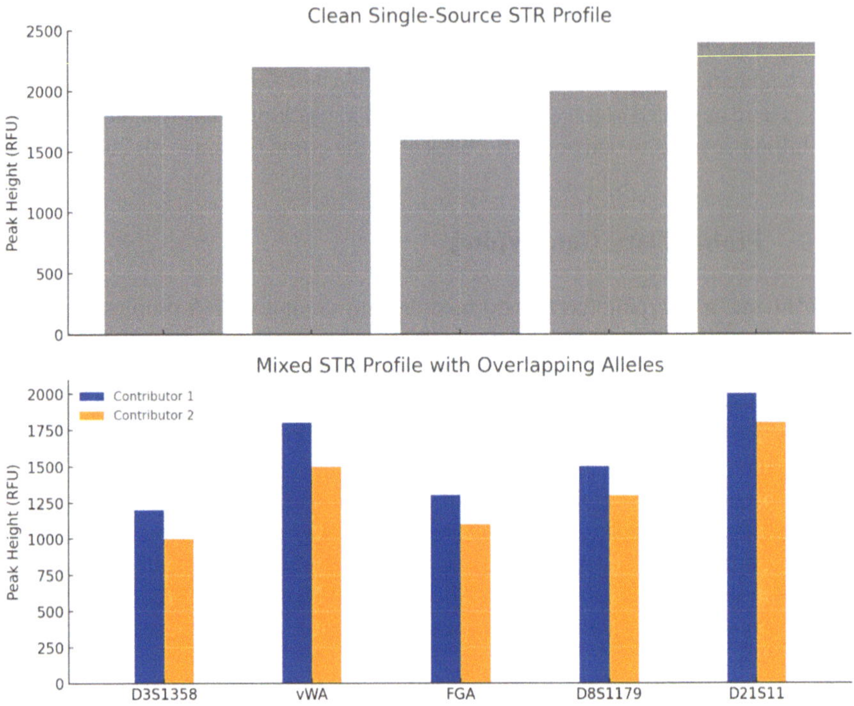

Fig. 14.2 STR profile example from blood evidence (clean vs. mixed). The *top chart* illustrates a clean single-source profile with distinct peaks for each STR locus. The *bottom chart* demonstrates a mixed profile from two contributors, showing overlapping alleles at each locus—common in forensic casework involving multiple individuals

Although mtDNA has lower discriminatory power—since maternal relatives share the same profile—it has been invaluable in disaster victim identification (DVI) and missing persons investigations. For example, mtDNA analysis was extensively used following the 2004 Indian Ocean tsunami and in historic identifications such as the remains of the Romanovs (James 2005).

3.6 Single Nucleotide Polymorphisms (SNPs)

Single nucleotide polymorphisms (SNPs) represent the most abundant form of genetic variation in the human genome. While less polymorphic than STRs individually, panels of SNPs can provide strong discriminatory power. Their small amplicon size also makes them well-suited for degraded DNA from old or compromised bloodstains.

Beyond identity testing, SNPs enable ancestry inference and phenotypic prediction. The HIrisPlex-S system, for example, can predict eye, hair, and skin color with high accuracy, providing investigative leads in cases lacking suspect profiles. This makes SNPs particularly relevant in "no suspect" cases or when blood evidence fails to match any existing database entries.

3.7 Next-Generation Sequencing (NGS)

Next-generation sequencing (NGS) represents a major technological leap in forensic genetics. Unlike traditional capillary electrophoresis-based STR typing, NGS simultaneously sequences multiple marker types—including STRs, SNPs, and mtDNA—in a single assay. This provides richer data, such as sequence variation within STR alleles that cannot be detected by length-based methods alone (Parson et al. 2016).

NGS has shown particular value in mixture deconvolution, ancestry estimation, and the analysis of highly degraded blood samples. Pilot studies using Illumina's MiSeq FGx platform demonstrated successful profiling from compromised and aged stains, suggesting NGS will become a standard in forensic practice as validation studies accumulate (Parson et al. 2016).

3.8 Epigenetic Markers

Emerging epigenetic markers extend forensic DNA analysis beyond identity. DNA methylation patterns can reveal the age of an individual or estimate the time since deposition of a bloodstain, offering investigators temporal information previously unattainable (Vidaki and Kayser 2018).

For example, specific CpG sites in blood DNA show consistent methylation changes with chronological age, allowing estimations within a margin of a few years. Similarly, RNA and microRNA signatures are being studied to determine the freshness of bloodstains, distinguishing between stains deposited hours, days, or months apart (Haas et al. 2009). These tools, though still experimental, represent the future of forensic inference.

4 Interpretation Challenges

4.1 Mixtures

Mixed blood samples pose unique interpretive challenges. When multiple individuals contribute DNA, the resulting profiles often overlap, making it harder to assign alleles to specific contributors. Analysts must carefully assess peak heights, stutter patterns, and contributor ratios to ensure accurate interpretation. Analysts typically start with a transparent set of assumptions: the number of contributors (NoC), whether to *condition on a known contributor* (e.g., a victim), and the propositions to be evaluated (e.g., *prosecution*: suspect contributed to the mixture; *defense*: an unknown person did) (Gill et al. 2012). Peak-height information, stutter modeling, and heterozygote balance help separate contributors, but these must be applied consistently and supported by validation data. Modern guidance favors a *likelihood-ratio (LR) framework* rather than categorical "match" language so that evaluative opinions remain reproducible and auditable (Gill et al. 2012; Willis et al. 2015). Open-source and validated tools (e.g., EuroForMix) and thoroughly validated commercial systems (e.g., STRmix™, TrueAllele®) estimate LRs by explicitly modeling mixture complexity and peak-height variability (Bright et al. 2017; Bleka et al. 2016).

Good practice in mixtures

- Use *quantitation and degradation indices* to inform expectations before interpretation (e.g., if high degradation is present, miniSTR/NGS may be prioritized).
- Consider *conditioning* on a known major contributor (e.g., the victim's blood) when appropriate—it can sharpen the LR for or against a putative minor contributor (Bright et al. 2017).
- Explore *alternative contributor counts* and report how conclusions vary as NoC changes; document sensitivity analyses (Willis et al. 2015; Butler et al. 2024).

4.2 PCR Inhibition

Blood contains *heme/hemoglobin* and other substances that *inhibit DNA polymerase*, causing drop-out, low peak heights, or locus-to-locus imbalance (Al-Soud and Rådström 2001). Analysts watch for classic signs—late Ct values in qPCR, locus-specific failures, and poor internal control performance—and respond with *clean-up chemistries* (silica columns, magnetic beads), *inhibitor-tolerant polymerases*, or additional purification (Opel et al. 2010; Sidstedt et al. 2020). Where inhibition cannot be fully relieved, *interpretation thresholds* and stochastic models should reflect the increased risk of drop-out (Gill et al. 2012; Butler et al. 2024).

4.3 Drop-in and Drop-out

At low template levels, *drop-out* (missing alleles) and *drop-in* (spurious small peaks) become more likely. Rather than excluding a person of interest due to a missing allele, probabilistic approaches *model the chance of drop-out* given the sample's quality and peak-height variance, producing an LR that incorporates this uncertainty (Bright et al. 2017; Gill et al. 2012). Replicate amplifications and consensus calling can reduce the impact of stochastic events, but the preferred solution is *formally modeling* the stochasticity (Butler et al. 2024).

4.4 Secondary Transfer of Blood DNA

Although *blood* transfers less readily than trace epithelial DNA, *secondary transfer* can occur—for example, a shoe steps in blood, and later the wearer touches another surface (Meakin and Jamieson 2013). Interpretation should stay tied to case circumstances: BPA may indicate whether deposition mechanisms match the narrative, and evaluative reporting should keep the alternative proposition (secondary transfer) explicit within the LR framework (Willis et al. 2015). Where feasible, analysts should also consider *activity-level propositions* (e.g., *how* DNA arrived), not just source-level questions (Aitken and Taroni 2004; Willis et al. 2015).

4.5 Degradation and Heat/UV Damage

Warm, humid environments and UV exposure *break DNA into shorter fragments*, stressing longer STR amplicons first. This produces a *"ski-slope" profile* (loss of longer loci) and heterozygote imbalance. Two practical responses are (i) *miniSTRs* that target shorter regions, and (ii) *NGS panels* that recover signal from very short fragments (Butler 2015). Quantitation kits that report a *degradation index* help set expectations before amplification, guiding the analyst toward the most suitable chemistry and interpretation plan.

4.6 Quality Controls and Contamination Management

Reagent blanks, extraction negatives, and elimination databases help detect or explain unexpected alleles. Any shared peaks between controls and evidence trigger a documented *contamination inquiry* and may constrain the propositions considered (Willis et al. 2015; Butler et al. 2024). When case facts allow, *conditioning on the victim* can simplify mixture interpretation in blood-heavy scenes while maintaining transparency about assumptions (Willis et al. 2015).

5 Case Application

Forensic scientists have used blood-based DNA profiling to solve crimes and correct miscarriages of justice. Cold cases dating back decades have been reopened with the advent of advanced genotyping, leading to convictions when partial bloodstains were successfully analyzed. Similarly, organizations such as the Innocence Project have leveraged blood DNA profiles to overturn wrongful convictions where earlier serological methods failed (Butler 2015).

One of the most significant impacts of blood DNA profiling has been in wrongful conviction cases. The Innocence Project reports that over 375 exonerations in the United States since 1989 were facilitated by DNA testing, with blood evidence playing a central role in many cases (Bowen 2024).

Cold cases have also been solved with modern methods. For example, re-analysis of blood evidence using STRmix™ in a 1990s homicide allowed deconvolution of a four-person mixture, resulting in a likelihood ratio exceeding 10^14, which upheld conviction despite initial legal challenges (Bright et al. 2017).

The role of blood DNA profiling in justice systems is exemplified by several landmark cases:

- *People v. Puckett (2010):* A cold case from the 1970s was solved using DNA profiling from a degraded blood sample, highlighting both the power and the limitations of low-template analysis (Butler 2015).
- *Pennsylvania Cold Case (2011):* Probabilistic genotyping of a decades-old bloodstain produced match probabilities exceeding one in a billion, resulting in conviction (Perlin et al. 2014).
- *Innocence Project Exonerations:* Blood DNA evidence has been central to post-conviction reviews, overturning wrongful convictions in over 375 U.S. cases since 1989 (Bowen 2024).

Blood DNA profiling has been instrumental in large-scale disaster victim identification (DVI). Following the 9/11 terrorist attacks, forensic teams analyzed thousands of fragmented and degraded blood and tissue samples, applying miniSTRs and mitochondrial DNA sequencing to provide identifications where conventional methods failed (Butler 2015). Similarly, after the 2004 Indian Ocean tsunami, blood and bone samples were analyzed using STRs and mtDNA to assist in the identification of thousands of victims across multiple countries (James 2005).

Regional case law has also shaped the admissibility of probabilistic genotyping from blood evidence. In the United States, *People v. Collins* (2015) marked one of the first cases where STRmix™-generated likelihood ratios were accepted as evidence, reinforcing probabilistic methods as legally defensible (Buckleton et al. 2019). In contrast, European jurisdictions have emphasized transparency and validation before full courtroom adoption, highlighting differences in international practice.

6 Ethical and Legal Considerations

While DNA profiling from blood evidence has transformed forensic science, its application raises profound ethical and legal questions. Unlike many other forensic markers, DNA contains not only identifiers for human identity but also information about ancestry, kinship, and potentially health-related traits. As such, the storage, use, and interpretation of DNA evidence extend beyond scientific considerations and touch directly upon *civil liberties, privacy rights, and the integrity of the justice system* (Kayser and De Knijff 2011).

6.1 Privacy and Data Protection

The most persistent ethical concern is the *retention of DNA profiles in national databases*. In several countries, profiles from suspects who are later acquitted or never charged remain in national repositories, creating privacy and discrimination risks. The landmark case *S. and Marper v. United Kingdom* (2008) before the European Court of Human Rights found that indefinite retention of nonconvicted individuals' DNA violated the right to privacy under Article 8 of the European Convention on Human Rights. This ruling forced the U.K. to adopt stricter time limits and removal policies for database entries.

By contrast, the United States has expanded its Combined DNA Index System (CODIS) through the *DNA Identification Act* and subsequent state legislation, permitting broader inclusion criteria, including some categories of arrestees. Proponents argue this enhances public safety, but critics contend it disproportionately affects marginalized groups and risks embedding systemic bias within genetic surveillance infrastructures (Greely 2020).

6.2 Familial Searching and Investigative Genetic Genealogy

The rise of *familial searching*—where partial matches in DNA databases point to relatives of potential suspects—raises additional ethical challenges. While powerful in identifying perpetrators, it places innocent family members under indirect genetic surveillance without their consent. A more controversial development is *Investigative Genetic Genealogy (IGG)*, in which law enforcement agencies search consumer ancestry databases to identify suspects through distant relatives. The Golden State Killer case famously relied on this method, demonstrating its investigative utility but also sparking debate about informed consent, data ownership, and the blurring of boundaries between recreational genetics and criminal justice (Greytak et al. 2019).

6.3 Transparency in Probabilistic Genotyping

Probabilistic genotyping systems, increasingly used to interpret complex blood DNA mixtures, also pose legal challenges. Courts demand that results be explained in accessible terms, typically through the *likelihood ratio framework*, rather than absolute or categorical claims of guilt (Gill et al. 2012). However, concerns have arisen about the *proprietary nature* of some algorithms, which limits defense access to source code for independent evaluation. Legal scholars argue that full transparency is necessary to uphold defendants' rights to challenge scientific evidence (Buckleton et al. 2019).

6.4 Global Disparities in Database Policies

International differences highlight the lack of consensus on how to balance the *probative value of DNA with privacy rights*. The U.K. maintains one of the largest forensic DNA databases in the world, while EU member states have introduced stricter retention rules influenced by human rights law. In contrast, the U.S. continues to prioritize expansion, with limited restrictions on retention. Organizations such as the European Network of Forensic Science Institutes (ENFSI) have called for *harmonization of standards* across borders to ensure compatibility in cross-national criminal investigations while respecting civil liberties (Doyle 2018).

6.5 Universal Databases Debate

The notion of a *universal DNA database*—including all citizens or all newborns—remains hotly contested. Advocates argue that it would eliminate racial or demographic bias by treating all individuals equally and would significantly reduce investigative bottlenecks. Critics, however, warn of *state overreach, surveillance abuse, and the erosion of privacy rights*. Ethical discourse increasingly emphasizes that the expansion of forensic capabilities must be paired with *robust oversight, independent auditing, and legal safeguards* to maintain public trust.

7 Future Directions

New breakthroughs in biology, software, and international coordination are reshaping how DNA evidence is handled worldwide. Blood evidence, traditionally analyzed through STR profiling, will increasingly be evaluated with *multilayered genomic and epigenomic approaches*, supported by artificial intelligence and harmonized international standards. These advances promise not only higher sensitivity and resolution but also more contextual insights, such as a donor's age, ancestry, or the timing of blood deposition.

7.1 Next-Generation Sequencing (NGS)

NGS technologies have already demonstrated their value in forensic applications. Unlike capillary electrophoresis, which measures allele length, NGS can capture *sequence-level variation within STR alleles*, thereby increasing discriminatory power (Parson et al. 2016). Importantly, NGS panels also permit simultaneous analysis of STRs, SNPs, and mtDNA, enabling investigators to extract *maximum information from minimal or degraded blood samples*. As costs decrease and validation studies accumulate, portable NGS instruments are likely to be deployed at crime scenes, accelerating turnaround times in major investigations (Butler 2015).

7.2 Epigenetics and Molecular Aging

Epigenetic research has expanded forensic possibilities beyond identification. *DNA methylation profiling* allows for the estimation of a donor's chronological age with a margin of a few years (Vidaki and Kayser 2018). Similarly, *RNA and microRNA (miRNA) expression patterns* can provide time-since-deposition (TSD) estimates for bloodstains, distinguishing whether they were deposited hours, days, or months earlier (Haas et al. 2009). These tools may prove invaluable in reconstructing crime scene timelines and verifying witness statements. Although still under development, epigenetic markers are moving steadily toward forensic validation and courtroom admissibility.

7.3 Artificial Intelligence and Robotics

Artificial intelligence (AI) is expected to *reshape DNA interpretation workflows*. Machine learning models are already being trained to automate STR peak calling, mixture deconvolution, and anomaly detection in electropherograms, minimizing human subjectivity (Luo et al. 2024). In combination with robotic sample processing, AI could streamline evidence handling and enable *real-time, on-site profiling*. For example, prototype devices now combine portable sequencing with AI-based interpretation, offering a future where investigators can generate interpretable DNA profiles directly at crime scenes without waiting for centralized laboratory processing.

7.4 Global Harmonization and Databasing

The globalization of crime and cross-border investigations highlights the need for *internationally harmonized DNA standards*. Organizations such as INTERPOL and ENFSI advocate for standardized collection, interpretation, and reporting frameworks to ensure compatibility across jurisdictions (Doyle 2018). Such harmonization is particularly critical in disaster victim identification (DVI), terrorism investigations, and organized crime cases, where blood evidence collected in one country may need to be analyzed or compared in another. Future frameworks may include not only standardized STR loci but also guidelines for epigenetic markers and AI-driven interpretation systems.

7.5 Ethical Frontiers in Emerging Technologies

With the expansion of capabilities, ethical considerations remain at the forefront. Real-time DNA sequencing at crime scenes, for example, raises questions about *data security, privacy, and potential misuse* if raw genomic data are stored or shared. Similarly, the application of epigenetic age prediction or phenotype inference must balance investigative utility with the risk of stigmatization or discrimination (Greely 2020). Building strong regulatory and oversight frameworks alongside scientific development will be essential to maintaining public trust.

8 Conclusion

DNA profiling from blood evidence continues to shape the landscape of forensic science, bridging biology, law, and investigative work. Its reliability in matching suspects, identifying victims, and uncovering wrongful convictions makes it one of the most transformative tools in modern criminal justice. Yet, the accuracy of these results hinges on meticulous sample handling—from collection to preservation, extraction, and interpretation.

Short Tandem Repeat (STR) analysis remains the gold standard for forensic DNA profiling, offering high discriminatory power. However, interpreting complex mixtures, handling low-template or degraded samples, and accounting for contamination require both technical expertise and contextual judgment. Advances in probabilistic genotyping, software-aided deconvolution, and new approaches like microhaplotypes and AI-based analysis are improving these capabilities. These innovations are expanding forensic possibilities, particularly in cold cases, mass disasters, and degraded evidence scenarios.

Nevertheless, as these technologies advance, so do the responsibilities of forensic scientists. DNA is not infallible; misinterpretation, bias, or procedural lapses can carry profound consequences—particularly in courtrooms. Ethical dilemmas arise when considering familial searching, databases, and long-term storage of genetic data. Practitioners must navigate not only scientific protocols but also the societal implications of their decisions.

Education, transparency, and continuous validation are essential. Legal professionals and forensic analysts alike must stay informed about the strengths and limitations of DNA evidence. Policymakers must also adapt regulations that protect individual rights while allowing room for scientific advancement. International cooperation and standardization—through bodies like ISFG and ENFSI—will further reinforce consistency and fairness in global forensic practices.

Ultimately, the promise of DNA profiling lies not just in its scientific precision, but in how responsibly it is applied. Ensuring justice requires more than technological power—it demands ethical awareness, critical thinking, and a steadfast commitment to accuracy. As DNA analysis continues to evolve, so too must the values and systems that guide its use.

References

Aitken C, Taroni F (2004) Statistics and the evaluation of evidence for forensic scientists, 2nd edn. Wiley

Alaeddini R (2012) Forensic implications of PCR inhibition—a review. Forensic Sci Int Genet 6(3):297–305

Al-Soud WA, Rådström P (2001) Purification and characterization of PCR-inhibitory components in blood cells. J Clin Microbiol 39(2):485–493. https://doi.org/10.1128/JCM.39.2.485-493.2001

Ballou SM, Kline MC, Stolorow MD, Taylor M, Williams S, Bamberger PS et al (2013) The biological evidence preservation handbook: best practices for evidence handlers. NIST

Bevel T, Gardner RM (2008) Bloodstain pattern analysis: with an introduction to crime scene reconstruction, 3rd edn. CRC Press/Taylor & Francis, Boca Raton, FL

Bleka Ø, Storvik G, Gill P (2016) EuroForMix: an open source software based on a continuous model to evaluate STR DNA profiles from a mixture of contributors with artefacts. Forensic Sci Int Genet 21:35–44

Bowen RT (2024) Unjust: wrongful convictions. In: Ethics and the practice of forensic science. CRC Press, pp 217–240

Bright JA, Taylor D, Gittelson S, Buckleton J (2017) The paradigm shift in DNA profile interpretation. Forensic Sci Int Genet 31:e24–e32

Buckleton JS, Bright JA, Gittelson S, Moretti TR, Onorato AJ, Bieber FR et al (2019) The probabilistic genotyping software STR mix: utility and evidence for its validity. J Forensic Sci 64(2):393–405

Butler JM (2015) Advanced topics in forensic DNA typing: interpretation. Academic Press. https://doi.org/10.1016/C2011-0-07495-1

Butler JM et al (2024) DNA mixture interpretation: a NIST scientific foundation review. National Institute of Standards and Technology, Gaithersburg, MD), NIST IR 8351. https://doi.org/10.6028/NIST.IR.8351

Doyle S (2018) Quality management in forensic science. Academic Press

Gill P, Gusmão L, Haned H, Mayr WR, Morling N, Parson W et al (2012) DNA commission of the International Society of Forensic Genetics: recommendations on the evaluation of STR typing results that may include drop-out and/or drop-in using probabilistic methods. Forensic Sci Int Genet 6(6):679–688

Greely HT (2020) The future of DTC genomics and the law. J Law Med Ethics 48(1):151–160

Greytak EM, Moore C, Armentrout SL (2019) Genetic genealogy for cold case and active investigations. Forensic Sci Int 299:103–113. https://doi.org/10.1016/j.forsciint.2019.03.039

Haas C, Klesser B, Maake C, Bär W, Kratzer AM (2009) mRNA profiling for body fluid identification by reverse transcription endpoint PCR and realtime PCR. Forensic Sci Int Genet 3(2):80–88

Hares DR (2015) Selection and implementation of expanded CODIS core loci in the United States. Forensic Sci Int Genet 17:33–34

James H (2005) Thai tsunami victim identification-overview to date. J Forensic Odontostomatol 23(1):1–18

Jeffreys AJ, Wilson V, Thein SL (1985) Individual-specific "fingerprints" of human DNA. Nature 316(6023):76–79

Kayser M, De Knijff P (2011) Improving human forensics through advances in genetics, genomics and molecular biology. Nat Rev Genet 12(3):179–192

Lee HC, Palmbach TM, Ambers A, Primorac D, Anđelinović Š (2023) Collection and preservation of physical evidence. In: Forensic DNA applications. CRC Press, pp 203–224

Linacre A, Tobe S (2013) Wildlife DNA analysis: applications in forensic science

Luo M, Yang W, Bai L, Zhang L, Huang JW, Cao Y et al (2024) Artificial intelligence for life sciences: a comprehensive guide and future trends. The Innovation Life 2(4):100105–100101

Meakin GE, Jamieson A (2013) DNA transfer: review and implications for casework. Forensic Sci Int Genet 7(4):434–443. https://doi.org/10.1016/j.fsigen.2013.03.015

Opel KL, Chung D, McCord BR (2010) A study of PCR inhibition mechanisms using real time PCR. J Forensic Sci 55(1):25–33

Parson W, Bandelt HJ (2007) Extended guidelines for mtDNA typing of population data in forensic science. Forensic Sci Int Genet 1(1):13–19. https://doi.org/10.1016/j.fsigen.2006.11.003

Parson W, Ballard D, Budowle B, Butler JM, Gettings KB, Gill P et al (2016) Massively parallel sequencing of forensic STRs: considerations of the DNA commission of the International

Society for Forensic Genetics (ISFG) on minimal nomenclature requirements. Forensic Sci Int Genet 22:54–63
Perlin MW, Dormer K, Hornyak J, Schiermeier-Wood L, Greenspoon S (2014) TrueAllele casework on Virginia DNA mixture evidence: computer and manual interpretation in 72 reported criminal cases. PLoS One 9(3):e92837
Purps J, Siegert S, Willuweit S, Nagy M, Alves C, Salazar R et al (2014) A global analysis of Y-chromosomal haplotype diversity for 23 STR loci. Forensic Sci Int Genet 12:12–23
Sidstedt M, Rådström P, Hedman J (2020) PCR inhibition in qPCR, dPCR, and MPS—mechanisms and solutions. Forensic Sci Int Genet 48:102233
Suchma KH, An FBI, Quantico V (2012) Handbook of forensic services. K. Waggoner (Ed.). FBI Laboratory Publication
Szibor R, Krawczak M, Hering S, Edelmann J, Kuhlisch E, Krause D (2003) Use of X-linked markers for forensic purposes. Int J Legal Med 117(2):67–74
Van Oorschot RA, Szkuta B, Meakin GE, Kokshoorn B, Goray M (2019) DNA transfer in forensic science: a review. Forensic Sci Int Genet 38:140–166
Vidaki A, Kayser M (2018) Recent progress, methods and perspectives in forensic epigenetics. Forensic Sci Int Genet 37:180–195. https://doi.org/10.1016/j.fsigen.2018.08.008
Virkler K, Lednev IK (2009) Raman spectroscopic signature of semen and its potential application to forensic body fluid identification. Forensic Sci Int 193(1–3):56–62
Willis SM, McKenna L, McDermott S, O'Donell G, Barrett A, Rasmusson B et al (2015) ENFSI guideline for evaluative reporting in forensic science. European Network of Forensic Science Institutes

Microbial Forensics: Analysis of Bloodstain Microbiome

15

Abstract

Microbial forensics is an emerging discipline that integrates microbiological and molecular approaches to support forensic investigations. The human microbiome—comprising diverse, individualized, and relatively stable microbial communities—offers promising potential as a biological identifier. Among its applications, bloodstain microbiome analysis provides a valuable avenue for cases where conventional DNA evidence is degraded, limited, or unavailable. This chapter explores the forensic relevance of bloodstain microbiome profiling, with emphasis on its utility in postmortem interval (PMI) estimation, source attribution, and individual discrimination. Recent advancements in next-generation sequencing (NGS), metagenomics, and bioinformatics are reviewed, along with standardized workflows for sample collection, DNA extraction, 16S rRNA sequencing, and computational analysis. Experimental studies simulating environmental and temporal influences on bloodstains are also discussed.

Bloodstains demonstrate characteristic microbial succession patterns that change over time and are shaped by environmental variables such as temperature, humidity, and substrate. These dynamic microbial shifts can be quantified to support PMI estimation, differentiate human from nonhuman blood sources, and associate individuals with specific environments. Despite encouraging progress, challenges related to standardization, contamination control, and database limitations remain. Bloodstain microbiome analysis holds significant promise for advancing microbial forensics by extending the evidentiary value of biological samples and offering novel parameters for forensic interpretation. Continued integration of multiomics technologies, AI-driven analytics, and robust reference databases will be essential for its transition into routine forensic practice.

H. R. Dash et al., *Forensic Investigation of Blood Evidence*,
https://doi.org/10.1007/978-981-95-9183-1_15

Keywords

Microbial Forensics • Bloodstain Microbiome • 16S rRNA Sequencing • Postmortem Interval • Metagenomics • Forensic Microbiology • DNA Evidence • Microbial Succession

1 Introduction

Forensic science applies scientific principles and analytical techniques to investigate crimes, identify victims or suspects, and reconstruct events at a crime scene. Traditionally, investigations have relied heavily on physical and biological evidence such as fingerprints, human DNA, blood, hair, or chemical residues. While these approaches have proven highly effective, they also face inherent limitations. Biological materials degrade quickly when exposed to heat, humidity, light, microbial activity, or contamination. In many cases, available evidence is minimal, compromised, or partially degraded, making it difficult to obtain reliable results (Houck and Siegel 2009).

To overcome these challenges, researchers have turned toward microbial forensics, an emerging discipline that utilizes microorganisms as forensic markers. Microbes are ubiquitous—on the skin, within the body, on objects we interact with, and throughout the environment. Each individual carries a unique and relatively stable microbial community, or microbiome, shaped by genetics, lifestyle, diet, health status, and environmental exposure. These microbial signatures can serve as individualized biological indicators. By analyzing microbial patterns, investigators can infer identity, trace interactions between individuals and objects, determine environmental context, and reconstruct events. Microbial forensics thus complements traditional forensic methods by integrating molecular biology with evidence interpretation (Breeze et al. 2005).

1.1 Overview of Microbial Forensics

Microbial forensics is a specialized branch of forensic science focused on identifying, characterizing, and interpreting microbial communities for investigative purposes. Unlike conventional techniques that rely primarily on human DNA, fingerprint analysis, or chemical evidence, microbial forensics explores the complex microbial ecosystems associated with humans, environments, and objects. These microbial communities can function as biological fingerprints, offering multilayered insights that link suspects, victims, and crime scenes (Budowle et al. 2017).

The field gained prominence following biocrime incidents, most notably the 2001 anthrax attacks in the United States, which underscored the importance of tracing pathogens to their source. Since then, microbial forensic approaches have extended to routine criminal investigations, including homicide, sexual assault, theft, and environmental crimes. By examining microbial diversity, abundance, and ecological succession, investigators can reconstruct sequences of events, identify contaminated areas, and uncover patterns that may not be visible through traditional methods.

A major strength of microbial forensics lies in its ability to complement and extend conventional evidence. While human DNA provides information about identity, microbial signatures reveal contextual details such as environmental exposure, surface contact, and temporal changes. Furthermore, microbial DNA is often more resilient than human DNA, surviving harsh environmental conditions including heat, UV radiation, moisture, and chemical exposure. Consequently, microbial evidence remains valuable even when conventional biological evidence is absent or degraded (De-Dios et al. 2020).

1.2 Relevance of Microbiome Analysis in Forensic Investigations

Advancements in next-generation sequencing (NGS), metagenomics, and bioinformatic analysis have significantly expanded the forensic applications of microbiome research (Metcalf et al. 2016; Javan et al. 2017). Microbial profiling is increasingly used in:

- *Postmortem Interval (PMI) Estimation:*
- Microbial communities change in predictable patterns during decomposition. These temporal shifts can improve PMI estimation, often with greater reliability than traditional morphological or entomological indicators.
- *Body Fluid Identification:*
- Distinct microbial signatures have been associated with blood, saliva, semen, and other biological fluids. Microbiome profiling can identify fluids even when chemical or protein markers have degraded.
- *Trace Evidence and Associative Analysis:*
- Microbial communities on surfaces, clothing, or personal items can link individuals to specific locations or objects, aiding crime scene reconstruction.
- *Environmental Stability:*
- Microbial DNA generally persists longer than human DNA under adverse conditions, extending the window for evidence recovery (Nodari et al. 2024).

These applications demonstrate the potential of microbiome analysis to strengthen forensic interpretation, especially in challenging or degraded evidence scenarios.

1.3 Concept of Bloodstain Microbiome

Among the various applications of microbial forensics, bloodstain microbiome analysis has emerged as a particularly promising area (Swayambhu et al. 2023;

Robinson et al. 2021). Although blood inside a healthy individual is largely sterile, exposure to the environment following deposition allows microorganisms from the air, surfaces, and the individual's skin to colonize the stain. These microbial communities evolve over time under the influence of environmental factors such as temperature, humidity, light, and the type of surface on which the blood is deposited.

Bloodstain microbiome analysis offers several forensic advantages:

- *Temporal Estimation:*
- Microbial succession within a bloodstain can help determine the age of the stain, providing insight into when an event occurred.
- *Environmental Interpretation:*
- The microbial composition of a bloodstain can indicate whether it was deposited indoors or outdoors and whether it was exposed to contaminants or specific environmental conditions.
- *Associative Evidence:*
- Unique microbial profiles may link a bloodstain to a particular individual, location, or object, even when human DNA is degraded or insufficient (Païssé et al. 2016).

Modern techniques—including NGS, metagenomics, and advanced bioinformatic pipelines—enable detailed characterization of these communities, significantly enhancing the evidentiary value of bloodstains and complementing traditional forensic analyses.

Microbial forensics represents a rapidly advancing field that strengthens and expands the capabilities of conventional forensic science. Bloodstain microbiome profiling, in particular, offers robust, individualized, and environmentally informative evidence. By incorporating microbial signatures into forensic workflows, investigators can overcome the limitations posed by degraded biological materials, improve PMI estimates, identify body fluids with greater accuracy, and establish connections between individuals, objects, and crime scenes. This chapter provides the foundational understanding of microbial forensics and highlights the growing relevance of bloodstain microbiome analysis in modern forensic investigations.

2 The Human Microbiome and Its Forensic Potential

The human body is not a single biological system but a complex, multispecies ecosystem composed of both human and microbial cells (Sender et al. 2016). Collectively known as the human microbiome, these microorganisms inhabit the skin, oral cavity, gastrointestinal tract, respiratory system, and other mucosal surfaces. They form dynamic, interactive communities essential for maintaining physiological balance and influencing overall health. In recent years, the study of the human microbiome has gained substantial relevance in forensic science. Because microbial communities differ between individuals and exhibit relative temporal stability, they can serve as biological signatures for personal identification, crime scene reconstruction, and postmortem interval estimation. This expanding field—microbial forensics—investigates how microbial patterns can supplement or replace traditional DNA-based methods, especially when human DNA is degraded, absent, or contaminated (Metcalf et al. 2017).

2.1 Structure and Diversity of the Human Microbiome

An estimated 10^{13}–10^{14} microorganisms inhabit the human body, including bacteria, fungi, viruses, and archaea. Each anatomical site provides a distinct ecological niche that shapes the community's composition and diversity. For example, the gut microbiome is dominated by *Firmicutes* and *Bacteroidetes*, the skin microbiome by *Staphylococcus* and *Corynebacterium*, and the oral microbiome by *Streptococcus* and *Veillonella* (Human Microbiome Project Consortium 2012).

Microbial composition varies significantly among individuals due to influences such as diet, hygiene practices, host genetics, lifestyle, health status, and geographic location. For instance, hunter-gatherer populations exhibit highly diverse gut microbiomes rich in *Prevotella*, *Proteobacteria*, and *Spirochaetes*, whereas urban populations tend to show less diverse communities dominated by *Bacteroides* and *Firmicutes* (Gupta et al. 2017). Understanding such inter-individual variability is crucial for forensic applications, as microbial profiles can act as distinguishing markers capable of differentiating one person from another.

2.2 Individuality and Temporal Stability of Microbial Signatures

Research consistently demonstrates that each individual possesses a unique and relatively stable microbial fingerprint (Franzosa et al. 2015; Schmedes et al. 2017). Although factors such as dietary changes, environmental exposure, and illness may introduce temporary fluctuations, a person's microbiome retains identifiable features over extended periods (Franzosa 2015).

Microbial signatures from the skin or oral cavity, for example, differ substantially between individuals and can therefore support personal identification (Schmedes et al. 2017; Metcalf et al. 2017). Studies have also shown that many microbial signatures persist for months on personal items such as mobile phones, keyboards, and wearable accessories. This stability enhances the forensic utility of microbiome profiling, enabling investigators to identify individuals even in situations where human DNA is absent or poorly preserved (Metcalf et al. 2017).

2.3 Microbial Transfer and Persistence at Crime Scenes

Humans constantly shed microorganisms into their surroundings through touch, respiration, and other daily activities. Approximately 36 million microbial cells per hour are released into the environment, forming an "invisible microbial cloud" around each individual (Karadayı et al. 2024). These microbes settle on surfaces—including door handles, clothing, furniture, or weapons—creating detectable biological traces that may persist for hours to days.

Experimental studies reveal that skin-associated microbes such as *Staphylococcus*, *Corynebacterium*, and *Propionibacterium* can be transferred to inanimate surfaces and remain detectable long after contact. For example, microbial DNA from a participant's hand was recovered from a knife handle several hours later, with 8.28% of bacterial species still detectable after 4 h and 6.95% after 24 h (Metcalf et al. 2017). Such persistence provides valuable forensic opportunities, particularly when human DNA is degraded, mixed, or entirely missing.

Unlike traditional DNA evidence, microbial signatures provide information not only about identity but also about environmental context, interactions, and temporal dynamics. Using advanced approaches such as metagenomic sequencing and 16S rRNA gene profiling, forensic scientists can compare microbial profiles between evidence and reference samples with high precision.

Microbial analysis of biological stains—including blood, saliva, or skin residues—can also offer insights into the source of the material, the time since deposition, and the possible number of contributors. With the integration of next-generation sequencing and machine learning algorithms, microbiome-based profiling is becoming increasingly accurate and represents a powerful complement to conventional forensic methods, particularly when standard DNA profiling is not feasible.

3 Bloodstain Microbiome: Origin and Characteristics

Bloodstains are among the most frequently encountered and forensically significant forms of biological evidence. Traditionally, their analysis has relied on serology and human DNA profiling. However, with advances in microbial forensics, attention has shifted toward the microbiome associated with bloodstains, revealing that blood—historically considered sterile—is in fact a dynamic microbial niche. The bloodstain microbiome, defined as the community of microorganisms present within or upon dried blood, can offer valuable information about the sample's origin, age, environmental exposure, and depositional history. These microbial patterns arise from both endogenous (host-derived) and exogenous (environmental) sources and evolve predictably over time. Understanding their composition, provenance, and temporal changes enhances forensic interpretation, especially when human DNA is degraded or insufficient (Rozenberga et al. 2024).

3.1 Microbial Communities in Fresh and Aged Bloodstains

Freshly deposited blood generally contains a low microbial load due to the immune-protected nature of the circulatory system. However, recent studies challenge the notion of complete blood sterility.

- *Study Type: 16S rRNA analysis of blood microbiome in healthy individuals* (Païssé et al. 2016)
- *Explanation:* Minute quantities of bacterial DNA are consistently detectable in healthy individuals. Dominant genera such as Pseudomonas, Staphylococcus, and Corynebacterium appear to originate from commensal sites like skin and mucosal surfaces. These findings suggest the existence of a core blood microbiome, relatively stable and shared across individuals.
- *Study Type: Longitudinal analysis of bloodstains under varying environmental conditions* (Cobaugh et al. 2015)
- *Explanation:* Time-series experiments show that microbial composition shifts significantly with humidity, temperature, and exposure duration. Early-stage stains are enriched with host-associated genera such as *Staphylococcus* and *Pseudomonas*, while older stains become dominated by environmental spore-formers like *Bacillus* and *Paenibacillus*. These predictable microbial succession patterns provide important clues for estimating the time since deposition (TSD) of bloodstains.
- *Study Type: Metagenomic sequencing of the 200-year-old bloodstain of Jean-Paul Marat* (De-Dios et al. 2020)
- *Explanation:* This exceptional study recovered bacterial, fungal, and even viral DNA from a centuries-old bloodstain. While the investigation focused on historical medical hypotheses, the preservation of trace microbial signatures demonstrates that elements of the original microbiome can persist for hundreds of years.

Collectively, these studies demonstrate that bloodstains undergo a transition from host-derived microbial communities to environmentally adapted taxa, making microbial profiling a powerful tool for forensic temporal and contextual analysis.

3.2 Endogenous Versus Exogenous Microbiota

The microbial composition of a bloodstain originates from two primary sources:

Endogenous Microbiota

Endogenous microbes originate from the host's bloodstream, skin, oral cavity, or mucosal surfaces. Rozenberga et al. (2024) reported that:

- 27.4% of bacterial ASVs detected in blood overlapped with other body sites—particularly skin.

- 25.3% of ASVs remained consistent across samples collected 3 months apart.

These findings indicate that blood carries a stable, individual-specific microbial background, enabling the possibility of microbial "fingerprints" for identity confirmation or exclusion (De-Dios et al. 2020).

Exogenous Microbiota
Exogenous microbes enter the bloodstain after deposition. These include airborne bacteria, soil microbes, and fungi that colonize the stain's surface. Studies such as De-Dios et al. 2020 demonstrated that:

- Distinct environmental microbes colonize bloodstains depending on exposure conditions.
- Historical samples exhibit environmental taxa atypical of human microbiomes, highlighting prolonged environmental interaction.

Bloodstains thus act as open microbial ecosystems, integrating both internal and external microbial inputs. Early stains are dominated by host-derived microbes, while older stains show increasing environmental influence. This balance between endogenous and exogenous signatures provides critical forensic indicators for reconstructing exposure duration, environmental setting, and staining history.

3.3 Factors Influencing Microbial Composition in Bloodstains

Microbial diversity and succession in bloodstains are shaped by a complex interplay of biological, environmental, and analytical factors.

Environmental Factors
Temperature, humidity, substrate type, and UV exposure profoundly influence microbial growth.

- High humidity and moderate temperatures support bacterial proliferation.
- Excessive dryness and UV light reduce microbial diversity.
- Porous substrates (fabric and wood) retain moisture and support longer microbial survival, unlike nonporous surfaces (glass and metal).

These factors explain why two bloodstains from the same donor may develop distinct microbial profiles under different environmental conditions.

Temporal Factors
Time since deposition is a major determinant of microbial succession.

- Early stains are dominated by host-associated taxa.
- Environmental microbes gradually increase with time.

Comparable temporal dynamics have been observed in seminal stains (Gürsoy et al. 2024), reinforcing the concept of time-dependent microbial change as a valuable forensic indicator.

Donor-Related Factors
The initial microbial composition depends on the donor's health, diet, lifestyle, and microbiome characteristics.

Chronic illnesses, immune disturbances, smoking, and medications can alter microbial signatures in the bloodstream. Such individual-specific variables must be considered during forensic interpretation.

Analytical and Preservation Factors
Sample handling and processing significantly affect microbial profiles.

- Delays in drying, contamination during collection, and inadequate storage can introduce artifacts.
- Advanced sequencing (metagenomics, 16S rRNA analysis) improves accuracy by distinguishing viable microorganisms from residual DNA.

Overall, the microbial composition of a bloodstain emerges from interactions between host-derived taxa, environmental inputs, and time-dependent ecological shifts. When systematically analyzed, these microbial signatures offer insights into the origin, age, and environmental history of blood evidence, reinforcing the value of microbial profiling as a powerful emerging tool in forensic science.

4 Analytical Techniques for Bloodstain Microbiome Analysis

The bloodstain microbiome has emerged as a valuable component of forensic biology, offering insights that complement traditional DNA and serological analyses. Advances in molecular biology, sequencing technologies, and bioinformatics now enable detailed characterization of microbial communities associated with bloodstains (Arriola et al. 2020). Together, these approaches—from sample collection and DNA extraction to sequencing and computational analysis—provide a robust framework for microbial identification and forensic interpretation (Hua et al. 2022).

4.1 Sample Collection and Preservation

Accurate microbiome analysis begins with proper sample collection and preservation. Because blood is a low-biomass sample prone to contamination, stringent precautions must be followed. Bloodstains are collected using sterile tools such as swabs, scalpels, or micropipettes, depending on whether the stain is wet or dried. The sampling environment and surfaces must be contamination-free, and sterile gloves and equipment are essential to prevent the introduction of exogenous microbes. For aged or absorbed stains, micro-sampling from the substrate (e.g., fabric, wood) is commonly performed (Yang and Watt 2005).

Once collected, samples must be preserved to maintain microbial DNA integrity. Air-drying minimizes microbial overgrowth, while storage at low temperatures in sterile, DNA-free containers prevents degradation. Some studies use fixatives such as ethanol or RNA later to stabilize nucleic acids. Preservation choice depends on downstream analyses—ethanol is typically suitable for long-term storage, whereas freezing at −20 °C or −80 °C is preferred for high-quality molecular work (Choo et al. 2015; Poulsen et al. 2021).

Environmental conditions at the scene (temperature, humidity, and substrate type) can influence microbial survival and must be documented, as they may affect community composition. Cross-contamination between biological samples must be strictly avoided through the use of separate sterile tools and controlled processing environments. Effective preservation ensures that the microbial community remains representative of the original biological context (Song et al. 2016).

4.2 DNA Extraction and Quantification

After preservation, the next critical step is microbial DNA extraction. This process isolates genetic material from both endogenous and exogenous microorganisms within the bloodstain. Efficient extraction is essential to capture low-abundance species and produce representative microbial profiles. Blood contains potential inhibitors such as hemoglobin and degraded proteins; therefore, extraction protocols must address these to avoid interference with enzymatic steps in sequencing (Howarth et al. 2022).

Most extraction methods combine mechanical lysis (e.g., bead beating) with chemical and enzymatic lysis to rupture microbial cell walls. Commercial extraction kits optimized for low-biomass forensic samples provide standardized, high-quality DNA suitable for sequencing.

Following extraction, DNA quantity and purity are assessed using spectrophotometric or fluorometric tools such as NanoDrop and Qubit. These analyses ensure that DNA is sufficiently concentrated and free from contaminants (Green and Sambrook 2012). Quantitative PCR (qPCR) may also be used to detect bacterial markers or assess DNA amplifiability before sequencing. Proper extraction and quantification are fundamental for generating reliable sequencing results and accurate microbial identification (Song et al. 2016). The generalized workflow of microbial metagenomics approach is given in Fig. 15.1.

4.3 16S rRNA Gene Sequencing and Metagenomics

The cornerstone of microbial community profiling is 16S rRNA gene sequencing, which targets conserved and variable regions within bacterial and archaeal genomes.

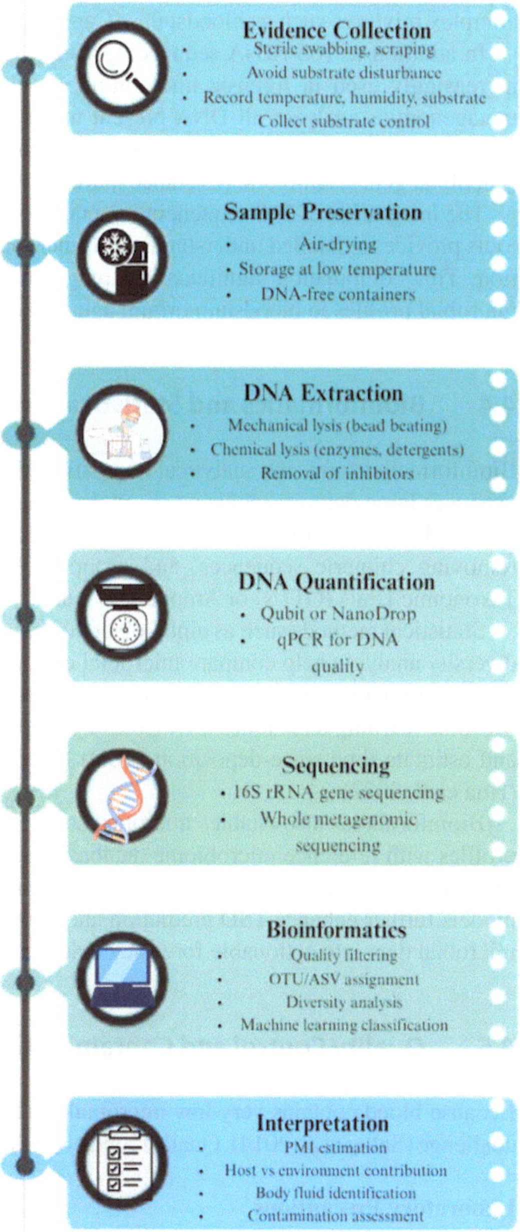

Fig. 15.1 Workflow for bloodstain microbiome analysis showing key steps from evidence collection to interpretive assessment of microbial profiles

This method enables taxonomic identification, often to the genus or species level, in complex mixtures such as bloodstains (Caporaso et al. 2012; Knight et al. 2018).

In addition to 16S rRNA sequencing, whole-genome shotgun metagenomics is increasingly used in forensic microbiome research. Unlike targeted sequencing, metagenomics analyzes all DNA present in the sample, providing comprehensive coverage of bacterial, fungal, and viral communities. This approach also reveals functional genes, antibiotic resistance markers, and metabolic pathways.

The integration of 16S sequencing, metagenomics, and advanced computational tools provides a detailed understanding of microbial community structure and function. This combination significantly strengthens forensic applications by linking microbial profiles to bloodstain origin, age, and environmental exposure.

4.4 Bioinformatics and Statistical Approaches

Bioinformatics forms the analytical foundation of microbiome research. Sequencing produces large datasets requiring computational workflows for filtering, classification, and interpretation. Initial preprocessing includes trimming low-quality reads, removing chimeric sequences, and grouping DNA sequences into Operational Taxonomic Units (OTUs) or Amplicon Sequence Variants (ASVs) (Amir 2021).

Statistical methods such as alpha-diversity (Shannon, Simpson indices) and beta-diversity analyses help compare microbial communities across samples and conditions. Ordination methods visualize community similarities and differences. Machine learning techniques are increasingly applied to classify unknown samples and estimate time-since-deposition (TSD) based on microbial succession patterns (Hua et al. 2022).

Bioinformatics also enables microbial source tracking by comparing bloodstain profiles with reference microbiome databases. This helps differentiate endogenous (host-derived) microbes from exogenous (environmental) contributors. AI-based models further enhance TSD prediction and contamination detection, transforming microbial data into actionable forensic evidence (Clarke et al. 2017).

4.5 Quality Control and Contamination Management

Because blood contains very low microbial biomass, contamination poses a major challenge (Salter et al. 2014). Quality control (QC) must be maintained at every stage:

Laboratory Precautions

- Use DNA-free consumables, sterile reagents, and dedicated pre-/post-PCR work areas.
- Include negative controls and blank extractions to detect contamination from reagents or laboratory environments.

Facility-Level Controls

- UV decontamination systems, HEPA-filtered air units, and strict aseptic techniques (gloves, masks, lab coats) help prevent introduction of human-associated microbes.

Data-Level Controls
Bioinformatic pipelines identify and remove contaminant sequences by comparing results to known reagent-associated taxa databases. Statistical analyses detect outlier patterns or cross-sample contamination.

Standardization and Reproducibility
Inter-laboratory consistency is ensured through initiatives such as the Microbiome Quality Control (MBQC) project, which promotes standardized protocols for extraction, sequencing, and data processing (Song et al. 2016).

Robust QC ensures that microbial signatures reflect the true biological origin of the bloodstain, enhancing their forensic reliability.

5 Forensic Applications of Bloodstain Microbiome Analysis

The forensic bloodstain microbiome provides a powerful set of investigative tools that complement traditional DNA, serology, and trace evidence analyses. Microbial signatures associated with blood can indicate time since deposition, identify species origin, support human identification, and help link suspects, victims, and crime scenes. These applications expand forensic capabilities, particularly in cases where biological material is degraded or insufficient for conventional testing.

5.1 Estimation of Postmortem Interval (PMI)

Accurately determining the postmortem interval (PMI) is essential for reconstructing events surrounding death. Traditional PMI indicators—such as rigor mortis, body temperature, biochemical degradation, or insect colonization—often become unreliable when the body is highly decomposed or when only trace evidence such as bloodstains remains (Madea 2016). In such scenarios, bloodstain microbiome analysis offers a promising alternative.

Microbial communities in bloodstains undergo predictable, time-dependent shifts after deposition (Hauther et al. 2015). These microbial successions function as a biological clock:

- *Early phase:* Dominance of bacteria such as *Staphylococcus* or *Pseudomonas*.

- *Later phase:* Appearance of genera such as *Bacillus* and *Clostridium*, which thrive during decomposition.

The presence or absence of these taxa, along with their relative abundances, helps estimate how long a bloodstain has been present (Tozzo et al. 2022).

Environmental conditions significantly influence microbial succession (Hyde et al. 2013):

- Warm, humid environments accelerate microbial growth and decomposition.
- Cold or dry environments slow down microbial activity and succession.

To ensure accuracy, PMI estimates derived from microbial evidence must always consider environmental factors such as temperature, humidity, oxygen exposure, and light. The combination of microbial profiling and contextual environmental data provides a more reliable PMI estimate when traditional indicators are absent or compromised.

5.2 Determination of Blood Origin and Human Identity

Bloodstain microbiome analysis offers new possibilities for determining the source of blood—both in terms of species and, potentially, individual identity (Aagaard et al. 2013).

Determining the Species of Origin

Each species possesses a characteristic microbial community shaped by its physiology, immune system, and environment. As a result:

- Human and animal blood differ in microbial composition.
- Bloodstain microbiome profiling can therefore distinguish human from nonhuman blood when traditional methods (e.g., precipitin tests and DNA analysis) fail due to degradation.

Human Identity Through Microbial Signatures

Even among humans, microbial communities are influenced by factors such as diet, lifestyle, exposure, and health. These differences create microbial fingerprints that may:

- Help associate a blood sample with an individual.
- Serve as supplementary evidence when nuclear or mitochondrial DNA is degraded or absent.

Advantages in Forensic Investigations

- Effective even with aged or degraded samples.
- Requires minimal biological material.
- Provides an additional layer of identification when traditional DNA methods yield limited results.

Thus, microbiome analysis enhances forensic interpretation by identifying species origin, supporting human identification, and complementing classical serological and genetic techniques.

5.3 Discrimination Between Human and Nonhuman Blood

Distinguishing human from nonhuman blood is critical in forensic casework, allowing investigators to focus on evidence relevant to the crime. While traditional methods (e.g., precipitin test, DNA barcoding) are effective, they may fail when samples are degraded, mixed, or minimal (Budowle et al. 2017).

Recent advances in microbiome profiling and spectroscopic techniques have improved species discrimination.

Raman Spectroscopy + PLS-DA
Bai et al. (2016) demonstrated the successful differentiation of human and animal blood using:

- *Raman spectroscopy:* A laser-based technique detecting subtle molecular variations.
- *Partial Least Squares Discriminant Analysis (PLS-DA):* A statistical method that classifies spectral patterns with high sensitivity.

Even in small or deteriorated samples, this approach produced clear distinctions between human and nonhuman blood, making it an accurate, rapid, and nondestructive forensic tool.

Forensic Importance

- *Crime scene screening:* Helps eliminate irrelevant samples.
- *Microbial signatures:* Provide species-specific patterns supporting identification.
- *Advanced analytical approaches:* Enable precise discrimination even in degraded samples.
- *Improved reliability:* Reduces the risk of misinterpretation in forensic casework.

5.4 Linking Suspects, Victims, and Crime Scenes Via Microbial Signatures

Microbial evidence can associate individuals with objects, environments, or each other—especially when conventional DNA evidence is absent, degraded, or mixed (Lax et al. 2015).

To illustrate the potential of microbial trace evidence, the following case study demonstrates how soil microbiome profiles can link objects to a specific location.

5.4.1 Case Study: Soil Microbiome Evidence Linking Objects to Crime Scenes

Introduction

When human DNA is not available or is degraded due to environmental exposure, microbial evidence offers an alternative means of establishing connections. Soil, skin, and environmental microbiomes vary significantly between locations and possess unique taxonomic signatures. This case study highlights how soil microbial profiling can link objects to a specific scene (Demanèche et al. 2017).

In a mock outdoor crime scene study, soil was collected from the scene itself, as well as from evidence items such as shoes and clothing that had been placed in contact with the soil. Soil samples from unrelated control locations were also collected to serve as comparisons. The objective was to determine whether microbial communities present on the evidence items matched those of the crime scene. Soil samples underwent standard DNA extraction protocols, followed by microbial profiling using 16S rRNA sequencing for bacteria and ITS sequencing for fungi. Data analysis included beta-diversity comparisons, cluster analysis, and microbial community profiling (Gouello et al. 2021). The results demonstrated strong evidence–scene linkage: microbial profiles of evidence items closely matched the crime scene soil. Notably, fungal communities provided better discrimination than bacterial communities, which showed moderate similarity across sites. Control soils differed sharply from the crime scene and evidence samples, confirming that the observed matches were not coincidental. Importantly, microbial evidence successfully linked the items to the scene even in the absence of human DNA, highlighting the potential of microbial profiling to support crime scene reconstruction.

Forensic Importance

- Provides an alternative evidence source when human DNA is compromised.
- Enhances trace evidence linkage between objects, suspects, and environments.
- Offers strong potential for scene reconstruction.
- Acts as a complementary—not replacement—tool alongside fingerprints, fibers, and DNA.

Limitations and Challenges

- *Environmental variability:* Microbial communities change with climate, moisture, sunlight, and substrate.

- *Transfer variability:* Microbial transfer depends on object surface, contact duration, and environmental conditions.
- Individual-level identification remains limited with current technology.
- Need for standardized protocols to ensure reproducibility and courtroom admissibility.

Soil microbial profiling provides a scientifically robust method for linking evidence to specific locations, even in the absence of human DNA. Although still developing, microbial forensics shows significant promise as a complementary tool in crime scene reconstruction and trace evidence interpretation. Continued research, methodological standardization, and validation studies will help integrate microbial profiling into routine forensic practice.

6 Environmental and Temporal Dynamics

6.1 Impact of Temperature, Humidity, and Substrate Type

Environmental conditions strongly influence the growth, survival, and infiltration of external microorganisms within bloodstain microbiomes (Fig. 15.2). Warm and humid environments accelerate bacterial metabolism, cellular replication, and the activity of host-associated taxa, resulting in higher total biomass. In contrast, rapid drying, desiccation, or UV exposure suppress growth, reduce taxonomic richness, and selectively favor desiccation-tolerant or UV-resistant taxa while eliminating moisture-dependent organisms (Eisenhofer et al. 2019).

Among all environmental variables, substrate type often has the greatest impact. Surfaces with dense environmental microbial communities rapidly inoculate the bloodstain upon deposition (Cobaugh et al. 2015). In a controlled study, substrate explained more variability in microbial composition than the type of bodily fluid itself. Bloodstains deposited on bedspreads or floors diverged significantly from those on sterile swabs, showing increased *Proteobacteri*a and *Cyanobacteria*, respectively. Low-biomass fluids such as peripheral blood or semen are particularly vulnerable; when deposited on microbially rich substrates, environmental taxa can dominate the recovered microbiome (Zhang et al. 2025).

Temperature, humidity, and substrate interact to shape microbial succession. Warm and moist conditions facilitate rapid colonization of substrate microbes, accelerating the loss of diagnostic host-associated taxa. Conversely, sterile or low-biomass substrates allow host signatures to persist longer, especially when environmental conditions limit microbial infiltration.

Best Practices

Best practices for bloodstain microbiome analysis emphasize careful sampling and contextual documentation. A substrate control should always be collected adjacent to the stain to account for background environmental microbes. Environmental parameters such as temperature, humidity, substrate type, and usage history at the

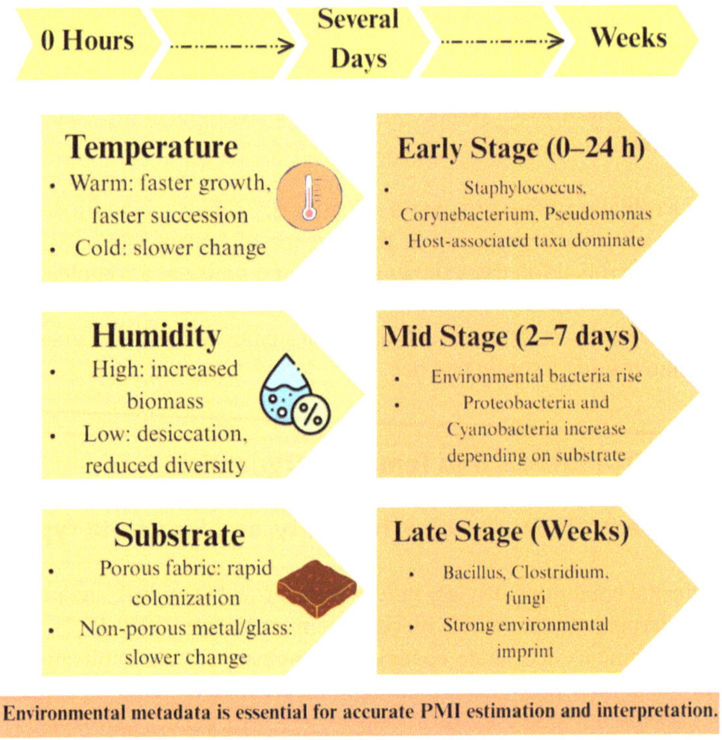

Fig. 15.2 Environmental and temporal factors influencing microbial succession in bloodstains, including temperature, humidity, and substrate effects

sampling site should be recorded, as these factors strongly influence microbial community composition. Computational tools, such as MicroDecon, can be employed to reduce background substrate noise, helping to isolate the stain-specific microbial signal; however, it is important to note that such tools cannot recover microbial signals that have already been lost or degraded.

6.2 Temporal Changes in Microbial Communities Postdeposition

After deposition, bloodstain microbiomes undergo predictable temporal shifts. Initially, community profiles resemble the donor's microbiota, containing taxa associated with skin or oral environments. Over subsequent days or weeks, environmental microorganisms gradually replace host-associated taxa (Zhang et al. 2025).

In a 7-day indoor exposure experiment, bloodstains on sterile swabs remained phylum-level similar to fresh samples. However, stains on nonsterile substrates showed substantial shifts toward substrate-associated taxa, illustrating that substrate often dictates the direction and magnitude of temporal change.

Early colonizers exploit the abundant labile nutrients of fresh blood. As nutrients deplete and conditions change, opportunistic secondary colonizers replace them. These successional patterns can be characterized using marker-gene sequencing or metagenomics. For instance, aged swabs have shown increasing proportions of Lactobacillus and decreasing Cyanobacteria and Proteobacteria, reflecting substrate-driven transitions.

The speed of microbial succession in biological fluids is strongly influenced by the initial microbial biomass. High-biomass fluids, such as saliva, tend to retain their characteristic microbial profiles for longer periods, whereas low-biomass fluids, including peripheral blood and semen, rapidly lose donor-specific taxa as environmental microbes outcompete the original host-associated communities. By combining temporal microbial trajectories with environmental metadata, researchers can estimate the relative time since deposition (TSD) of a sample. To model these patterns robustly, it is essential to conduct longitudinal controls under conditions that mimic the crime scene, apply microbial source-tracking algorithms to distinguish host versus environmental contributions, and convert observed community shifts into statistical metrics that support defensible postmortem interval (PMI) estimates.

6.3 Microbial Succession Patterns in Bloodstains

Microbial succession in bloodstains exhibits reproducible shifts in dominant taxa that function as molecular clocks. Early stages typically contain donor-associated taxa from skin, oral, or endogenous sources. Within days, environmental colonizers disrupt these profiles, with specific patterns depending on substrate and environment. Controlled studies have shown increases in Proteobacteria on fabric and Cyanobacteria on floor surfaces. Saliva and menstrual blood stains frequently retain Firmicutes as dominant taxa (Mazzarelli et al. 2023).

Successional patterns should be interpreted using suites of taxa, not single indicators. Taxon-specific variation necessitates integrating abundance, presence/absence, and diversity metrics into molecular clock models. Models must be trained across diverse substrates, temperatures, and humidities to ensure reliable time-since-deposition estimates. One study reported that substrate contributed more variation than time (via PERMANOVA), although decomposition stage still served as a useful temporal indicator—highlighting the importance of combining both factors (Cobaugh et al. 2015).

Use source-tracking tools such as FEAST to partition microbial contributions from host, substrate, and unknown sources (Shenhav et al. 2019). Decontamination methods improve body-fluid clustering by removing substrate reads. In casework, collecting substrate controls is essential for accurate deconvolution.

Best-Practice Reporting Requirements

Best-practice reporting in forensic microbiome studies requires comprehensive documentation of environmental metadata, including temperature, humidity, and substrate type, as these factors strongly influence microbial succession patterns. Researchers should report sequencing depth and include all negative and substrate controls to ensure data quality and reproducibility. Additionally, uncertainty estimates should accompany temporal inferences to provide transparency regarding the confidence of time-since-deposition (TSD) estimates. Models used to infer TSD should be validated across multiple ecological conditions before being presented as forensic evidence in court. Following these guidelines maximizes the forensic utility of microbial succession data and ensures that interpretations are scientifically robust and defensible.

7 Case Studies and Experimental Investigations

7.1 Laboratory Simulations of Bloodstain Microbiome Evolution

Laboratory simulations allow researchers to observe the evolution of bloodstain microbiomes under controlled conditions that approximate real crime scenes. By manipulating variables such as temperature, humidity, substrate type, and light exposure, investigators can systematically study microbial succession from the moment blood contacts a surface. This controlled environment eliminates external variability and produces reliable data supporting the use of microbial communities for PMI estimation (Sacco et al. 2024).

Early-stage laboratory samples consistently show the donor's microbial signature, with taxa resembling skin- or oral-associated communities. As time progresses, especially under elevated temperature or humidity, these communities shift as environmental microbes colonize the stain. Predictably, taxa already present on the nearby substrate—whether fabric, soil, or flooring—gradually outcompete the low-biomass microbial community of the blood itself. This effect is particularly pronounced when the bloodstain is deposited on surfaces bearing high environmental microbial loads, echoing findings from multiple forensic microbiome studies (Wahab et al. 2025).

Laboratory simulations also enable researchers to observe how bloodstains respond to controlled patterns of heat, desiccation, UV exposure, and prolonged oxygen exposure. Repeated sampling over days or weeks yields quantitative temporal data, supporting the development of microbial succession models for PMI estimation (Swayambhu et al. 2023). These experiments further reveal how long donor-associated taxa persist and the rate at which environmental taxa dominate.

Additionally, simulations allow refinement of forensic workflows, including evaluation of DNA extraction kits, marker gene targets, sequencing depth, contamination risks, and the performance of decontamination protocols. The results emphasize the importance of strict chain of custody, method validation, and reproducible workflows for forensic admissibility.

Substrate comparisons further demonstrate how different surfaces influence microbial trajectories. Cloth, soil, and indoor flooring each promote distinctive succession patterns. Evaluating these interactions clarifies when substrate effects overshadow the biological signal and informs interpretation in real forensic casework.

In summary, laboratory simulations show that bloodstain microbiomes predictably colonize and transform over time. These patterns provide a foundation for interpreting aged or degraded samples and reinforce the forensic relevance of microbial succession.

7.2 Reported Forensic Cases Utilizing Microbial Analysis

Microbial forensics has been successfully applied in multiple real-world investigations. Though the field is still maturing, case studies demonstrate the power of microbial signatures for source attribution, transmission tracking, and evidentiary linkage. The referenced PDF documents two instructive cases.

7.2.1 Case 1: Shiga Toxin–Producing *Escherichia coli* O104:H4 Outbreak (Germany and France) (Muniesa et al. 2012)

Whole-genome sequencing (WGS) was used to compare isolates recovered from two countries. Genetic distinctions between strains informed the reconstruction of outbreak history, demonstrating how microbial profiling supports attribution in public health investigations. Although not a criminal case, this example highlights the precision of microbial analyses in linking related events across geographic boundaries.

7.2.2 Case 2: Deliberate HIV Transmission by a Physician (Louisiana) (Schmidt 1998)

In this criminal case, investigators followed strict biosafety and chain-of-custody protocols before sequencing viral genes from both the victim and the physician's patient. Phylogenetic analysis revealed that the viral strain infecting the victim clustered closely with the strain present in the physician's patient. Courts accepted this microbial evidence, marking a landmark case demonstrating that microbial genetic signatures can meet forensic and legal standards when supported by validated methods.

These investigations show that microbial evidence can:

- Determine human vs. nonhuman origins of biological material
- Reconstruct transmission pathways or intentional introduction
- Identify shared sources or donors
- Support interpretations about whether a stain was moved or deposited under unusual conditions
- Contribute to PMI estimation via microbial succession patterns

As microbial forensics expands, community-level or strain-level microbial signatures may increasingly complement—and in some cases challenge—the weight of DNA-based evidence.

7.3 Comparative Studies with Other Body Fluid Microbiomes

Comparative microbiome studies help differentiate blood from other major body fluids such as saliva, semen, and vaginal secretions (Dobay et al. 2019). Understanding these differences enhances the ability to identify fluid type, resolve mixed stains, and strengthen forensic conclusions.

Blood is characteristically low in microbial biomass, making it more vulnerable to rapid environmental overprinting. In contrast, saliva contains a high microbial load dominated by *Streptococcus*, *Prevotella*, and other oral taxa. Vaginal fluid typically exhibits a Lactobacillus-dominated community, while semen contains taxa associated with the male urogenital tract. These distinct microbial profiles generally do not overlap, allowing effective discrimination when sequencing depth is sufficient.

The referenced literature emphasizes that low-biomass fluids, especially blood, are most susceptible to being overtaken by substrate-associated microbes (Table 15.1). This must be considered when comparing fluids across different surface types (clothing, soil, indoor flooring) or interpreting aged stains.

Comparative studies of body-fluid microbiomes reveal distinct patterns of fluid-specific succession. Saliva, with its high microbial biomass, tends to retain oral-dominant taxa for extended periods, whereas vaginal fluids experience rapid community shifts influenced by pH and oxygen exposure. Blood, in contrast, quickly loses donor-specific taxa and is rapidly overprinted by environmental microbes. Monitoring these successional trajectories can provide critical forensic insights, including the identification of which fluid was deposited first in a mixed stain, whether multiple fluids were deposited simultaneously or at different times, and the expected degradation patterns for complex or degraded evidence.

Microbial profiling is especially valuable when traditional presumptive or confirmatory tests fail due to chemical treatment, degradation, or loss of protein/DNA targets. High-quality sequencing datasets feed machine-learning classifiers that differentiate fluids based on microbial community structure. As more controlled and casework-derived datasets become available, these classifiers grow more accurate, enabling more confident fluid identification even in highly degraded or mixed samples (Wohlfahrt et al. 2023).

Table 15.1 Comparison of microbial characteristics across major body fluids relevant to forensic identification and discrimination

Body fluid	Biomass level	Dominant taxa	Stability over time	Forensic use	References
Blood	Very low microbial biomass (nearly sterile in healthy individuals)	*Staphylococcus*, *Streptococcus*, skin-associated and transient environmental microbes	Low stability; microbial profile largely influenced by contamination during sampling	Useful for confirming human origin; not reliable for individual discrimination based on microbiota	Dobay et al. (2019)
Saliva	High microbial biomass	*Streptococcus*, *Veillonella*, *Prevotella*, *Haemophilus*	Moderately stable; changes with diet, hygiene, and circadian rhythm	Strongly useful for body-fluid identification due to distinct oral microbiome	Dobay et al. (2019)
Semen	Moderate microbial biomass	*Lactobacillus*, *Staphylococcus*, *Corynebacterium*, *Gardnerella*	Moderately stable; influenced by sexual activity and genital health	Useful for semen confirmation and potential individualization through seminal microbiome signatures	López et al. (2022)
Vaginal Fluid	High microbial biomass	*Lactobacillus* spp. dominant; *Gardnerella*, *Atopobium* in dysbiosis	Stability varies; highly dependent on hormonal state and menstrual cycle	Highly useful for identifying vaginal origin and discriminating between normal vs. dysbiotic states	Wohlfahrt et al. (2023)
Skin/Touch DNA (for comparison)	Variable; usually low	*Staphylococcus*, *Corynebacterium*, *Propionibacterium*	Varies widely; sensitive to washing, environment	Useful for activity-level inference and trace evidence; inter-individual variability can assist discrimination	López et al. (2022)

8 Challenges and Limitations

8.1 Sample Degradation and Contamination Risks

One of the primary challenges in analyzing microbial evidence from bloodstains is the extensive degradation that occurs once the stain dries. DNA rapidly deteriorates under environmental stressors such as heat, light, oxygen exposure, humidity, and microbial enzymatic activity (Butler 2015). Microbial cells also degrade, and taxa with fragile or sensitive cell envelopes disappear first, further distorting the original community structure. The loss of these taxa weakens the true microbial signature and limits recovery of potential donor-associated organisms.

At the same time, additional background noise is introduced through airborne microbes, surface-associated microbiota, and laboratory contaminants—all of which can generate false signals or shift community composition away from the original biological fingerprint (Neckovic et al. 2020).

The referenced literature (Eisenhofer et al. 2019) highlights that low-biomass samples, such as bloodstains, are particularly susceptible to contamination throughout sequencing workflows. Reagent contamination may artificially inflate environmental taxa, skewing relative abundance profiles and misleading interpretation. Even in marker gene (amplicon) studies, kit contaminants can comprise a significant portion of reads when biological DNA is low. This effect becomes even more pronounced in aged or degraded stains, where donor DNA declines faster than background contaminants.

Substrate contamination introduces additional complexity. All surfaces—floors, fabrics, furniture, soil, or indoor materials—harbor their own microbial communities. Once blood contacts a substrate, rapid microbial acquisition occurs, and substrate-associated taxa often outcompete the original blood microbiota. Without appropriate substrate controls, these environmental taxa may be mistaken for temporally driven succession rather than contamination.

Contamination can also arise downstream, during pipetting, extraction, PCR amplification, and library preparation. Reagents and plasticware may carry detectable levels of microbial DNA (Eisenhofer et al. 2019). As discussed in the referenced studies, negative controls enable detection and computational removal of contaminants; however, no decontamination pipeline can fully restore the original biological profile once contamination dominates. Thus, prevention remains the most effective strategy.

Careful sampling practices can mitigate many of these risks. This includes:

- Using sterile tools and avoiding contact with surrounding surfaces
- Packaging samples promptly to limit exposure
- Freezing samples when possible to delay degradation
- Incorporating multiple negative and process controls at each analytical step
- Recording batch effects to distinguish biological variation from workflow-induced artifacts
- Employing validated cleaning and bioinformatic pipelines

Despite unavoidable degradation and contamination, meaningful microbial patterns can still be recovered when evidence is collected meticulously, controls are rigorously implemented, and analytical workflows are standardized. With proper documentation and quality assurance, bloodstain microbiome profiles can yield scientifically defensible observations. Various key challenges of bloodstain microbiome analysis and their suggestive mitigation strategies has been highlighted in Table 15.2.

8.2 Interpretation Complexity and Lack of Standardization

Interpreting microbial evidence requires a cautious and highly contextual approach. Microbial communities are extraordinarily diverse, dynamic, and sensitive to environmental influence. This complexity is compounded by the absence of standardized protocols for sampling, sequencing, and data analysis across forensic laboratories. As highlighted in the referenced literature, microbial forensics still lacks uniformity in methodology, making cross-study comparisons difficult and sometimes unreliable (Eisenhofer et al. 2019; Salter et al. 2014).

Different laboratories employ various DNA extraction kits, primer sets, and sequencing depths, each of which can significantly alter the microbial groups detected in a sample. For example, one primer set may amplify a particular bacterial group while another fails to detect it entirely. Similarly, bioinformatic variability—including choices of filtering thresholds, clustering algorithms, and taxonomic databases—can reshape diversity estimates and change community profiles. These inconsistencies hinder the development of universal classification tools and reduce the comparability of results across laboratories.

Interpretation is further complicated by the need to separate host-associated microbial signals from those originating in the surrounding environment. Although succession models describe broad trends, these trajectories must be adjusted for temperature, humidity, substrate type, sample age, and other contextual variables.

Table 15.2 Key challenges in bloodstain microbiome analysis with recommended mitigation strategies

Challenge	Description	Mitigation Strategy
Sample Degradation	Heat, UV, and oxygen degrade microbial DNA	Rapid collection, freezing, and low-temperature storage
Contamination	Substrate microbes overpower donor taxa	Use substrate controls, sterile tools, and negative controls
Low Biomass	Blood carries fewer microbes than other fluids	Deep sequencing and optimized extraction
Lack of Standardization	Variation in primers, kits, and sequencing pipelines	Adopt uniform protocols and comprehensive metadata reporting
Interpretation Complexity	Hard to separate host signals from environmental signals	Use source-tracking tools and multiomics approaches
Legal Challenges	Lack of error rates and absence of formal statistical models	Transparent reporting and validated workflows

Differences in biological materials also influence interpretation; for instance, blood typically has much lower biomass compared to saliva or vaginal fluid, leading to distinct succession patterns and greater susceptibility to contamination.

A major unresolved issue concerns the weight of microbial evidence in legal contexts. Unlike human DNA profiling, microbial communities do not follow Mendelian inheritance, lack population frequency databases, and are influenced by random environmental variability. Courts require transparent explanations of how microbial patterns are interpreted, yet the absence of validated statistical frameworks introduces uncertainty and may weaken evidentiary value (Dobay et al. 2019; Wohlfahrt et al. 2023).

Mitigating these limitations requires comprehensive standardization. Consistent protocols are needed for sampling, storage, sequencing workflows, metadata reporting, and analysis. Reference databases documenting bloodstain microbial succession under diverse conditions would allow more reliable comparisons between case samples and established profiles. Interpretation becomes more robust when microbial data are integrated with conventional evidence—such as environmental measurements, substrate characteristics, and known timelines.

Given the interpretive complexity, forensic practitioners must remain precise, transparent, and conservative in their conclusions. Clear documentation of analytical steps and decision-making criteria enhances scientific defensibility and reduces uncertainty in forensic evaluations.

8.3 Ethical and Legal Considerations in Microbial Evidence

The use of microbial evidence raises important ethical and legal questions. Microbial signatures may reveal sensitive information related to an individual's health, lifestyle, medical conditions, behaviors, or environmental exposures. Even a bloodstain microbiome can inadvertently disclose clinically relevant data, necessitating careful consideration of responsible data use and privacy protections.

The referenced literature emphasizes that microbial forensics does not yet fit neatly within existing legal frameworks. Microbial patterns may link individuals to objects or locations with greater resolution than traditional DNA profiling, yet the field lacks standardized error rates, validated interpretive models, and codified admissibility standards (Castro and De Ungria 2022). Courts, therefore, require clear documentation, rigorous chain-of-custody practices, and explicit descriptions of analytical limitations.

Informed consent is another critical issue. Microbiome sampling often captures far more information than conventional DNA typing. Individuals may be unaware that their microbial data could reveal health status or other personal traits. Ethical practice requires explicit policies outlining what data are collected, how they will be stored, who may access them, and how long they will be retained. Without such safeguards, investigators risk violating privacy and eroding public trust.

Legal challenges also arise from contamination risks. If environmental DNA or substrate-associated microbes infiltrate a sample, false associations may occur.

Forensic experts must be able to demonstrate rigorous contamination control and provide comprehensive metadata—including handling conditions, storage temperatures, batch history, and full documentation of laboratory and bioinformatic procedures.

Additional legal complexities involve cross-jurisdictional variability in privacy laws, data protection regulations, evidence storage requirements, and admissibility standards. International or inter-state collaborations may require formal agreements defining microbial data handling, access rights, and transfer protocols.

Responsible application of microbial forensics relies on transparency, proportionality, and respect for privacy. Practitioners must clearly communicate both the strengths and limitations of microbial evidence, avoiding overstatement of its significance. Ethical stewardship is essential not only for safeguarding individual rights but also for maintaining scientific integrity and public confidence in forensic practice.

9 Future Perspectives

9.1 Integration of Multiomics Approaches

The future trajectory of microbial forensics is closely tied to the adoption of multiomics platforms—including metagenomics, transcriptomics, proteomics, metabolomics, and epigenomics—which offer higher-resolution microbial signatures and help overcome the inherent limitations of single-omic datasets (Maric et al. 2023; Nielsen et al. 2023). Current workflows rely primarily on DNA-based massive parallel sequencing (MPS), a highly informative but imperfect approach. Challenges associated with MPS—such as short read lengths, large storage requirements, and complex bioinformatics—have prompted interest in incorporating multiomics to improve resolution and functional interpretation (Wohlfahrt et al. 2023).

Multiomics integration can reveal functional, metabolic, and regulatory states of microbial communities. For example, combining metagenomic profiles with metabolomic fingerprints may provide insight into microbial metabolic activity, responses to environmental stress, and substrate-specific biochemical transformations (Maric et al. 2023; Nielsen et al. 2023). Such insights may improve postmortem interval (PMI) estimation or enhance geolocation accuracy by linking microbial functions to local environmental signatures.

Similarly, multiomics datasets may improve discrimination between closely related strains and strengthen microbial source attribution where geographically adjacent sites share similar taxonomic compositions (Wohlfahrt et al. 2023). These approaches can also enhance understanding of environmental variability, microbial stability, and population-level diversity—key constraints in current forensic analyses.

To be effective, multiomics-based workflows will require harmonized protocols, interoperable databases, and multidisciplinary collaboration. As bioinformatics tools continue to evolve, standardized cross-omics analytical frameworks will be essential to maintain consistency and avoid interpretive variability (Nielsen et al. 2023). Ultimately, multiomics strategies will make microbial evidence more precise, ecologically contextualized, and legally defensible, expanding the forensic toolkit.

9.2 Development of Microbial Databases for Forensic Use

One of the most urgent needs in microbial forensics is the development of comprehensive, standardized, and geographically representative microbial reference databases (Dobay et al. 2019; Maric et al. 2023; Wohlfahrt et al. 2023). Current limitations—including the absence of robust reference genomes, inconsistent metadata, and incomplete microbial diversity coverage—significantly constrain the evidentiary power of microbial analysis.

Effective databases must include strain-level genomic resolution, since forensic interpretation often depends on detecting subtle differences between genetically similar organisms. Soil forensics, for instance, requires detailed microbial biogeography maps due to extreme spatial specificity. Similarly, human-associated microbiome databases must account for population-level variation, enabling baseline frequency estimates and informed assessments of probative value.

Standardization is equally critical. Different taxonomic databases and annotation pipelines often produce inconsistent identifications. Beyond the well-documented limitations of 16S rRNA-based profiling, inconsistencies across reference libraries continue to present classification challenges. A validated microbial forensic database system must include:

- Controlled and versioned taxonomic annotation
- Documented algorithmic parameters
- Cross-jurisdiction compatibility
- Transparent reporting practices

Future databases must also incorporate rich environmental metadata—temperature, humidity, substrate type, geolocation, exposure duration—because microbial communities are acutely sensitive to environmental conditions. Such contextual data are foundational for interpreting microbial evidence at the activity level, particularly for evaluating persistence, transfer likelihood, and contamination risk.

Recent developments in national-level microbial forensic infrastructures demonstrate the value of centralized, well-funded systems. These frameworks support biosurveillance, outbreak attribution, and the integration of microbial evidence into forensic casework, strengthening the reliability and interpretability of microbial signatures.

9.3 Machine Learning and AI in Microbiome Data Interpretation

Machine learning (ML) and artificial intelligence (AI) are poised to transform forensic microbiome interpretation by enabling the analysis of high-dimensional, sparse, and compositional datasets—characteristics that pose substantial challenges to traditional statistical methods (Eisenhofer et al. 2019; Knights et al. 2011). Microbiome data commonly contain numerous zeroes, exhibit batch effects, and reflect complex ecological interdependencies, making ML methods particularly suitable.

Recent research has explored the use of:

- Random forest classifiers for predicting sample contributors or differentiating microbiomes from individuals living in the same environment. These classifiers show promise but exhibit increasing error rates in mixed-origin samples, indicating caution for forensic applications (Dobay et al. 2019; Maric et al. 2023).
- Bayesian tools such as SourceTracker, which can identify microbial contributions in mixed samples. Although effective in specific contexts (e.g., sexual contact studies), they require a minimum microbial contribution threshold (~10%), limiting applicability in low-transfer scenarios (Knights et al. 2011).
- ML-based contaminant detection, which may help identify and remove substrate-derived or laboratory-introduced taxa before interpretation (Eisenhofer et al. 2019).

Beyond classification, AI-driven pattern recognition could integrate multiomics data to identify taxonomic, functional, and metabolic signatures that are difficult to detect using conventional pipelines. As bioinformatics tools remain fragmented and frequently updated, AI systems may help stabilize analytical frameworks, automate quality control, and reduce reliance on heuristic software chains.

Despite these advantages, forensic implementation requires validated, interpretable, and standardized models. Error rates must be quantifiable, algorithms transparent, and workflow reproducible across laboratories. While AI offers immense potential, it will not become a routine forensic tool without careful validation and regulatory acceptance.

9.4 Potential for Routine Forensic Implementation

Transitioning microbial forensics into routine forensic practice requires coordinated progress in standardization, validation, infrastructure, and policy development. Currently, a major barrier is the lack of standardized protocols for sample collection, preservation, laboratory processing, sequencing, and analysis. These workflows must be validated for sensitivity, specificity, reproducibility, and limits of detection before they can gain widespread judicial acceptance.

Routine implementation also requires rigorous management of contamination risks, substrate effects, and the complexity of microbial transfer dynamics. Studies have repeatedly shown that substrate contamination can alter microbial profiles, emphasizing the need for validated collection procedures, stringent laboratory workflows, and robust computational decontamination methods (Salter et al. 2014; Eisenhofer et al. 2019). Batch effects, index hopping, and inconsistent pipelines highlight the need for controlled processing environments.

At an organizational level, national microbial forensic systems—integrated with public health surveillance, specialized laboratories, and trained personnel—can provide the infrastructure needed for regular forensic use. Such frameworks have already strengthened responses to biocrimes, environmental contamination, and outbreak investigations (Oliveira and Amorim 2018).

For microbial profiling to become standard forensic practice, accreditation bodies must develop microbial-specific quality standards analogous to those used for DNA profiling. Emerging links between microbial forensics and bioterrorism investigations suggest increasing institutional recognition of the field. Nonetheless, routine implementation will depend on unified advancements in method standardization, laboratory capacity, validated analytics, and national-level policy frameworks.

10 Conclusion

Bloodstain microbiome analysis expands the forensic toolkit by providing information when human DNA is degraded, limited, or unavailable. Microbial communities in bloodstains reflect both donor and environmental inputs, evolve predictably, and respond to temperature, humidity, and substrate, offering temporal and contextual indicators that complement traditional evidence.

Controlled laboratory studies and case investigations show that microbial succession can be modeled, enabling source attribution and temporal reconstruction. However, challenges such as low biomass, substrate overprint, contamination, and evolving bioinformatic pipelines highlight the need for caution and rigorous documentation.

Standardization is essential for reliable and admissible microbial evidence. Consistent protocols, comprehensive reference databases, and harmonized reporting improve reproducibility, transparency, and legal defensibility. Multiomics and machine learning approaches further enhance temporal, functional, and spatial inference, but require thorough validation and clearly defined error rates before use in court.

Ethical and legal considerations—including privacy protection, informed consent, chain-of-custody, and contamination control—must parallel technological advances. Integrating microbial profiles with conventional evidence strengthens investigations, supporting fair, scientifically grounded conclusions. When underpinned by robust methods and databases, bloodstain microbiome analysis can transition from research to routine forensic application, extending analytical capabilities without replacing human DNA profiling.

References

Aagaard K, Riehle K, Ma J, Segata N, Mistretta TA, Coarfa C et al (2013) The human microbiome project: metagenomic analysis of microbial communities in human samples. Nature 486(7402):207–214

Amir A (2021) Microbiome analysis using 16S amplicon sequencing: from samples to ASVs. In: Deep sequencing data analysis. Springer US, New York, pp 123–141

Arriola LA, Cooper A, Weyrich LS (2020) Palaeomicrobiology: application of ancient DNA sequencing to better understand bacterial genome evolution and adaptation. Front Ecol Evol 8:40

Bai P, Wang J, Yin H, Tian Y, Yao W, Gao J (2016) Discrimination of human and nonhuman blood by Raman spectroscopy and partial least squares discriminant analysis. Anal Lett 50(2):379–388. https://doi.org/10.1080/00032719.2016.1176033

Breeze RG, Budowle B, Schutzer SE (2005) Microbial forensics. Elsevier Academic Press. https://doi.org/10.1086/498524

Budowle B, Schmedes SE, Wendt FR (2017) Increasing the reach of forensic genetics with massively parallel sequencing. Forensic Sci Med Pathol 13(3):342–349

Butler, J. M. (2015). Advanced topics in forensic DNA typing: Interpretationy. Academic Press. San Diego, California

Caporaso JG, Lauber CL, Walters WA, Berg-Lyons D, Huntley J, Fierer N et al (2012) Ultra-high-throughput microbial community analysis on the Illumina HiSeq and MiSeq platforms. ISME J 6(8):1621–1624

Castro AE, De Ungria MCA (2022) Methods used in microbial forensics and epidemiological investigations for stronger health systems. Forens Sci Res 7(4):650–661. https://doi.org/10.1080/20961790.2021.2023272

Choo JM, Leong LE, Rogers GB (2015) Sample storage conditions significantly influence faecal microbiome profiles. Sci Rep 5(1):16350

Clarke TH, Gomez A, Singh H, Nelson KE, Brinkac LM (2017) Integrating the microbiome as a resource in the forensics toolkit. U.S. Department of Justice, National Institute of Justice. (NCJ 252050)

Cobaugh KL, Schaeffer SM, DeBruyn JM (2015) Functional and structural succession of soil microbial communities below decomposing human cadavers. PLoS One 10(6):e0130201

De-Dios T, van Dorp L, Charlier P, Morfopoulou S, Lizano E, Bon C et al (2020) Metagenomic analysis of a blood stain from the French revolutionary Jean-Paul Marat (1743–1793). Infect Genet Evol 80:104209

Demanèche S, Schauser L, Dawson L, Franqueville L, Simonet P (2017) Microbial soil community analyses for forensic science: application to a blind test. Forensic Sci Int 270:153–158. https://doi.org/10.1016/j.forsciint.2016.12.004

Dobay L, Hanssen E et al (2019) Microbiome-based body fluid identification of samples exposed to indoor conditions. Forensic Sci Int Genet 42:102172

Eisenhofer R, Minich JJ, Marotz C, Cooper A, Knight R, Weyrich LS (2019) Contamination in low microbial biomass microbiome studies: issues and recommendations. Trends Microbiol 27(2):105–117

Franzosa EA, Huang K, Meadow JF et al (2015) Identifying personal microbiomes using metagenomic codes. Proc Natl Acad Sci 112(22):E2930–E2938. https://doi.org/10.1073/pnas.1423854112

Gouello A, Dunyach-Remy C, Siatka C, Lavigne J-P (2021) Analysis of microbial communities: an emerging tool in forensic sciences. Diagnostics 12(1):Article 1. https://doi.org/10.3390/diagnostics12010001

Green MR, Sambrook J (2012) Molecular cloning. A laboratory manual, 4th edn. Cold Spring Harbor Laboratory Press, New York, p 448

Gupta VK, Paul S, Dutta C (2017) Geography, ethnicity or subsistence-specific variations in human microbiome composition and diversity. Front Microbiol 8:1162. https://doi.org/10.3389/fmicb.2017.01162

Gürsoy N, Karadayı S, Akmayan İ, Karadayı B, Özbek T (2024) Time-dependent change in the microbiota structure of seminal stains exposed to indoor environment. Int J Legal Med 138(2):591–602. https://doi.org/10.1007/s00414-023-03108-9

Hauther KA, Cobaugh KL, Jantz LM, Sparer TE, DeBruyn JM (2015) Forensic analysis of microbial succession in human cadavers. Appl Environ Microbiol 81(21):7439–7448

Houck MM, Siegel JA (2009) Fundamentals of forensic science. Academic Press (Elsevier). Burlington, MA/San Diego, CA

Howarth A, Drummond B, Wasef S, Matheson CD (2022) An assessment of DNA extraction methods from blood-stained soil in forensic science. Forensic Sci Int 341:111502

Hua H, Meydan C, Afshin EE, Lili LN, D'Adamo CR, Rickard N, Dudley JT, Price ND, Zhang B, Mason CE (2022) A wipe-based stool collection and preservation kit for microbiome community profiling. Front Immunol 13:889702. https://doi.org/10.3389/fimmu.2022.889702

Human Microbiome Project Consortium (2012) Structure, function and diversity of the healthy human microbiome. Nature 486:207–214. https://doi.org/10.1038/nature11234

Hyde ER, Haarmann D, Lynne A, Bucheli SR, Petrosino JF (2013) The living dead: bacterial community structure of a cadaver at the onset and end of decomposition. PLoS One 8(10):e77733

Javan GT, Finley SJ, Smith T, Miller J, Wilkinson JE (2017) Cadaver thanatomicrobiome signatures: the ubiquitous nature of Clostridium species in human decomposition. Front Microbiol 8:2096

Karadayı S, Yılmaz İ, Özbek T, Karadayı B (2024) Transfer and persistence of microbiota markers from the human hand to the knife: a preliminary study. J Forensic Leg Med 107:102757. https://doi.org/10.1016/j.jflm.2024.102757

Knight R, Vrbanac A, Taylor BC, Aksenov A, Callewaert C, Debelius J et al (2018) Best practices for analysing microbiomes. Nat Rev Microbiol 16(7):410–422

Knights D, Costello EK, Knight R (2011) Supervised classification of human microbiota. FEMS Microbiol Rev 35(2):343–359

Lax S, Smith DP, Hampton-Marcell J, Owens SM, Handley KM, Scott NM, Gibbons SM, Larsen P, Shogan BD, Weiss S, Metcalf JL, Ursell LK, Vazquez-Baeza Y, Van Treuren W, Hasan NA, Gibson MK, Colwell R, Dantas G, Knight R (2015) Longitudinal analysis of microbial interaction between humans and the indoor environment. Science 345(6200):1048–1052

López C, Hanssen E, Dobay L et al (2022) Advances in microbial metagenomics and artificial intelligence analysis in forensic identification. Front Microbiol 13:1046733

Madea B (2016) Methods for determining time of death. Forensic Sci Med Pathol 12(4):451–485

Maric T, Brecx M et al (2023) A review on microbial species for forensic body fluid identification in healthy and diseased humans. Curr Microbiol 80:299

Mazzarelli G, Pezzuto F, Cappetti B et al (2023) Changes in the microbial community after vaginal fluid exposure in different simulated forensic situations. Forensic Sci Int 345:111831

Metcalf JL, Xu ZZ, Weiss S, Lax S, Van Treuren W, Hyde ER et al (2016) Microbial community assembly and metabolic function during mammalian corpse decomposition. Science 351(6269):158–162

Metcalf JL, Xu ZZ, Bouslimani A, Dorrestein P, Carter DO, Knight R (2017) Microbiome tools for forensic science. Trends Biotechnol 35(9):814–823. https://doi.org/10.1016/j.tibtech.2017.03.006

Muniesa M, Hammerl JA, Hertwig S, Appel B, Brüssow H (2012) Shiga toxin-producing Escherichia coli O104: H4: a new challenge for microbiology. Appl Environ Microbiol 78(12):4065–4073

Neckovic A, van Oorschot RAH, Szkuta B, Durdle A (2020) Challenges in human skin microbial profiling for forensic science: a review. Genes 11(9):Article 1015. https://doi.org/10.3390/genes11091015

Nielsen B et al (2023) Metatranscriptomic characterization of six types of forensic samples and its potential application to body fluid/tissue identification. Forensic Sci Int Genet 65:102888

Nodari R, Arghittu M, Bailo P, Cattaneo C, Creti R, D'Aleo F, Saegeman V, Franceschetti L, Novati S, Fernández-Rodríguez A (2024) Forensic microbiology: when, where and how. Microorganisms 12(5):988. https://doi.org/10.3390/microorganisms12050988

Oliveira M, Amorim A (2018) Microbial forensics: new breakthroughs and future prospects. Appl Microbiol Biotechnol 102(24):10377–10391. https://doi.org/10.1007/s00253-018-9414-6

Païssé S, Valle C, Servant F, Courtney M, Burcelin R, Amar J, Lelouvier B (2016) Comprehensive description of blood microbiome from healthy donors assessed by 16 S targeted metagenomic sequencing. Transfusion 56(5):1138–1147

Poulsen CS, Kaas RS, Aarestrup FM, Pamp SJ (2021) Standard sample storage conditions have an impact on inferred microbiome composition and antimicrobial resistance patterns. Microbiol Spectrum 9(2):e01387–e01321

Robinson JM, Pasternak Z, Mason CE, Elhaik E (2021) Forensic applications of microbiomics: a review. Front Microbiol 11:608101

Rozenberga M, Saksis R, Elbere I, Birzniece L, Briviba M, Konrade I, Klovins J (2024) Analysis of the microbial community on the surface of bloodstains under different environmental conditions in areas with minimal human interference. Curr Microbiol 81(10):307. https://doi.org/10.1007/s00284-024-03833-3

Sacco MA, Gualtieri S, Tarallo AP, Calanna L, La Russa R, Aquila I (2024) The role of molecular investigations in estimating the time since deposition (TSD) of bloodstains: a systematic review of the literature. Int J Mol Sci 25(13):7469

Salter SJ, Cox MJ, Turek EM, Calus ST, Cookson WO, Moffatt MF, Turner P, Parkhill J, Loman NJ, Walker AW (2014) Reagent and laboratory contamination can critically impact sequence-based microbiome analyses. BMC Biol 12:87

Schmedes SE, Woerner AE, Budowle B (2017) Forensic human identification using skin microbiomes. Appl Environ Microbiol 83(22):e01672–e01617. https://doi.org/10.1128/AEM.01672-17

Schmidt RJ (1998) Criminal case involving HIV-1 transmission by injection —. FindLaw, Louisiana

Sender R, Fuchs S, Milo R (2016) Revised estimates for the number of human and bacteria cells in the body. PLoS Biol 14(8):e1002533

Shenhav L, Thompson M, Joseph TA, Briscoe L, Furman O, Bogumil D, Mizrahi I, Halperin E (2019) FEAST: fast expectation-maximization for microbial source tracking. Nat Methods 16(7):627–632

Song SJ, Amir A, Metcalf JL, Amato KR, Xu ZZ, Humphrey G, Knight R (2016) Preservation methods differ in fecal microbiome stability, affecting suitability for field studies. mSystems 1(3):e00021–e00016. https://doi.org/10.1128/mSystems.00021-16

Swayambhu M, Kümmerli R, Arora N (2023) Microbiome-based stain analyses in crime scenes. Appl Environ Microbiol 89(1):e01325–e01322

Tozzo P, Amico I, Delicati A, Toselli F, Caenazzo L (2022) Post-mortem interval and microbiome analysis through 16S rRNA sequencing: a systematic review. Diagnostics 12(11):2641. https://doi.org/10.3390/diagnostics12112641

Wahab A, Menon S, Jacob AA (2025) Evaluation of microbial growth in dried blood stains. Int J Innov Sci Res Technol 10(4):3750–3753

Wohlfahrt D, Tan-Torres AL, Green R, Brim K, Bradley N, Brand A et al (2023) A bacterial signature-based method for the identification of seven forensically relevant human body fluids. Forensic Sci Int Genet 63:102895

Yang DY, Watt K (2005) Contamination controls when preparing archaeological remains for ancient DNA analysis. J Archaeol Sci 32(3):331–336

Zhang J, Yu D, Zhang L, Wang T, Yan J (2025) Environmental microbiota from substrate may interfere with microbiome-based identification of forensically relevant body fluids: a pilot study. Forensic Sci Int Genet 74:103170. https://doi.org/10.1016/j.fsigen.2024.103170

Artificial Intelligence and Emerging Technologies in Blood Forensics

16

Abstract

Blood evidence has long been central to forensic investigations, but traditional methods are constrained by human subjectivity. Emerging technologies, particularly artificial intelligence (AI), are beginning to change this. They introduce greater precision, reproducibility, and efficiency into blood forensic analysis. This paper explores the integration of AI across four major domains of blood forensics: bloodstain pattern analysis, age estimation of bloodstains, three-dimensional (3D) crime scene reconstruction, and the development of cloud-based forensic databases. In pattern analysis, convolutional neural networks (CNNs) and other machine learning models have achieved classification accuracies surpassing human performance, reducing bias and enhancing consistency. In age estimation, hyperspectral imaging and spectroscopy paired with AI enable nondestructive, substrate-independent assessments of bloodstain timelines, advancing investigative accuracy. AI-driven simulations and immersive technologies such as virtual and augmented reality are redefining 3D reconstructions, providing investigators and jurors with intuitive visualizations of complex events. Meanwhile, forensic databases enhanced with computer vision and natural language processing (NLP) facilitate real-time case linkage and contextual analysis across jurisdictions. Despite these benefits, challenges remain concerning explainability, dataset bias, standardization, and the admissibility of AI-generated evidence under legal standards like the Daubert criteria. This paper argues that while AI will not replace forensic expertise, it will augment it by providing reproducible insights, rapid analysis, and scalable infrastructure. Ultimately, the responsible integration of AI and emerging technologies in blood forensics offers a pathway toward more objective science and stronger foundations for justice, provided that ethical safeguards, transparency, and human oversight remain central to their deployment.

H. R. Dash et al., *Forensic Investigation of Blood Evidence*,
https://doi.org/10.1007/978-981-95-9183-1_16

Keywords

Forensic Science · Bloodstain Analysis · Artificial Intelligence · Hyperspectral Imaging · Crime Scene Reconstruction.

1 Introduction

Blood evidence has consistently played a central role in forensic investigations, offering insights into both the nature and sequence of criminal acts. From early criminal cases in the nineteenth century to modern forensic laboratories, it has often provided compelling narratives—though not always without controversy. Yet despite its enduring value, blood evidence has also presented a longstanding challenge: the interpretation of bloodstain patterns is not always straightforward. Subtle differences in droplet size, distribution, and trajectory may suggest different scenarios, and the reliance on human experts introduces subjectivity into the analysis process (Sirisathitkul and Sirisathitkul 2025). As a result, while bloodstain pattern analysis (BPA) has been influential, it has historically been viewed with caution regarding its reproducibility and courtroom reliability.

Traditionally, BPA has been performed by forensic analysts trained to classify stains into categories such as passive drops, impact spatters, or transfer marks. Analysts also attempt to infer the mechanism of injury, point of origin, or even the sequence of events. However, studies over the past two decades have documented significant *inter-observer variability* in bloodstain interpretations. Different experts analyzing the same crime scene photographs may reach divergent conclusions about what occurred, raising questions about the admissibility of BPA testimony under standards such as the Daubert or Frye criteria in U.S. courts (Sirisathitkul and Sirisathitkul 2025). This lack of consistency reflects not only the inherent complexity of fluid dynamics in bloodstain formation but also the cognitive biases that can influence human interpretation.

The advent of *artificial intelligence (AI)* and *emerging technologies* in forensic science marks a turning point in addressing these limitations. With the rapid development of computer vision, machine learning, and high-resolution imaging, researchers have begun to apply computational methods to blood forensics with the goal of making analysis more objective, replicable, and scientifically defensible (Sirisathitkul and Sirisathitkul 2025). AI-driven approaches allow for the systematic examination of thousands of stains in a way that no human analyst could achieve within the same timeframe. Moreover, the ability of machine learning models to recognize patterns invisible to the human eye has opened possibilities for new forms of evidence interpretation.

One of the earliest areas where AI has shown promise is in *bloodstain pattern classification*. By training convolutional neural networks (CNNs) on large image datasets, researchers have been able to achieve classification accuracies exceeding 90%, surpassing many traditional approaches (Chen et al. 2025). Such algorithms not only distinguish between spatter categories but also detect confounding artifacts such as insect activity or cleaning attempts, which have historically complicated manual analysis. Importantly, these computational models also generate reproducible outputs, meaning that two analysts using the same trained model will arrive at the same result, thereby reducing subjective variability.

Another critical application of AI in blood forensics is *bloodstain age estimation*, an area that has long challenged forensic scientists. Determining how long a bloodstain has been at a crime scene is crucial for establishing timelines, corroborating witness statements, and distinguishing between relevant and irrelevant stains. Traditional biochemical methods often involve destructive sampling and can be highly sensitive to environmental conditions such as temperature, humidity, and light exposure. By contrast, emerging technologies such as hyperspectral imaging and infrared spectroscopy, when paired with machine learning algorithms, enable *nondestructive, image-based estimation of bloodstain age* with considerably improved accuracy (Hassan et al. 2024). These approaches use AI models to learn degradation patterns in hemoglobin and other biomolecules, producing results that are both rapid and adaptable to real-world forensic settings.

Beyond classification and age estimation, AI technologies have also enhanced *3D reconstruction of crime scenes involving blood evidence*. Physics-based simulations combined with AI algorithms can reconstruct the trajectory of blood droplets, estimate angles of impact, and create immersive virtual models of crime scenes. Such reconstructions allow investigators, jurors, and judges to visualize events more intuitively, potentially increasing the clarity and impact of forensic testimony (Acampora and Vitiello 2024). While early reconstructions required extensive manual input, automated systems now use deep learning to streamline the process, reducing both time and error.

Importantly, these technological advancements are not confined to research laboratories. Several forensic laboratories worldwide are beginning to explore AI-based tools for operational use, integrating them into digital workflows and cloud-based forensic databases. Automated systems that can compare new stains with existing case data are being developed to support cross-case analysis, potentially identifying links between cases or uncovering patterns suggestive of serial offenses (Ketsekioulafis et al. 2024). This represents a significant step toward a more networked, data-driven forensic science ecosystem.

However, the introduction of AI into blood forensics also raises significant *ethical and legal concerns*. Unlike human experts, AI systems often function as "black boxes," producing highly accurate predictions without transparent reasoning. This lack of explainability challenges the principle of evidentiary transparency in court, where judges and juries must be able to understand how conclusions were reached (Pappachan et al. 2024). Furthermore, the reliability of AI depends on the quality and representativeness of training datasets. If training data are biased or incomplete—for example, if they underrepresent stains formed in certain environmental conditions—then the resulting models may yield skewed or misleading results. This raises concerns about fairness and reliability in judicial outcomes.

Despite these challenges, there is widespread recognition that the *integration of AI and emerging technologies into blood forensics is both inevitable and necessary*. The increasing complexity of criminal cases, combined with public demands for scientific rigor and objectivity, compels forensic science to adopt methods that minimize error and maximize transparency. Researchers are therefore calling for a balanced approach: one that embraces the power of AI while also ensuring rigorous validation, standardized protocols, and robust oversight mechanisms.

In summary, the field of blood forensics is undergoing a profound transformation. Where once the interpretation of bloodstains was limited to expert judgment, today it is being augmented by powerful computational tools capable of rapid, reproducible, and precise analysis. As this shift unfolds, it is essential for forensic scientists, legal professionals, and policymakers to critically engage with the opportunities and risks that AI presents. This paper explores these developments in greater depth, examining how artificial intelligence and emerging technologies are reshaping blood forensics, what benefits they offer, and what challenges must be addressed to ensure their responsible application in the justice system.

2 AI in Bloodstain Pattern Analysis (BPA)

Traditionally, bloodstain pattern analysis (BPA) relied on the careful visual inspection of stain morphology, size, and distribution. While this expert judgment has helped solve many cases, it has also been criticized as prone to subjectivity. Although expert judgment has contributed to solving countless cases, it is increasingly recognized that this method is prone to subjectivity and cognitive bias. Analysts may disagree on whether a stain was created by impact, transfer, or passive mechanisms, leading to concerns about reliability and reproducibility in court (Sirisathitkul and Sirisathitkul 2025). In recent years, however, artificial intelligence (AI) and machine learning (ML) have emerged as powerful tools to address these limitations, offering quantitative, replicable, and scalable approaches to bloodstain classification.

2.1 From Visual Inspection to Machine Learning Models

The integration of computer vision and pattern recognition into BPA began with relatively simple machine learning classifiers such as support vector machines (SVMs), decision trees (DTs), and random forests (RFs). These models were trained on hand-crafted features such as droplet size, circularity, and edge sharpness to distinguish between categories like impact spatters, drip stains, and transfer marks. While these early models showed moderate accuracy, they represented an important step toward removing subjective judgment from BPA (Zulfiqar et al. 2021).

Subsequent research demonstrated that these models could be optimized using feature extraction techniques from hyperspectral imaging, allowing classifiers to access information beyond the visible spectrum. For example, Giulietti et al. (2022) developed neural networks to correct spectral distortions caused by different substrates, which significantly improved classification accuracy. Similarly, Wei et al. (2024) combined ATR-FTIR spectroscopy with SVMs to identify bloodstains on various substrates, showing that machine learning can generalize across complex crime scene environments.

2.2 The Rise of Deep Learning in BPA

While traditional ML models rely heavily on manually defined features, deep learning, particularly convolutional neural networks (CNNs), has revolutionized BPA by enabling automated feature learning directly from raw image data. CNNs excel at identifying subtle spatial patterns, making them ideal for complex stain morphology.

Sirisathitkul and Sirisathitkul (2025) reported that CNN models achieved over 95% classification accuracy when distinguishing between passive, projected, and transfer stains. More advanced architectures such as Inception v3 and residual neural networks (ResNets) have been tested, with Chen et al. (2025) demonstrating their ability to differentiate true impact spatters from fly artifacts, a distinction that has long troubled forensic analysts. These models not only increased accuracy but also reduced the time required for large-scale stain analysis, offering scalability that human analysts could never achieve.

Furthermore, CNN-based BPA systems have been applied in 3D reconstructions of spatter trajectories. Murugan et al. integrated Inception v3 with geometric modeling to reconstruct angles of impact, while Jung et al. (2024) introduced hybrid frameworks that combine CNNs with Decision Trees to improve interpretability. These hybrid approaches address one of the major criticisms of AI in forensic contexts: the "black box" problem, where models generate results without transparent reasoning.

2.3 Enhancements Through Hyperspectral Imaging

A particularly promising frontier in BPA involves the combination of hyperspectral imaging (HSI) with machine learning. Unlike standard photography, HSI captures data across hundreds of spectral bands, revealing chemical and structural details not visible to the human eye. Zulfiqar et al. (2021) demonstrated that integrating HSI with classifiers such as CNNs, SVMs, and RFs could identify bloodstains even under degraded or contaminated conditions. Recent work by Garapati et al. (2025) reviewed how hyperspectral data, paired with CNNs and Decision Trees, offers higher robustness against environmental variations such as lighting, substrate interference, or partial cleaning.

These developments suggest that HSI-enhanced BPA could become a standard tool in forensic laboratories, particularly when analysts are faced with ambiguous or degraded evidence.

2.4 Interdisciplinary Innovations

AI in BPA does not exist in isolation; it is increasingly integrated with other forensic tools. For instance, Ojo et al. (2023) used CNNs and SVMs for predicting crime scene events with drone-collected imagery, enabling investigators to combine aerial scene data with close-up BPA evidence. Similarly, Abraham et al. (2021) proposed a "Tree-CNN" hybrid framework to automatically classify broader categories of crime scene images, demonstrating how BPA can be integrated into multimodal forensic intelligence systems.

Additionally, the growing field of explainable AI (XAI) seeks to bridge the gap between computational power and courtroom admissibility. Researchers argue that hybrid models combining interpretable algorithms (like Decision Trees) with deep learning can generate both accurate and legally defensible outcomes (Thakkar et al. 2023).

Table 16.1 compares common AI models applied to BPA, highlighting their relative strengths, reported accuracy, and key references.

Figure 16.1 visualizes the reported accuracy rates of major AI models, illustrating the superior performance of CNNs and ResNets compared to earlier approaches.

2.5 Challenges and Limitations

Despite these advances, several challenges remain before AI-based BPA can be widely adopted in forensic practice. One major concern is dataset bias. Most machine learning models are trained on controlled laboratory-generated stains, which may not reflect the complexity of real-world crime scenes where environmental contamination, cleaning attempts, or mixed blood sources complicate analysis. As Jung et al. (2024) noted, CNNs require vast amounts of high-quality labeled data, and obtaining such datasets for BPA remains resource-intensive.

Another limitation is the explainability problem. Judges, jurors, and even forensic practitioners may be hesitant to accept results generated by models that cannot provide clear reasoning. Although hybrid approaches are being explored, the tension between accuracy and interpretability remains a key research question.

Table 16.1 Comparison of AI models in bloodstain pattern analysis

AI Model	Application	Reported Accuracy	Key Reference
Support vector machines (SVM)	Basic stain classification	~80%	Zulfiqar et al. (2021)
Decision Trees	Interpretable decision-making	~85%	Thakkar et al. (2023)
Convolutional neural networks (CNN)	Automated feature extraction, high accuracy	90â€“95%+	Sirisathitkul and Sirisathitkul (2025)
Residual neural networks (ResNets)	Distinguishing impact vs. fly artifacts	95%+	Chen et al. (2025)

Source: Author's compilation based on Sirisathitkul and Sirisathitkul (2025), Chen et al. (2025)

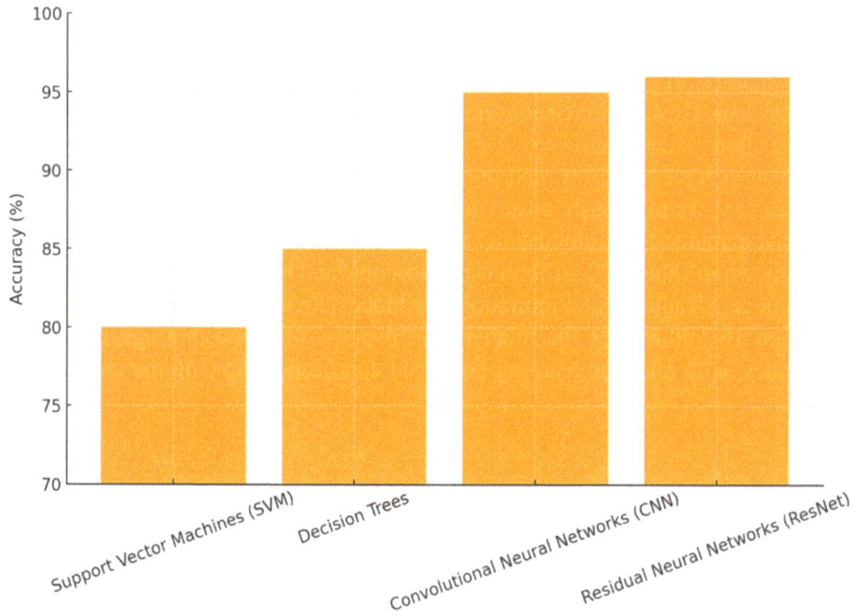

Fig. 16.1 Reported accuracy of AI models in BPA. (Source: Author's compilation based on Chen et al. (2025))

2.6 The Road Ahead

Looking forward, BPA is likely to evolve into a multilayered analytical process where AI complements, rather than replaces, human expertise. Analysts will increasingly act as supervisors who interpret AI outputs, validate results, and provide courtroom testimony that integrates both computational and human reasoning. Collaborative databases and cloud-based BPA platforms will enable global sharing of AI models, ensuring continuous improvement and standardization across jurisdictions.

Ultimately, AI is transforming BPA from a subjective craft into a quantifiable science, improving both the accuracy and credibility of forensic evidence. While challenges of data quality, interpretability, and legal acceptance remain, the trajectory of research strongly suggests that AI-*assisted BPA will become a cornerstone of forensic science in the next decade.*

3 Age Estimation of Bloodstains

Establishing the time since deposition (TSD) of bloodstains is one of the most critical yet challenging tasks in forensic science. A reliable estimate of how long a bloodstain has been present at a scene can help investigators reconstruct timelines, confirm or challenge alibis, and distinguish between relevant and irrelevant biological traces. Traditionally, forensic scientists have relied on biochemical and morphological assessments to determine bloodstain age. These include observing visible changes in color due to hemoglobin degradation or employing destructive chemical tests. However, such conventional methods suffer from several drawbacks: they are often inaccurate, destructive, and highly sensitive to environmental conditions such as light, humidity, and temperature (Kumar et al. 2020). Consequently, there has been a strong push toward developing more objective, nondestructive, and quantifiable methods for bloodstain age estimation.

As shown in Table 16.2, different spectroscopic and machine learning techniques offer unique advantages and limitations for bloodstain age estimation.

Figure 16.2 illustrates the relative effectiveness of bloodstain age estimation techniques, with machine learning-enhanced approaches showing the most robust results.

3.1 Traditional Approaches and Their Limitations

Bloodstains undergo a predictable yet complex biochemical transformation after deposition. Hemoglobin deoxygenates, oxidizes, and eventually breaks down into biliverdin and bilirubin, causing visible changes in color from bright red to brownish-black. Early forensic methods relied heavily on these *colorimetric observations*, but such techniques are inherently subjective. Even instrumental methods, such as infrared or Raman spectroscopy, while more precise, have often required destructive sampling or provided inconsistent results in uncontrolled conditions (Thanakiatkrai et al. 2013).

Moreover, traditional approaches struggle to incorporate *environmental variability*. A bloodstain exposed to sunlight may degrade more quickly than one in the shade, while humidity and substrate material can also significantly alter spectral characteristics (Fonseca et al. 2025). Without accounting for these factors, conventional models often fail to generalize, undermining their forensic reliability.

Table 16.2 Techniques for bloodstain age estimation

Technique	Strength	Limitation	Key Reference
Infrared spectroscopy	Accurate chemometric models	Destructive, sensitive to environment	Kumar et al. (2020)
Hyperspectral imaging (HSI)	Substrate-independent, detailed spectra	Expensive equipment	Giulietti et al. (2023)
Near-infrared (NIR) spectroscopy	Portable, on-site analysis	Dependent on calibration	Fonseca et al. (2025)
Random forest regression	Handles temporal variability in color	Requires large training data	Seki et al. (2024)

Source: Author's compilation based on key references

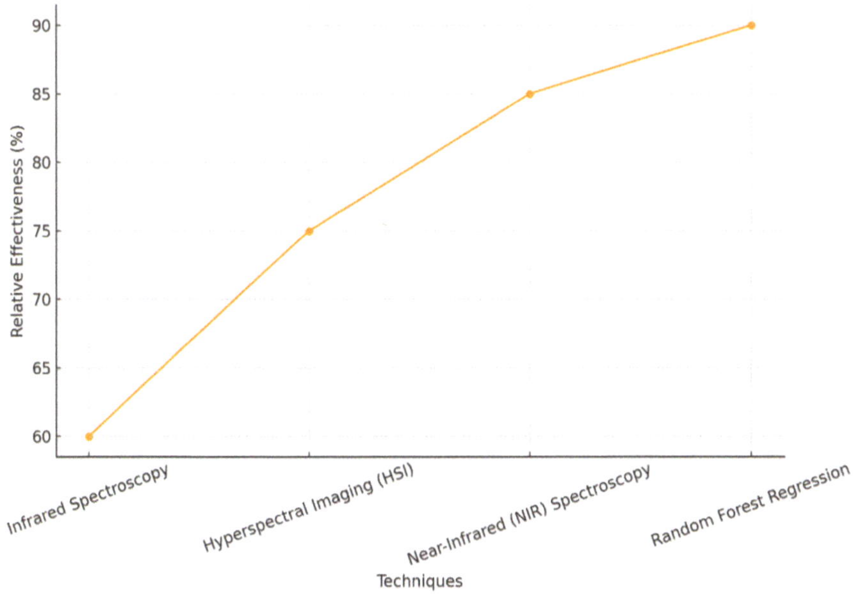

Fig. 16.2 Effectiveness of techniques in bloodstain age estimation. (Source: Author's compilation based on Giulietti et al. (2023), Seki et al. (2024))

3.2 Hyperspectral Imaging (HSI) as a Game Changer

Hyperspectral imaging (HSI) has emerged as one of the most promising techniques for noninvasive bloodstain analysis. Unlike ordinary photography, HSI captures information across dozens or even hundreds of spectral bands, revealing details of hemoglobin degradation invisible to the naked eye. Each pixel in an HSI image contains a complete reflectance spectrum, allowing forensic scientists to monitor biochemical changes with high sensitivity (Giulietti et al. 2022).

AI models such as *convolutional neural networks (CNNs)* can then be trained on these spectral signatures to predict bloodstain age. For example, Giulietti et al. (2023) developed a neural network–based HSI approach that corrected substrate distortions, enabling accurate TSD estimation regardless of the surface material. Similarly, Butt et al. (2022) introduced a hybrid 2D–3D CNN model that processed both spatial and spectral information, demonstrating improved performance over traditional chemometric models. These advances indicate that HSI, when combined with machine learning, offers a *substrate-independent* and *environmentally robust* method of bloodstain aging.

3.3 Machine Learning and Spectroscopy

In addition to HSI, other spectroscopic techniques such as *attenuated total reflectance–Fourier transform infrared (ATR-FTIR)*, *near-infrared (NIR) spectroscopy*, and *Raman spectroscopy* have been paired with machine learning to estimate bloodstain age. Kumar et al. (2020) used ATR-FTIR with chemometric models to estimate bloodstain ages ranging from 1 to 175 days, reporting significantly higher accuracy than traditional biochemical tests.

Recent studies have extended these methods using more advanced ML techniques. For instance, Seki et al. (2024) developed a *random forest regression model* using temporal colorimetric data, demonstrating reliable predictions even under varying lighting conditions. Similarly, Fonseca et al. (2025) used handheld NIR spectroscopy devices coupled with chemometric analysis to estimate the age of bloodstains on different substrates, paving the way for portable, on-scene applications. These methods show the versatility of combining *spectral data with AI*, allowing for more rapid and accurate forensic assessments.

3.4 Incorporating Environmental Variables

One of the most significant advantages of AI models in bloodstain age estimation is their ability to incorporate *environmental variability*. Traditional models often failed when stains were exposed to nonstandard conditions, but machine learning algorithms can learn from diverse datasets that include variables such as *temperature, humidity, substrate, and light exposure*. For example, Cooney et al. (2024) demonstrated that HSI-based deep learning models could discriminate between human and animal bloodstains while simultaneously accounting for environmental noise.

Such approaches not only improve reliability but also reduce false conclusions in forensic practice. By learning from a broad spectrum of real-world conditions, AI-based models can provide *more consistent and legally defensible results*, thereby increasing their admissibility in court.

3.5 Emerging Trends and Applications

Beyond laboratory validation, researchers are now developing *portable and real-time systems* for bloodstain age estimation. Advances in miniaturized HSI cameras and handheld NIR devices mean that investigators could soon perform age estimation *directly at crime scenes* without the need to transport samples to a lab (Fonseca et al. 2025). AI algorithms embedded into these devices could provide immediate TSD estimates, dramatically speeding up investigations.

Additionally, cross-disciplinary efforts are expanding the potential applications of these tools. For example, Devassy and George (2021) applied HSI to beverage stains, demonstrating that similar methods could distinguish between biological and nonbiological stains—a critical task in ensuring only relevant evidence is processed. These innovations suggest that AI-based spectral methods will soon form part of a *standard toolkit in forensic practice.*

3.6 Challenges and Ethical Considerations

Despite the promise of AI-enhanced bloodstain age estimation, challenges remain. First, *dataset quality and diversity* are crucial. Most current studies use laboratory-prepared stains, which may not fully capture the complexity of real crime scenes where blood mixes with dirt, cleaning agents, or other contaminants. Expanding datasets to include such variability is essential for model robustness (Giulietti et al. 2023).

Second, forensic science faces the ongoing issue of *explainability*. Judges and juries must be able to understand how AI models arrive at their predictions. Black-box models, while accurate, may face skepticism in court if their reasoning cannot be clearly articulated (Sirisathitkul and Sirisathitkul 2025). Efforts to integrate explainable AI (XAI) methods are therefore critical to bridging the gap between computational accuracy and judicial acceptance.

Finally, there are ethical concerns about over-reliance on AI in high-stakes legal contexts. While AI can augment human expertise, it should not replace forensic judgment. Instead, AI-based age estimation should be viewed as a *complementary tool* that provides objective support to expert testimony.

4 3D Crime Scene Reconstruction

The reconstruction of crime scenes has long been a central practice in forensic science, helping investigators visualize events and present findings to courts in a clear and persuasive manner. Among the most challenging aspects of reconstruction is accurately tracing the trajectory and distribution of blood spatters, which can provide vital clues about the location of victims and suspects, the number of blows or gunshots, and the type of weapon used. Traditionally, investigators relied on stringing methods, trigonometric calculations, and two-dimensional photographs to approximate the origin of bloodstains. While these methods provided useful insights, they were limited by observer bias, measurement errors, and the inability to capture the full complexity of three-dimensional (3D) dynamics (Raneri 2018).

The emergence of artificial intelligence (AI) and physics-based simulation tools has transformed this area of forensic practice. Today, forensic scientists are increasingly using advanced imaging, computer vision, and 3D modeling systems to create accurate, immersive reconstructions of crime scenes. By integrating video footage, laser scanning, geospatial data, and fluid dynamic models of blood droplet motion, these tools generate highly detailed virtual environments where events can be replayed and examined from multiple perspectives (Villa et al. 2023). This technological shift not only improves investigative accuracy but also strengthens the ability to communicate findings to juries in a compelling, understandable format.

4.1 From Stringing to Simulation

Earlier techniques such as *stringing*—in which elastic strings were physically attached to bloodstains to estimate trajectories—offered crude visualizations but lacked precision. Modern systems instead rely on *3D scanning technologies* such as structured light, photogrammetry, and laser scanning to capture the entire crime scene in digital form. These scans can then be processed by AI algorithms that identify stains, classify them, and compute their likely origins based on droplet size, angle, and velocity (Wang et al. 2019).

For example, Home et al. (2024) demonstrated that virtual reality (VR) reconstructions based on 3D scans helped jurors better understand bloodstain evidence than traditional two-dimensional exhibits. Such immersive presentations significantly reduced misinterpretation, underscoring the courtroom value of AI-assisted 3D models.

4.2 AI and Physics-Based Models

One of the most powerful applications of AI in 3D reconstruction is the use of *fluid dynamics simulations* combined with machine learning to replicate how blood behaves under different forces. By training on datasets of experimentally generated spatters, AI systems can estimate droplet trajectories with remarkable accuracy. These models can then be integrated into 3D digital environments, enabling investigators to replay events and test alternative hypotheses (Sirisathitkul and Sirisathitkul 2025).

Recent studies highlight the growing precision of these methods. Sacco et al. (2025) applied BPA-based simulations to reconstruct a staged crime scene, showing how AI models could distinguish genuine impact spatters from deliberately fabricated ones. Similarly, Patsiouras et al. (2024) combined computer vision with deep learning to reconstruct both ballistic and bloodstain evidence in virtual 3D environments, illustrating the potential of multimodal AI frameworks in forensic reconstruction.

4.3 Virtual and Augmented Reality Applications

The integration of *virtual reality (VR)* and *augmented reality (AR)* into forensic reconstruction is expanding rapidly. VR systems allow investigators and juries to enter immersive, reconstructed crime scenes where blood trajectories and points of origin can be visualized in real time (Maneli and Isafiade 2022). This approach has been shown to enhance comprehension in nonexpert audiences, who often struggle with abstract forensic testimony.

Augmented reality, by contrast, overlays digital reconstructions onto the physical crime scene, enabling investigators to view predicted trajectories and potential impact points while still on site (Sachdeva et al. 2024). This fusion of digital and physical evidence improves both the speed and accuracy of investigations.

4.4 Geospatial and Multimodal Reconstructions

Recent advancements also emphasize the *integration of geospatial data*. Razali et al. (2025) demonstrated how drone-based aerial imagery combined with ground-level scans could reconstruct crime scenes on both macro and micro levels, including the mapping of blood spatters across large areas. Such multimodal datasets allow for more holistic reconstructions, combining body positioning, weapon trajectory, and spatter distribution into a single, unified model.

Similarly, Ma (2024) introduced reinforcement learning models into crime scene simulations, enabling AI systems to adaptively improve their reconstructions by testing multiple event hypotheses against observed data. This type of *adaptive simulation* is particularly valuable in complex cases where multiple plausible scenarios must be explored.

4.5 Legal and Ethical Considerations

While 3D reconstruction holds clear benefits, it also raises challenges. Courts must evaluate whether AI-driven reconstructions meet evidentiary standards such as the *Daubert test*, which requires methods to be scientifically valid and generally accepted. Critics argue that AI models—particularly those relying on deep learning—may operate as "black boxes" whose internal reasoning is not transparent to juries or judges (Arthanari et al. 2025).

Another concern involves the *risk of persuasive bias*. Highly realistic VR reconstructions may unduly influence jurors by presenting a particular narrative as fact, even if the underlying assumptions are contested. To mitigate these risks, experts stress the need for transparency about model parameters, error rates, and the limitations of reconstructions (Sessa et al. 2025).

4.6 Future Directions

The trajectory of current research suggests that AI-based 3D reconstruction will soon become a standard component of forensic investigations. Advances in *explainable AI (XAI)* may improve the transparency of blood spatter models, while the increasing availability of portable scanning equipment will enable on-site reconstruction in real time. Furthermore, integrated platforms may allow investigators to upload case data into cloud-based systems, where AI can compare reconstructions against large forensic databases, potentially identifying patterns across cases or linking crimes that share ballistic or spatter characteristics.

Ultimately, the integration of AI into 3D crime scene reconstruction represents more than just a technological upgrade—it reflects a fundamental shift toward *data-driven, visually immersive forensic practice*. By combining computational accuracy with human oversight, these tools promise to enhance both investigative rigor and courtroom communication.

5 Integration with Forensic Databases

One of the most significant recent advances in forensic science is the integration of *artificial intelligence (AI)* with *cloud-based forensic databases*. Historically, forensic analysis has been siloed, with individual laboratories or jurisdictions managing their own datasets of crime scene images, case files, and laboratory reports. This fragmentation often limited the ability of investigators to identify connections between seemingly unrelated cases. With the rise of digital forensic infrastructures and AI-powered analytics, however, it has become possible to conduct *real-time comparisons of blood patterns, DNA profiles, and case notes across multiple databases*, allowing investigators to identify serial crimes or uncover hidden linkages (Sirisathitkul and Sirisathitkul 2025).

5.1 Evolution of Forensic Databases

Forensic databases have traditionally been associated with DNA profiling systems such as CODIS in the United States or NDNAD in the United Kingdom. These platforms revolutionized criminal investigations by enabling automated matches between crime scene samples and known offender profiles. More recently, similar approaches have been applied to *pattern evidence*, including bloodstain patterns, fingerprints, and even shoeprints (Price 2004). However, unlike DNA profiles, which are relatively straightforward to encode, bloodstain evidence is complex, multidimensional, and context-dependent. This complexity makes *AI-driven methods essential* for meaningful database integration.

5.2 AI-Powered Blood Pattern Comparison

Cloud-based platforms now incorporate *computer vision and deep learning algorithms* capable of analyzing thousands of bloodstain images simultaneously. These systems automatically extract morphological features such as size, distribution, angle, and velocity indicators, and compare them against reference libraries of experimentally generated patterns.

For instance, Chen et al. (2025) demonstrated how residual neural networks (ResNets) can distinguish between impact spatters and fly artifacts. When integrated into forensic databases, such models allow investigators to query whether a blood pattern found at one scene resembles patterns from other cases. This has the potential to uncover serial crimes or link cases that might otherwise appear unrelated.

Additionally, *probabilistic databases* can store AI-generated likelihood scores rather than binary classifications, providing investigators with nuanced insights into how closely two patterns match (Garapati et al. 2025). This adds statistical rigor and transparency, increasing the admissibility of database comparisons in court.

5.3 Natural Language Processing in Forensic Reports

Another frontier in forensic database integration is the use of *Natural Language Processing (NLP)* to analyze unstructured textual data, such as forensic reports, witness statements, and crime scene notes. While bloodstain images provide visual evidence, textual records often contain crucial contextual details about lighting conditions, substrate materials, and investigative observations. Manually reviewing such reports is time-consuming and error-prone, especially in large-scale investigations spanning multiple cases or jurisdictions.

AI-powered NLP models are now being developed to *automatically extract key entities, events, and conditions* from forensic reports, linking them to visual evidence stored in databases (Thakkar et al. 2023). For example, an NLP system might identify that a report mentions "arterial spurting" and link that descriptor to corresponding blood pattern images, enabling cross-case searches based on both text and imagery. This multimodal integration significantly enhances investigative efficiency and consistency.

5.4 Linking Cases Across Jurisdictions

One of the major benefits of database integration is the ability to link cases across jurisdictions. Razali et al. (2025) emphasized the role of *geospatially driven forensic databases*, where location-tagged spatter reconstructions are stored alongside case files. Investigators in different regions can then query whether similar blood patterns have been observed elsewhere, potentially uncovering serial offenders.

Moreover, Ojo et al. (2023) demonstrated that drone-collected imagery, when integrated into forensic databases, could support large-scale spatter pattern mapping across outdoor environments. When coupled with AI algorithms, such data integration enhances the ability to trace offender movement patterns and cross-reference them with existing case histories.

5.5 Challenges and Limitations

Despite their promise, AI-integrated forensic databases face several challenges:

1. *Standardization of Data*: Different labs may capture images at varying resolutions, under different lighting conditions, or with inconsistent labeling practices. Without standardized protocols, AI models may produce biased or unreliable results (Jung et al. 2024).
2. *Data Privacy and Security*: Forensic data often includes sensitive personal information. Cloud-based systems must therefore comply with strict data protection regulations, raising questions about *ownership, access, and cross-border data sharing* (Arthanari et al. 2025).
3. *Explainability of AI Models*: Courts require transparent reasoning. If a database links two cases based on an opaque deep learning model, judges and juries may struggle to evaluate its reliability. Efforts in *explainable AI (XAI)* are essential to ensure forensic databases remain legally defensible (Sessa et al. 2025).
4. *Over-Reliance Risk*: Investigators may over-rely on automated matches, overlooking alternative explanations. Human oversight remains critical to prevent wrongful linkages.

5.6 Future Directions

The future of AI-enhanced forensic databases lies in *multimodal integration*, where bloodstain images, spectral data, textual reports, and geospatial information are combined into unified platforms. Such systems could provide investigators with *holistic case comparisons*, moving beyond isolated evidence types.

Emerging research also points toward *blockchain-enabled forensic databases*, ensuring tamper-proof audit trails for all case entries and matches (Ma 2024). This would enhance trust in database integrity and reduce the risk of manipulation. Furthermore, ongoing advancements in *federated learning*—where AI models are trained across multiple decentralized datasets without sharing raw data—could help address privacy concerns while still improving model performance across jurisdictions.

6 Ethical and Legal Considerations

While AI offers significant benefits, it also raises concerns about transparency, explainability, and admissibility in court. Many AI models operate as "black boxes," providing predictions without clear reasoning about how decisions are reached. This opacity complicates their acceptance under legal standards such as the Daubert criteria, which emphasize testability, known error rates, and general scientific acceptance (Koellner 2025). Scholars argue that unless AI evidence is interpretable, courts may hesitate to admit it, especially in high-stakes criminal trials (Montgomery

2025). Beyond transparency, the risk of ***algorithmic bias*** is another pressing issue. If training datasets underrepresent certain conditions, environments, or demographics, outcomes may inadvertently favor or disadvantage specific groups, mirroring the inequities already present in the justice system (Garrett and Rudin 2023). Forensic AI must therefore be subject to rigorous validation and cross-jurisdictional standardization to ensure fairness and reliability. Experts increasingly call for the development of ***explainable AI (XAI)*** frameworks and independent audit mechanisms, which would make algorithmic reasoning accessible to judges, juries, and defense counsel (Williams 2025). Without these safeguards, there is a danger that AI could reinforce existing biases rather than correct them, undermining both the scientific integrity of forensic evidence and the legitimacy of judicial outcomes.

7 Conclusion

The integration of artificial intelligence (AI) and emerging technologies into blood forensics represents a fundamental turning point in forensic science. For decades, blood evidence has been recognized as one of the most compelling forms of physical evidence in criminal investigations, offering insights into the dynamics of violent encounters, the positions of victims and suspects, and the sequence of events. Yet despite its value, traditional approaches to analyzing blood evidence have long been limited by subjectivity and variability. Experts, even when highly trained, often disagree about how stains were created or what sequence of actions produced them, and courts have frequently expressed concern about the reproducibility of bloodstain pattern analysis. The recent adoption of computational methods, particularly machine learning and computer vision, is gradually reshaping this landscape, moving the discipline toward a more standardized, quantifiable, and scientifically defensible practice.

One of the clearest areas of progress has been in bloodstain pattern analysis (BPA). Historically, analysts classified stains by visually comparing their morphology and distribution, a process that was prone to bias and error. Machine learning models such as convolutional neural networks (CNNs) now offer a more objective alternative. Studies demonstrate that CNNs can classify stains into categories such as passive, projected, or transfer with accuracy rates above 90 percent, often exceeding human performance (Acampora and Vitiello 2024; Chen et al. 2025). These models are not susceptible to fatigue or cognitive bias and, importantly, produce consistent results that can be reproduced across analysts and laboratories. This shift transforms BPA from a largely interpretive art into a computational science, laying the groundwork for greater acceptance in the courtroom.

Equally transformative is the ability of AI to estimate the age of bloodstains. Establishing how long a stain has been present is essential for reconstructing timelines, confirming or challenging alibis, and distinguishing relevant from irrelevant stains. Traditional biochemical methods have been destructive and unreliable, particularly under varying environmental conditions. Advances in hyperspectral imaging (HSI) and spectroscopy, combined with machine learning models, have significantly improved accuracy. Neural network approaches have been shown to track hemoglobin degradation across substrates and lighting conditions, reducing error margins to mere hours (Giulietti et al. 2023; Fonseca et al. 2025). The emergence of portable devices, such as handheld near-infrared spectrometers, further suggests that such analysis will soon be possible directly at crime scenes, eliminating delays and reducing reliance on laboratory testing (Seki et al. 2024). The result is a technology that not only accelerates investigations but also strengthens the evidentiary foundation of forensic timelines.

The ability to reconstruct crime scenes in three dimensions has also been revolutionized by AI and simulation technologies. Investigators once relied on rudimentary methods such as stringing techniques and static photographs, which provided only approximations of spatter trajectories. Today, AI models integrated with physics-based simulations can create virtual environments that depict blood droplet motion and impact angles with striking fidelity. Immersive technologies such as virtual reality (VR) and augmented reality (AR) allow jurors, judges, and investigators to walk through reconstructed scenes, visualize trajectories, and better understand the spatial dynamics of violent events (Home et al. 2024; Villa et al. 2023). These tools not only improve investigative accuracy but also enhance courtroom communication by translating complex scientific data into visual experiences that nonexperts can grasp. While concerns remain about the potential for overly persuasive visuals, the benefits for clarity and transparency are undeniable.

The development of cloud-based forensic databases further underscores the transformative potential of AI. Historically, forensic work has been fragmented, with each laboratory or jurisdiction maintaining its own records. AI-enabled databases now allow investigators to compare blood patterns across thousands of cases, revealing connections that might otherwise remain hidden. Machine learning models can assign probabilistic scores to pattern similarities, improving reliability while maintaining transparency about error margins (Razali et al. 2025). Natural Language Processing (NLP) adds another dimension, extracting key details from unstructured forensic reports and linking them to visual evidence. This integration makes it possible to search across both text and imagery, accelerating the investigative process and enabling a more holistic view of case evidence (Thakkar et al. 2023). In effect, these platforms create a living library of forensic knowledge that grows with each new case, expanding the collective intelligence of the field.

Despite these advances, significant challenges remain. Perhaps the most persistent obstacle is the issue of explainability. Many of the most accurate AI models function as "black boxes," producing predictions without transparent reasoning. Courts, however, require that evidence be explainable under standards such as the Daubert criteria, which emphasize testability, error rates, and general scientific acceptance (Koellner 2025). Without interpretability, even highly accurate AI outputs may be deemed inadmissible. Equally concerning is the risk of bias in training datasets. If models are trained primarily on laboratory-prepared stains under controlled conditions, they may fail to generalize to the variability of real-world crime scenes. This raises the risk that forensic AI could reproduce or even amplify existing inequities in the justice system (Garrett and Rudin 2023). To address these issues, scholars and practitioners increasingly advocate for explainable AI frameworks, standardized protocols, and independent audits of forensic algorithms (Williams 2025). Such measures are essential if AI is to achieve not only scientific rigor but also public and legal legitimacy.

Looking ahead, the future of forensic AI appears both promising and complex. Analysts are unlikely to be replaced; instead, their roles will evolve. Human expertise will remain vital in interpreting AI outputs, contextualizing them within the broader framework of an investigation, and presenting them in court. What AI offers is augmentation: the ability to process vast datasets, identify subtle patterns invisible to the human eye, and provide reproducible insights that strengthen expert testimony. As DNA analysis redefined forensic science in the late twentieth century, AI now promises to do the same in the twenty-first, bringing speed, accuracy, and objectivity to disciplines that have long been criticized for their subjectivity.

The integration of AI into blood forensics also underscores the importance of cross-disciplinary collaboration. Computer scientists, forensic practitioners, ethicists, and legal scholars must work together to ensure that these tools are developed responsibly. This includes not only improving model accuracy but also addressing issues of privacy, security, and fairness. Cloud-based forensic databases, for example, must navigate questions of data ownership and consent, while immersive reconstructions must be presented with appropriate safeguards against misleading jurors. Only through sustained collaboration can forensic AI fulfill its promise while maintaining the integrity of the justice system.

In closing, the incorporation of AI and emerging technologies into blood forensics offers a path toward greater reliability, transparency, and efficiency in criminal investigations. It has the potential to transform subjective practices into quantifiable sciences, to accelerate the pace of analysis, and to enhance the clarity of courtroom communication. At the same time, these tools bring new challenges that demand careful consideration: the risk of bias, the opacity of black-box models, the need for standardized protocols, and the ethical dilemmas of deploying powerful computational systems in matters of life and liberty. The future of forensic science will not be defined solely by technological innovation but by the choices society makes in balancing accuracy with fairness, speed with accountability, and automation with human judgment. If approached thoughtfully, AI in blood forensics can become not a replacement for human expertise but a partner to it—one that enhances the pursuit of truth and strengthens the foundations of justice.

References

Abraham J, Ng R, Morelato M, Tahtouh M, Roux C (2021) Automatically classifying crime scene images using machine learning methodologies. Forens Sci Int Digit Investig 39:301273

Acampora G, Vitiello A (2024) A comparison of deep learning methods for multi-class classification of bloodstain patterns. In: Innovations and applications of artificial intelligence. Springer, pp 325–342. https://doi.org/10.1007/978-981-96-4596-1_21

Arthanari A, Raj SS, Ravindran V (2025) A narrative review in application of artificial intelligence in forensic science: enhancing accuracy in crime scene analysis and evidence interpretation. J Int Oral Health 17(1):45–53. https://journals.lww.com/jioh/fulltext/2025/01000/a_narrative_review_in_application_of_artificial.2.aspx

Butt MHF, Ayaz H, Ahmad M, Li JP (2022) A fast and compact hybrid CNN for hyperspectral imaging-based bloodstain classification. In: Proceedings of the IEEE congress on evolutionary computation. IEEE. https://doi.org/10.1109/CEC.2022.9870277

Chen L, Zhu Y, Ma C, Lyu Z (2025) Automated identification of impact spatters and fly spots with a residual neural network. Forensic Sci Int 380:111625. https://doi.org/10.1016/j.forsciint.2024.111625

Cooney GS, Köhler H, Chalopin C, Babian C (2024) Discrimination of human and animal bloodstains using hyperspectral imaging. Forensic Sci Med Pathol 20(2):490–499

Devassy BM, George S (2021) Forensic analysis of beverage stains using hyperspectral imaging. Sci Rep 11(1):7325. https://doi.org/10.1038/s41598-021-85737-x

Fonseca AC, Pereira JF, Honorato RS, Pimentel MF (2025) Age estimation of bloodstains deposited on different floor tiles using handheld NIR spectrometer and chemometrics. Microchem J 213:113900. https://doi.org/10.1016/j.microc.2025.113900

Garapati K, Maram SS, Manikandan VM (2025) Hyperspectral imaging in forensic science: an overview, application areas and recent advancements. In: Computational intelligence in emerging technologies. Springer, pp 89–108. https://doi.org/10.1007/978-3-031-83127-0_6

Garrett BL, Rudin C (2023) The right to a glass box: rethinking the use of artificial intelligence in criminal justice. Cornell Law Rev 109(3):621–670. https://scholarship.law.duke.edu/faculty_scholarship/6996

Giulietti N, Discepolo S, Castellini P, Martarelli M (2022) Correction of substrate spectral distortion in hyperspectral imaging by neural networks for bloodstain characterization. Sensors 22(19):7311. https://doi.org/10.3390/s22197311

Giulietti N, Discepolo S, Castellini P, Bini A (2023) Neural network–based hyperspectral imaging for substrate-independent bloodstain age estimation. Forensic Sci Int 349:111589. https://doi.org/10.1016/j.forsciint.2023.111589

Hassan HAM, Abouelwafa MM et al (2024) Automating forensic analysis: integrating machine learning algorithms for forensic evidence and serial killer profiling. In: Proceedings of the 2024 international conference on artificial intelligence and forensics. IEEE. https://doi.org/10.1109/IAF.2024.10783645

Home PH, Norman DG, Wade K, Spearing E, Williams MA (2024) 3D scanning a crime scene to enhance juror understanding of Bloodstain Pattern Analysis evidence. Sci Justice 64(3):333–338

Jung H, Jo YS, Ahn Y, Jeong J, Lim SK (2024) A first step towards a machine learning-based framework for bloodstain classification in forensic science. Forensic Sci Int 365:112278

Ketsekioulafis I, Filandrianos G, Katsos K, Thomas K, Spiliopoulou C, Stamou G, Sakelliadis EI (2024) Artificial intelligence in forensic sciences: a systematic review of past and current applications and future perspectives. Cureus 16(9):e70363

Koellner E (2025) Black box justice? Legal evidence, digital democracy, and the risks of AI in hyper-specialized courts [Preprint]. SSRN. https://papers.ssrn.com/sol3/papers.cfm?abstract_id=5215622

Kumar R, Sharma K, Sharma V (2020) Bloodstain age estimation through infrared spectroscopy and chemometric models. Sci Justice 60(5):413–421. https://doi.org/10.1016/j.scijus.2020.07.004

Ma J (2024) Smart tech meets forensics: enhancing crime scene investigation with digital simulations. Forensic Sci Int 349:111622. https://doi.org/10.1016/j.forsciint.2024.111622

Maneli MA, Isafiade OE (2022) 3D forensic crime scene reconstruction involving immersive technology: a systematic literature review. IEEE Access 10:120456–120475. https://doi.org/10.1109/ACCESS.2022.9858116

Montgomery RM (2025) Augmenting forensic science through AI: the next leap in multidisciplinary approaches [Preprint]. ResearchGate. https://www.researchgate.net/publication/388456617_Augmenting_Forensic_Science_Through_AI

Ojo TP, Chi H, Hilliard E, Yan J (2023) Crime scene prediction for unmanned aerial vehicles investigation via machine learning. In: Proceedings of the 8th international conference on computational intelligence and applications. IEEE. https://doi.org/10.1109/ICCIA.2023.10208630

Pappachan P, Adi NS, Firmansyah G, Rahaman M (2024) Deep learning-based forensics and anti-forensics. In: Digital forensics and cyber crime investigation. CRC Press, Boca Raton, FL, pp 211–240

Patsiouras E, Vasileiou SK, Gkekas K (2024) Integrating AI and computer vision for ballistic and bloodstain analysis in 3D digital forensics. IEEE Trans Artif Intell 5(4):765–777. https://doi.org/10.1109/TAI.2024.10796512

Price C (2004) Blood pattern analysis. In: Emes A, Price C (eds) From crime scene to court: the essentials of forensic science. Routledge, London, pp 115–141

Raneri D (2018) Enhancing forensic investigation through the use of modern three-dimensional (3D) imaging technologies for crime scene reconstruction. Aust J Forensic Sci 50(2):123–139. https://doi.org/10.1080/00450618.2018.1424245

Razali AF, Edward KL, Ariff MFM, Jasni NS (2025) Innovative forensic features and blood spatter 3D reconstruction from geospatial-driven datasets. Eng Technol Appl Sci Res 15(1):27–35. https://www.etasr.com/index.php/ETASR/article/view/10327

Sacco MA, Gualtieri S, Princi A, Raffaele R, La Russa R (2025) Application of bloodstain pattern analysis (BPA) to reconstruct a staged crime scene in a complex forensic case. Forensic Sci Int Rep 15:100465. https://doi.org/10.1016/j.fsir.2025.100465

Sachdeva S, Garg P, Sharma N, Sivaperumal P (2024) Artificial intelligence in crime scene reconstruction. In: Artificial intelligence in forensic science. CRC Press, pp 15–29

Seki T, Hsiao YY, Ishizawa F, Sugano Y, Takahashi Y (2024) Establishment of a random forest regression model to estimate the age of bloodstains based on temporal colorimetric analysis. Legal Med 63:102231. https://doi.org/10.1016/j.legalmed.2023.102231

Sessa F, Chisari M, Esposito M, Guardo E, Mauro LD (2025) Advancing diagnostic tools in forensic science: the role of artificial intelligence in gunshot wound investigation—a systematic review. Forensic Sci 5(3):30. https://doi.org/10.3390/forensicsci5030030

Sirisathitkul C, Sirisathitkul Y (2025) Bloodstain in forensics: From visual inspections to AI-assisted pattern analysis and age estimation. History of science and technology. Retrieved from https://hst-journal.com/index.php/hst/article/view/688

Thakkar P, Patel D, Hirpara I, Jagani J, Patel S (2023) A comprehensive review on computer vision and fuzzy logic in forensic science applications. J Data Sci 21(4):445–463. https://doi.org/10.1007/s40745-022-00408-6

Thanakiatkrai P, Yaodam A, Kitpipit T (2013) Age estimation of bloodstains using smartphones and digital image analysis. Forensic Sci Int 232(1–3):68–75. https://doi.org/10.1016/j.forsciint.2013.09.027

Villa C, Lynnerup N, Jacobsen C (2023) A virtual, 3D multimodal approach to victim and crime scene reconstruction. Diagnostics 13(17):2764. https://doi.org/10.3390/diagnostics13172764

Wang J, Li Z, Hu W, Shao Y, Wang L, Wu R (2019) Virtual reality and integrated crime scene scanning for immersive and heterogeneous crime scene reconstruction. Forensic Sci Int 302:109859. https://doi.org/10.1016/j.forsciint.2019.109859

Wei CT, You JL, Weng SK, Jian SY, Lee JCL, Chiang TL (2024) Enhancing forensic investigations: identifying bloodstains on various substrates through ATR-FTIR spectroscopy combined with machine learning algorithms. Spectrochim Acta A Mol Biomol Spectrosc 308:123755

Williams B (2025) AI and the future of digital evidence in law and journalism [Preprint]. ResearchGate. https://www.researchgate.net/profile/Bruce-William-2/publication/393412672_AI_AND_THE_FUTURE_OF_DIGITAL_EVIDENCE_IN_LAW_AND_JOURNALISM/links/68683a3307b3253fd1cc4fed/AI-AND-THE-FUTURE-OF-DIGITAL-EVIDENCE-IN-LAW-AND-JOURNALISM.pdf

Zulfiqar M, Ahmad M, Sohaib A, Mazzara M, Distefano S (2021) Hyperspectral imaging for bloodstain identification Sensors 21(9):3045

Part IV

Applications and Case Studies

17 Crime Scene Reconstruction Using Blood Evidence

Abstract

Blood evidence is among the most compelling forms of forensic material encountered at crime scenes. Its value extends far beyond identification through DNA; blood also functions as a silent witness, recording movements, forces, and sequences of events through its spatial distribution and microscopic characteristics. Crime scene reconstruction using blood evidence enables investigators to determine the positioning of victims and suspects, the type of weapon used, and whether a scene has been manipulated or staged. Bloodstain pattern analysis (BPA) remains a cornerstone of this process, as it interprets the size, shape, and direction of stains to reveal underlying actions. BPA has proven particularly useful in differentiating homicides from suicides and in uncovering inconsistencies within staged crime scenes. Recent developments in molecular forensics have expanded the possibilities further. Techniques such as RNA sequencing allow for estimating the time of deposition, while DNA profiling has shown resilience even under conditions of environmental stress. These biological insights complement digital innovations such as hyperspectral imaging, three-dimensional modeling, and artificial intelligence, which improve both accuracy and communication of findings in legal settings. Despite progress, challenges persist. Environmental degradation, methodological subjectivity, and the complexity of conveying technical details to juries remain significant obstacles to courtroom miscommunication or degraded evidence. Nevertheless, by combining traditional analysis with molecular biology and digital technologies, the role of blood evidence in reconstruction continues to grow. As forensic science evolves, blood remains central to reconstructing events with precision and ensuring that truth is revealed in the pursuit of justice.

Keywords

Bloodstain pattern analysis · Crime scene reconstruction · Forensic science · DNA evidence · Hyperspectral imaging · Staged crime scenes

H. R. Dash et al., *Forensic Investigation of Blood Evidence*,
https://doi.org/10.1007/978-981-95-9183-1_17

1 Introduction

Blood has long been regarded as one of the most profound forms of forensic evidence, offering investigators not just a biological identifier but also a detailed physical account of events. In forensic science, few materials carry as much narrative weight as blood. Its presence at a crime scene can immediately alter the trajectory of an investigation, prompting questions about who was involved, what occurred, how it happened, and in what sequence. This is because blood is more than a biological trace—it behaves according to physical principles, forming patterns that reflect the actions and forces that generated them (Mokwena and Matlou 2025).

Unlike many other forms of trace evidence, blood possesses a dual significance. At the molecular level, DNA extracted from blood can reliably link individuals to a scene. At the macroscopic level, however, the distribution, shape, volume, and microscopic features of bloodstains can provide investigators with essential information about the position, movement, and interactions of victims and perpetrators. This dual functionality elevates blood from a simple indicator of violence to a dynamic record of that violence. For example, a pool of blood beneath a victim might suggest a collapse in place, while a trail of small, spattered droplets may indicate motion after injury (Zuidema Agius 2024).

One of the most critical aspects of blood evidence lies in its ability to tell a story that is not easily manipulated. While perpetrators can attempt to clean or alter a scene, genuine blood patterns are difficult to replicate convincingly without leaving inconsistencies. In cases where the truth is obscured—such as staged suicides, accidental misdirection, or attempts to destroy evidence—blood patterns often remain as the most reliable witness (Nagwanshi et al. 2024). It is this inherent resistance to deception that makes blood evidence invaluable in crime scene reconstruction.

Reconstructing a crime scene is not merely about analyzing what happened but understanding how, when, and why it occurred. Blood evidence, through the discipline of bloodstain pattern analysis (BPA), allows investigators to piece together a sequence of actions. BPA considers the size, shape, location, and dispersion of bloodstains, drawing on principles of physics to interpret directionality, velocity, and angles of impact. These interpretations, in turn, help forensic experts infer the type of weapon used, the number of blows struck, whether a victim was moving, and the possible location of assailants and victims at different stages of the encounter (Mokwena and Matlou 2025; Zuidema Agius 2024).

As forensic science has evolved, so too has the application of BPA. Once based primarily on observational expertise, modern BPA now incorporates technologies such as three-dimensional modeling, digital trajectory analysis, and hyperspectral imaging. These tools increase the precision of pattern interpretation, reduce subjectivity, and allow reconstructions to be presented more effectively in courtrooms. For instance, Razali et al. (2025) describe how 3D modeling based on geospatial datasets enhances accuracy in determining blood trajectories, while Sheth et al. (2025) demonstrate how hyperspectral imaging with artificial intelligence can distinguish true blood from substances like paint or rust that often mimic its appearance.

The integration of BPA with molecular biology represents a key advancement in recent forensic methodologies. Traditionally, BPA could not determine the age of a bloodstain, limiting its utility in establishing timelines. This gap is now being addressed through RNA sequencing and protein-based analyses. Gosch et al. (2025) introduced the use of RNA biomarkers tied to circadian rhythms to estimate the time of deposition, providing a temporal context that complements physical pattern data. Similarly, Hou et al. (2025) used advanced molecular spectroscopy to distinguish genuine blood from counterfeit stains, thereby enhancing the reliability of blood evidence in staged scenes.

These molecular methods, however, are not without limitations. RNA, while informative, degrades rapidly in uncontrolled environments. DNA, though generally stable, can be compromised by heat, humidity, and some presumptive chemical tests used during initial evidence collection (Di et al. 2025). Such fragility underscores the importance of forensic protocol and careful sequencing of analytical procedures to preserve both the biological and physical integrity of samples. In complex scenes where environmental factors may degrade or obscure stains, a multidisciplinary approach—combining BPA, molecular biology, and digital tools—becomes essential.

Despite its scientific rigor, one of the greatest challenges associated with blood evidence lies in how it is communicated. The courtroom setting introduces an audience—judges, jurors, and attorneys—who may not be versed in the technical language or visual interpretation of forensic science. If blood evidence is to play its proper role in the justice process, it must be explained with clarity and supported by demonstrable methodology. This is where visual technologies, such as 3D scene reconstructions and animated stain modeling, have a vital role. They not only help experts articulate their findings but also ensure that the jury can comprehend the significance of what they are being shown (Muqadas et al. 2025; Razali et al. 2025).

The interpretation of blood evidence is further complicated by subjectivity. While efforts have been made to standardize BPA, some conclusions remain dependent on expert interpretation. Zuidema Agius (2024) pointed out the variability that can emerge when different analysts examine the same set of stains. This has fueled debate regarding the admissibility of BPA in courts and emphasized the need for rigorous training, certification, and ongoing calibration of forensic practitioners. AI-based tools are currently being developed to address this issue by standardizing pattern classification and reducing human bias (Butt et al. 2025).

At the core of all these developments is a recognition that blood evidence is not merely a reactive tool but a proactive one. It does not just validate a theory—it helps build it. In cases where witness statements conflict, where physical evidence is sparse, or where the motive remains unclear, bloodstains can serve as a silent but articulate witness. Whether through passive pooling, arterial spurting, transfer impressions, or high-velocity spatter, each stain contributes a fragment of the larger narrative.

The goal of this paper is to explore the evolving landscape of blood evidence interpretation, focusing on how traditional BPA is being transformed by digital technologies and molecular innovations. It will also examine how interdisciplinary collaboration is essential to addressing existing limitations, from environmental degradation to legal admissibility. Drawing on research from leading forensic studies, including those by Mokwena and Matlou (2025), Gosch et al. (2025), Sheth et al. (2025), and others, the paper seeks to offer a comprehensive view of how blood evidence is used not only to solve crimes but to construct a coherent and scientifically defensible account of what occurred.

Ultimately, blood remains central to forensic science because of its versatility, reliability, and narrative power. It connects biology to behavior, substance to sequence, and action to consequence. In a field where the stakes involve justice, liberty, and truth, the importance of accurate and nuanced interpretation of blood evidence cannot be overstated.

2 Bloodstain Pattern Analysis (BPA)

Bloodstain pattern analysis (BPA) is one of the most enduring and essential techniques in forensic science. It provides investigators with a scientific framework for interpreting the physical characteristics of blood at crime scenes and linking those findings to the actions that may have caused them. Blood is unique in its behavior; its fluid properties, response to gravity, and interaction with surfaces make it possible to reconstruct not only the events of a crime but also the positioning of individuals and the sequence of their movements. Because of this, BPA is often described as the "language of blood," one that requires both scientific precision and investigative intuition to translate (Mokwena and Matlou 2025).

2.1 Foundations of BPA

At its core, BPA examines the size, shape, distribution, and directionality of bloodstains. These characteristics are determined by a combination of physical forces, including gravity, velocity, and the angle of impact. By observing these traits, forensic experts can classify stains into categories such as passive drops, projected blood, transfer patterns, and cast-off stains (Mokwena and Matlou 2025). Each type of pattern conveys distinct information:

- *Passive stains* are formed when blood drips due to gravity, often reflecting stationary bleeding.
- *Projected patterns* arise from external force applied to a blood source, such as arterial spurts caused by a breached artery.
- *Transfer patterns* occur when a bloodied surface contacts another surface, leaving impressions such as handprints or footwear marks.
- *Cast-off patterns* result from blood flung from a moving object, like a weapon swung repeatedly.

Table 17.1 Types of bloodstain patterns and their forensic significance

Pattern type	Description	Forensic use
Passive	Formed by the force of gravity alone, e.g., drips, pools	Indicates stationary bleeding
Projected	Result from blood propelled by external force, e.g., spurts	Reveals force direction and source type
Transfer	Created when a bloody object contacts another surface	Links objects or individuals to the scene
Cast-off	Generated when blood is flung from a moving object	Estimates weapon movement and number of blows

These distinctions allow analysts to infer not only the type of violence inflicted but also the approximate number of blows, the position of participants, and even whether a victim was moving during the attack (Zuidema Agius 2024).

These stain types are systematically categorized in Table 17.1, providing a concise overview of their physical characteristics and forensic implications.

This table summarizes the four principal categories of bloodstains—passive, projected, transfer, and cast-off—highlighting their physical causes and investigative value.

2.2 Application in Crime Scene Reconstruction

One of the primary contributions of BPA is its ability to situate events within a broader narrative of a crime scene. By reconstructing trajectories, experts can establish where an assailant and victim were located at different points in the incident. For example, elongated bloodstains with pointed ends indicate the direction of travel, which can help confirm whether a victim was standing, crouching, or lying down when injured (Mokwena and Matlou 2025). This information can be critical in cases where witness statements conflict with physical evidence.

BPA has also proven particularly valuable in identifying staged crime scenes. Nagwanshi et al. (2024) demonstrated that when offenders attempt to disguise homicides as suicides, blood patterns often reveal inconsistencies. For instance, a supposed self-inflicted gunshot may be contradicted by spatter patterns inconsistent with the victim's position or by stains suggesting movement that could not have been self-generated. Similarly, Sacco et al. (2025) analyzed staged cases and found that BPA could expose manipulation by highlighting anomalies in stain distribution. In these situations, BPA serves not only as a tool of reconstruction but also as a safeguard against deception.

2.3 Integration with Other Forensic Disciplines

While BPA is powerful in its own right, it rarely stands alone. Instead, it is often integrated with other forensic methods such as DNA analysis, toxicology, or digital reconstructions to strengthen investigative conclusions. For example, BPA may identify the location of a blood source, while DNA testing confirms whose blood it is. Likewise, toxicology results may explain why a victim was incapacitated in a position consistent with certain blood patterns. In interdisciplinary approaches, BPA becomes a critical bridge between physical evidence and biological identification (Razali et al. 2025).

Digital technologies have amplified the potential of BPA by reducing reliance on subjective interpretation. Three-dimensional modeling of bloodstains enables investigators to visualize trajectories more accurately and present findings in court with greater clarity. Razali et al. (2025) highlighted how geospatially driven 3D reconstructions improved precision in interpreting stain angles and distances. Similarly, Sheth et al. (2025) applied hyperspectral imaging with artificial intelligence to classify stains, making it easier to distinguish true blood from visually similar materials like rust or paint. These integrations help minimize analyst bias and enhance the credibility of BPA in legal contexts.

2.4 Advances and Innovations

Recent advances are pushing BPA beyond traditional boundaries. Butt et al. (2025) introduced a spatial–spectral transformer model capable of differentiating blood from confounding substances such as beetroot juice, which has historically led to false positives. These innovations highlight how BPA is adapting to challenges of contamination and deception.

Moreover, studies like Gosch et al. (2025) illustrate how BPA can be enriched by molecular biology. RNA sequencing allows forensic experts to estimate the time of blood deposition, providing an additional layer of reconstruction that complements physical stain analysis. This integration of molecular and physical approaches signals a future where BPA is not only about what happened but also about when it happened.

2.5 Limitations and Critiques

Despite its strengths, BPA is not without criticism. One of the most pressing concerns is subjectivity. Analysts may draw different conclusions from the same set of stains, raising questions about consistency and reliability. Zuidema Agius (2024) argued that while BPA is grounded in science, its interpretation often relies on expert judgment, making rigorous training essential.

Environmental factors also pose challenges. Heat, humidity, cleaning efforts, or absorbent surfaces can distort bloodstains, complicating reconstruction. For example, porous materials like fabric or wood can cause stains to spread irregularly, reducing the accuracy of angle-of-impact calculations (Mathews 2025). Additionally, degraded or partial stains may lead to incomplete reconstructions.

Another limitation lies in courtroom communication. Jurors may struggle to understand the technical nuances of stain analysis, particularly when interpretations differ between experts. This has fueled debate about the admissibility of BPA in legal systems and underscored the importance of combining it with digital reconstructions or molecular evidence for stronger evidentiary value (Muqadas et al. 2025).

2.6 The Future of BPA

The future of BPA lies in greater integration with technology and interdisciplinary science. Automated systems powered by artificial intelligence may help reduce subjectivity by standardizing classifications. Digital tools such as 3D modeling and hyperspectral imaging will continue to enhance both accuracy and accessibility. Furthermore, the merging of BPA with molecular forensics promises to provide deeper temporal and biological context, making blood evidence more robust in both investigations and courtrooms (Sheth et al. 2025).

BPA's enduring significance lies in its ability to transform stains into stories. While challenges remain, its adaptability ensures that it will continue to evolve as a critical component of forensic science. When combined with emerging innovations, BPA allows investigators not just to identify what occurred but to reconstruct the unfolding of events in a way that brings clarity to complex and often chaotic crime scenes.

3 Biological and Molecular Advances

Blood evidence is not only a visual marker but also a rich biological material that retains information at the molecular level. While bloodstain pattern analysis (BPA) provides critical physical insights, it has limitations in terms of subjectivity and environmental degradation. To overcome these challenges, forensic science has increasingly turned to molecular biology and advanced chemical analyses to extract deeper layers of information from bloodstains. These methods allow investigators to determine not just *how* blood was deposited, but also *when* and *under what conditions*. The following subsections explore the most significant molecular advances that have transformed crime scene reconstruction.

3.1 DNA Profiling and Its Forensic Reliability

Deoxyribonucleic acid (DNA) analysis remains the cornerstone of biological forensics. Since the late twentieth century, DNA typing from blood has been the gold standard for personal identification, linking suspects and victims to crime scenes with high accuracy. However, modern research has focused not only on identification but also on enhancing DNA recovery under challenging conditions.

Di et al. (2025) investigated the resilience of DNA in blood exposed to thermal environments, such as those often encountered in arson cases. Their results showed that despite significant degradation, enough genetic material could frequently be retrieved to create partial or complete profiles. This demonstrates that even compromised blood evidence may remain probative, expanding its utility in reconstructions where offenders attempt to destroy evidence.

Another aspect concerns the effects of presumptive blood tests. While chemical reagents such as luminol and phenolphthalein are valuable in detecting blood traces, some reagents can degrade nucleic acids, complicating later DNA analysis (Di et al. 2025; Zuidema Agius 2024). This finding highlights the importance of strategic sequencing of forensic procedures—prioritizing sample preservation when DNA testing is expected.

3.2 RNA-Based Analysis and Temporal Estimation

One of the most groundbreaking molecular developments is the use of ribonucleic acid (RNA) to estimate the *time of deposition* of bloodstains. Traditional BPA cannot determine when a stain was created, which often leaves critical gaps in reconstruction.

Gosch et al. (2025) advanced this field by identifying RNA biomarkers whose expression patterns fluctuate with circadian rhythms. By targeting these markers, their study provided a method to approximate whether blood was deposited in the morning, afternoon, or evening. This temporal dimension enhances crime scene reconstruction, enabling investigators to test alibis or reconstruct the sequence of events across hours rather than merely determining "before" or "after."

RNA analysis has additional applications beyond time-of-day estimation. Specific RNA signatures can indicate tissue of origin, distinguishing between venous blood, menstrual blood, or other body fluids. This increases contextual accuracy in crime reconstruction, especially in complex scenes involving multiple fluid types.

A comparative overview of these molecular techniques, their primary uses, and limitations is shown in Table 17.2.

3.3 Protein and Metabolomic Profiling

Proteins and metabolites within blood are also becoming critical markers in forensic science. Unlike DNA and RNA, which focus on genetic and temporal information, proteomic and metabolomic analyses provide chemical signatures that reflect environmental exposure, health status, and even cause of death.

Table 17.2 Molecular techniques and their role in blood evidence analysis

Technique	Purpose	Limitations
DNA profiling	Identify individuals from blood samples	Degraded in extreme heat or with chemicals
RNA analysis	Estimate time of blood deposition	Highly sensitive to environmental degradation
Protein profiling	Differentiate real and fake bloodstains	Requires sophisticated lab equipment
Metabolomics	Analyze environmental exposure and stain age	Still emerging in forensic standardization

Hou et al. (2025) applied *time-of-flight secondary ion mass spectrometry (ToF-SIMS)* to differentiate genuine blood from counterfeit stains. Their method analyzed the molecular composition of stains at a microscopic level, exposing staged scenes where substances were used to imitate blood. Such counterfeit detection is vital because staged evidence can otherwise mislead reconstructions, especially when perpetrators attempt to confuse investigators by planting false stains.

Metabolomics has also emerged as a complementary field. Swayambhu (2025) introduced the *FORMics* approach, which examines the metabolomic and microbiome signatures of body fluids. This approach helps not only in identifying the fluid type but also in determining post-depositional changes, such as microbial growth in bloodstains. These changes can offer insights into how long a stain has been exposed to the environment, indirectly providing timeline information.

The table outlines emerging biological techniques—DNA, RNA, protein, and metabolomics—alongside their forensic applications and current limitations.

3.4 Bloodstain Aging and Degradation Studies

A significant challenge in crime reconstruction is determining how long blood has been at a scene. While RNA and metabolomics provide promising avenues, other studies focus directly on physical and chemical degradation.

Forensic scientists have investigated hemoglobin oxidation, color changes, and spectral properties of blood as indicators of age. Research suggests that ultraviolet-visible spectroscopy and hyperspectral imaging can track these chemical transitions, offering objective metrics for stain aging (Sheth et al. 2025). Combining such techniques with traditional BPA allows investigators to not only reconstruct the mechanics of bloodshed but also assess whether stains are consistent with the alleged time of events.

3.5 Integration with Digital Technologies

Molecular findings gain greater forensic utility when paired with digital and imaging technologies. For example, hyperspectral imaging can reveal stains invisible to the naked eye while simultaneously analyzing chemical signatures. Butt et al. (2025) introduced a transformer-based model that integrates hyperspectral data with artificial intelligence, effectively distinguishing true blood from other visually similar substances. When combined with RNA or DNA profiling, this dual-layered analysis strengthens conclusions by cross-validating physical and molecular characteristics.

Similarly, Razali et al. (2025) incorporated molecular insights into 3D reconstruction models, ensuring that digital simulations reflect not only geometric accuracy but also the biological state of stains. This holistic approach supports both investigative interpretation and courtroom communication, where jurors benefit from visually clear and scientifically grounded presentations.

3.6 Challenges and Limitations of Molecular Approaches

Despite their promise, molecular techniques are not without challenges. RNA is notoriously unstable, degrading rapidly under environmental stress such as heat, humidity, or ultraviolet light (Gosch et al. 2025). This limits the reliability of RNA-based temporal estimates in uncontrolled settings.

Proteomic and metabolomic analyses require sophisticated instruments and expertise, raising issues of accessibility for resource-limited forensic laboratories (Hou et al. 2025). Additionally, interpreting microbial and metabolic signatures can be complex, and standard forensic protocols are still being developed for courtroom admissibility.

Finally, contamination remains a persistent risk. Blood evidence collected from uncontrolled scenes may contain environmental or human contaminants that alter molecular results. Therefore, strict chain-of-custody and preservation practices are essential to ensure reliability.

3.7 The Future of Molecular Blood Analysis

Looking ahead, the integration of DNA, RNA, proteomic, and metabolomic approaches represents the next frontier in forensic blood evidence analysis. Rather than focusing solely on identity, future reconstructions will incorporate molecular timelines, fluid-type differentiation, and chemical degradation patterns to build comprehensive narratives of crime scenes.

Emerging technologies such as *nanopore sequencing* may eventually allow on-site molecular profiling, reducing delays in laboratory testing. Similarly, portable mass spectrometers could make field-level counterfeit detection feasible. As these innovations develop, molecular methods will increasingly complement traditional BPA, reducing subjectivity and enhancing scientific credibility.

4 Digital and Imaging Technologies

The introduction of digital forensics and advanced imaging has dramatically transformed the way blood evidence is analyzed and presented in forensic investigations. While traditional methods relied heavily on visual inspection and two-dimensional photographs, the integration of digital technologies allows experts to reconstruct bloodstains with greater precision, objectivity, and reproducibility. These innovations are not only improving accuracy during investigations but also enhancing how findings are communicated in legal settings, where clear visual evidence is critical for juror comprehension.

4.1 Transition from Traditional Photography to Digital Forensics

Historically, blood evidence was documented through standard photography, sketches, and manual measurements. Although effective in capturing static details, these methods were limited in their ability to convey spatial relationships or account for complex three-dimensional patterns (Muqadas et al. 2025). The advent of digital photography and computer-based modeling has expanded forensic capabilities, enabling investigators to generate reconstructions that preserve scenes virtually, allowing evidence to be reviewed repeatedly without relying solely on physical samples.

4.2 Three-Dimensional (3D) Bloodstain Reconstruction

One of the most significant advancements in digital forensics is the use of 3D reconstruction. Razali et al. (2025) demonstrated how geospatial datasets can be combined with digital modeling to recreate blood spatter patterns with a high degree of accuracy. Unlike flat photographs, 3D models capture the true spatial geometry of stains, allowing analysts to calculate impact angles, trajectories, and point of origin with precision.

This innovation is particularly useful in cases involving multiple victims or complex crime scenes. By modeling blood trajectories in three dimensions, investigators can determine the relative positions of individuals during violent encounters. Furthermore, these models can be used in courtrooms as interactive visual aids, allowing juries to observe the dynamics of bloodshed in a way that static images cannot convey.

4.3 Hyperspectral Imaging for Bloodstain Identification

A longstanding challenge in forensic analysis is differentiating blood from visually similar substances such as rust, paint, or fruit juices. Hyperspectral imaging has emerged as a solution to this problem by capturing data across a wide spectrum of wavelengths, beyond what the human eye can perceive.

Sheth et al. (2025) applied hyperspectral imaging in combination with artificial intelligence (AI) classification algorithms, developing a system that could distinguish genuine blood from false positives with high reliability. This is significant for outdoor crime scenes, where environmental contaminants often obscure or mimic blood evidence. By integrating AI, hyperspectral systems can rapidly scan large areas, flagging potential bloodstains for further confirmatory testing.

4.4 Artificial Intelligence and Machine Learning Applications

Artificial intelligence has become a powerful tool in digital forensics, especially in pattern recognition and classification tasks. Butt et al. (2025) introduced a transformer-based spatial–spectral model designed to analyze hyperspectral forensic images. Their system successfully identified bloodstains even when they were visually indistinguishable from other fluids, such as beetroot juice.

Machine learning models not only enhance stain classification but also help automate tasks traditionally performed manually, such as calculating angles of impact or classifying stain categories. By reducing human subjectivity, AI applications in BPA and digital imaging improve reproducibility and scientific reliability—two critical aspects for courtroom admissibility (Zuidema Agius 2024).

4.5 Integration of Digital Imaging with Traditional BPA

Digital imaging technologies are not replacing traditional BPA but rather augmenting it. Forensic experts still analyze stain morphology and trajectory based on physical principles; however, digital tools allow for cross-validation of findings. For example, a cast-off pattern documented digitally can be mathematically modeled to confirm whether it supports a proposed narrative of events.

Razali et al. (2025) showed that integrating 3D reconstructions with BPA provided a more holistic understanding of crime scenes. This combination allowed investigators to reconcile discrepancies between witness testimonies and stain trajectories, strengthening case outcomes.

4.6 Challenges and Limitations of Digital Methods

Despite their advantages, digital and imaging technologies face several challenges. High-resolution imaging and 3D modeling require specialized equipment and software, which may be costly for resource-limited forensic laboratories (Muqadas et al. 2025). Additionally, hyperspectral and AI-driven systems require extensive datasets for training, raising questions about bias and generalizability when applied across diverse environmental contexts.

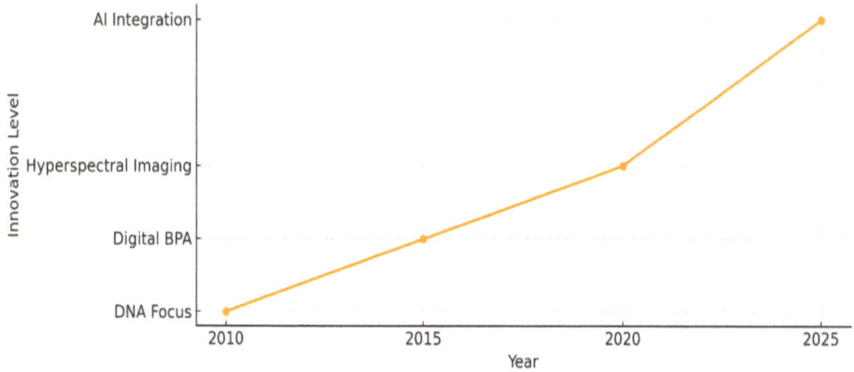

Fig. 17.1 Timeline of technological innovations in blood evidence analysis

Another issue concerns legal admissibility. Courts often demand established standards and validation for scientific techniques. While BPA has long-standing precedent, newer digital tools are still in the process of achieving broad legal acceptance. This underscores the need for peer-reviewed validation studies and standardized protocols.

As illustrated in Fig. 17.1, the evolution of blood evidence analysis has paralleled advancements in imaging, artificial intelligence, and molecular biology.

This figure tracks major forensic milestones from traditional DNA profiling to current AI-integrated hyperspectral imaging models.

5 Interdisciplinary Applications

Blood evidence rarely stands alone in the investigative process. Instead, it frequently interacts with other forms of forensic evidence, and when interpreted in conjunction with them, it enriches the reconstruction of crime scenes. This interdisciplinary perspective has grown in importance as forensic science becomes increasingly collaborative and technologically integrated.

5.1 Blood and Footwear Impression Evidence

One of the most direct intersections is with footwear impression analysis. Blood-soaked footwear impressions can provide powerful evidence linking suspects to crime scenes. Mathews (2025) examined how outsole impressions in blood develop on dark, porous surfaces and demonstrated that these impressions retain identifiable characteristics even when deposited on challenging substrates. In practice, this means that investigators can connect specific shoes to specific stains, effectively placing an individual at the scene and within the timeline of events.

The significance of such findings lies not only in linking a suspect to a location but also in reconstructing movement. A trail of bloody footprints can show where a suspect walked, whether they approached or fled, and whether they carried or dragged an object. Combined with BPA, these impressions provide both directional and positional evidence that deepens the reconstruction narrative.

5.2 Blood and Biochemical/Microbial Signatures

Another emerging interdisciplinary frontier is the integration of blood analysis with microbiome and metabolome studies. Traditional DNA and RNA methods sometimes fail when samples are degraded or heavily contaminated. In such cases, microbial and metabolic signatures offer alternative pathways for identification.

Swayambhu (2025) introduced the *FORMics* approach, which leverages the unique microbial communities and chemical metabolites found in blood and other body fluids. These signatures not only help distinguish between fluid types but can also indicate the age and condition of stains. For instance, microbial succession in a bloodstain exposed to open air may follow predictable patterns, offering indirect timelines similar to those developed in entomology for decomposing bodies.

This integration underscores the interdisciplinary ethos of modern forensic science: where biology, chemistry, and digital analysis converge to fill gaps that a single discipline alone could not address.

6 Challenges and Limitations

Despite these advances, significant challenges continue to hinder the analysis and interpretation of blood evidence in forensic reconstruction. These challenges highlight the delicate balance between scientific promise and practical reality.

6.1 Environmental Degradation

Environmental factors pose one of the greatest threats to the reliability of blood evidence. Heat, humidity, exposure to ultraviolet light, and cleaning attempts all accelerate the degradation of DNA and RNA within stains, making recovery of intact genetic material difficult (Zuidema Agius 2024). Similarly, physical patterns may be distorted when blood soaks into porous surfaces or is smeared during attempts at removal.

Such degradation complicates not only biological analysis but also the physical interpretation of stains. For example, elongated stains may no longer indicate the original angle of impact once environmental factors alter their boundaries. To mitigate these issues, forensic teams must prioritize immediate documentation and collection, often relying on digital imaging to preserve stains before they deteriorate.

6.2 Subjectivity in Interpretation

Another persistent challenge is the subjectivity associated with BPA. Although grounded in physics, BPA still relies on the analyst's expertise to interpret patterns. Two experts may examine the same stains yet reach slightly different conclusions, a problem that undermines consistency and raises questions of admissibility in court (Muqadas et al. 2025).

The integration of digital tools, such as 3D modeling and AI-based classification, is helping reduce this subjectivity by standardizing measurements and offering replicable computational results. However, these technologies are not yet universally available, particularly in resource-constrained jurisdictions. Until broader adoption occurs, BPA's reliance on human interpretation remains a limitation.

6.3 Legal Admissibility and Communication

The courtroom presents a unique challenge for blood evidence. Judges and jurors may struggle to comprehend complex forensic testimony, particularly when it involves molecular biology or digital reconstructions. Without clear communication, even scientifically valid evidence can lose persuasive power.

As Mokwena and Matlou (2025) note, forensic experts must bridge this communication gap by presenting evidence in accessible formats. This often means supplementing testimony with visual aids such as 3D reconstructions or simplified charts that help jurors grasp technical concepts without losing accuracy. Ensuring that advanced forensic methods meet standards of reliability and understandability remains an ongoing concern.

7 Conclusion

Crime scene reconstruction using blood evidence has progressed from a discipline rooted in visual inspection to a complex science that integrates molecular biology, digital imaging, and interdisciplinary approaches. Traditional BPA continues to provide the foundation for interpreting stain patterns, yet it is increasingly complemented by advances in RNA sequencing, proteomics, metabolomics, and hyperspectral imaging. Together, these innovations allow investigators not only to determine *how* events occurred but also to estimate *when* they happened, providing richer timelines for crime scene reconstruction (Gosch et al. 2025).

The field's interdisciplinary applications further strengthen its impact. Footwear impressions in blood extend the narrative of stain patterns by linking suspects directly to crime scenes, while microbiome and metabolome studies offer identification methods when DNA is degraded or unavailable (Mathews 2025; Swayambhu 2025). These contributions demonstrate that blood evidence is most powerful when viewed not in isolation but in collaboration with other forensic disciplines.

Nonetheless, challenges remain. Environmental degradation can compromise stains, subjectivity still influences BPA, and courtroom communication continues to be a hurdle (Zuidema Agius 2024; Muqadas et al. 2025). Yet these limitations are being addressed through technological innovation, rigorous training, and interdisciplinary cooperation.

Ultimately, blood evidence endures as one of the most reliable and versatile forms of forensic material. Its patterns, molecules, and interdisciplinary connections together form a multidimensional record of violent events. When analyzed with scientific rigor and interpreted in context, blood becomes more than a trace of violence—it becomes a silent but eloquent witness, guiding investigators toward truth and justice.

References

Butt MHF, Peng B, Manzoor M, Tabraiz S (2025) Spatial–spectral transformer with gated local and spectral self-attention for hyperspectral forensic imaging. SSRN. https://papers.ssrn.com/sol3/papers.cfm?abstract_id=5334108

Di J, Jin J, Zhang J, Xu X, Li C, Guo K, Shi C (2025) Research on correlation between DNA typing and trace characteristics of blood after thermal exposure. Forensic Sci Int Genet 76:102039. https://www.sciencedirect.com/science/article/pii/S1872497324001686

Gosch A, Sendel S, Caliebe A, Courts C (2025) TrACES of time: towards estimating time-of-day of bloodstain deposition by targeted RNA sequencing. Forensic Sci Int Genet 77:102102. https://www.sciencedirect.com/science/article/pii/S1872497325000675

Hou Y, Wu Y, Guo C, Chen B, Zhao YB, Li Z (2025) A new insight into the distinction between genuine and faux blood fingermarks using ToF-SIMS. Microchem J *199*:108779. https://www.sciencedirect.com/science/article/pii/S0026265X25022210

Mathews SA (2025) The development of outsole impressions in blood on dark, porous surfaces [Doctoral dissertation, Murdoch University]. Murdoch Research Repository. https://research-portal.murdoch.edu.au/esploro/outputs/graduate/The-Development-of-Outsole-Impressions-in/991005787483607891

Mokwena RJ, Matlou TV (2025) The relevance of understanding bloodstain patterns in crime reconstruction. OIDA Int J Sustain Dev 18(4):45–57. https://papers.ssrn.com/sol3/papers.cfm?abstract_id=5172105

Muqadas JN, Aien NU, Fatima T (2025) Crime scene to judgment: identifying gaps in crime scene investigation. Forensic Insights Res 2(1):1–12. https://ammanif.com/journal/fi/index.php/home/article/view/147

Nagwanshi BK, Kosariya SS, Mishra R, Chandra TL, Chakraborty A (2024) Distinguishing homicide from suicide: a forensic investigation into a staged crime scene. IJAR 10(10):199–202

Razali AF, Edward KL, Ariff MFM, Jasni NS (2025) Innovative forensic features and blood spatter 3D reconstruction from geospatial datasets. Eng Technol Appl Sci Res 15(2):987–993. https://www.etasr.com/index.php/ETASR/article/view/10327

Sacco MA, Gualtieri S, Princi A, Raffaele R (2025) Application of bloodstain pattern analysis to reconstruct a staged crime scene. Forensic Sci Int Rep 15:101052. https://www.sciencedirect.com/science/article/pii/S2589871X25000531

Sheth TB, Shah MS, Patel SH (2025) A hybrid inception network for hyperspectral imaging-based bloodstain classification. Int Res J Med Sci 3(4):12–20. https://www.irjms.com/wp-content/uploads/2025/04/Manuscript_IRJMS_03634_WS.pdf

Swayambhu M (2025) FORMics: Forensic microbiome and metabolome analysis for body fluid identification [Doctoral dissertation, University of Zurich]. ZORA. https://www.zora.uzh.ch/id/eprint/277711

Zuidema Agius B (2024) The use of blood pattern analysis to strengthen forensic evidence in solving crimes [Master's thesis, University of Malta]. University of Malta Repository. https://www.um.edu.mt/library/oar/handle/123456789/128168

Case Studies in Blood Evidence Interpretation

18

Abstract

Blood evidence interpretation has transitioned from a largely observational art to a multidisciplinary science that integrates biology, mathematics, fluid dynamics, and genomics. Once limited to basic visual assessments, bloodstain pattern analysis (BPA) has matured into a rigorous field supported by experimental validation, standardized protocols, and advanced technology. This chapter provides a comprehensive review of BPA's evolution, scientific foundations, and its integration with modern DNA profiling techniques, including next-generation sequencing (NGS) and forensic genealogy. To demonstrate its practical application, we examine ten landmark case studies showcasing BPA's ability to reconstruct events, expose false narratives, solve decades-old cold cases, and even overturn wrongful convictions. These cases illustrate the powerful narrative value of blood evidence, which uniquely combines spatial reconstruction with genetic identification. The discussion emphasizes how BPA and DNA interpretation are increasingly interdependent, highlighting innovations such as probabilistic genotyping, digital scene modeling, and artificial intelligence–driven classification. Legal and ethical considerations, including the privacy implications of forensic genealogy and the dangers of expert bias, are also examined. As forensic science undergoes continuous transformation, BPA remains one of its most compelling tools—offering investigators not only proof of identity but also a scientific narrative of crime scenes that withstands judicial scrutiny. This article underscores the importance of rigorous training, certification, and peer review to ensure that blood evidence interpretation continues to advance justice and maintain public confidence.

Keywords

Bloodstain pattern analysis · Forensic DNA profiling · Crime scene reconstruction · Next-generation sequencing · Forensic science

H. R. Dash et al., *Forensic Investigation of Blood Evidence*,
https://doi.org/10.1007/978-981-95-9183-1_18

1 Introduction

Few types of forensic evidence carry the weight and credibility of blood. Unlike trace evidence such as fibers or fingerprints, blood evidence has a dual role: it can link a suspect or victim to a crime through genetic profiling while also revealing how the crime occurred through its spatial distribution. Each droplet, smear, or cast-off arc holds a narrative embedded in its geometry, trajectory, and placement, providing investigators with invaluable context (Bevel et.al, 2025).

For much of the twentieth century, crime scene interpretation relied heavily on experience and observational skills. Early practitioners, such as French scientist Victor Balthazard, pioneered the idea that bloodstains obeyed mathematical principles. In 1939, Balthazard and colleagues published groundbreaking research showing that stain length-to-width ratios could reveal impact angles (Balthazard et al. 1939). This represented one of the earliest examples of forensic science integrating physics into investigative practice.

BPA gained further credibility through the work of Herbert Leon MacDonell, whose seminal publication *Flight Characteristics of Human Blood and Stain Patterns* (1971) introduced experimental methodologies to law enforcement training. MacDonell's work moved BPA from anecdotal interpretation to systematic science, creating a framework for measurement, replication, and peer review.

The DNA revolution of the 1980s added another layer of sophistication. Alec Jeffreys' discovery of DNA fingerprinting in 1985 transformed forensic science, enabling unparalleled individualization of biological evidence (Wambaugh 1995). Combined with BPA, DNA evidence allowed investigators to reconstruct entire crime scenes with both scientific precision and human identification. Over time, RFLP analysis gave way to PCR amplification, STR typing, and, more recently, NGS technologies that can analyze even highly degraded or minute samples (Butler 2014).

Despite its power, BPA has faced criticism for its reliance on expert interpretation. The 2009 National Research Council report identified BPA as a discipline that, while rooted in science, required more rigorous validation and error-rate studies (National Research Council 2009). Wrongful convictions tied to flawed BPA testimony, such as those of David Camm and Julie Rea Harper, revealed the potential consequences of overconfidence and poor training. These cases sparked reforms, leading to greater emphasis on accreditation, proficiency testing, and peer-reviewed methodologies (Wolson 2024).

Today, BPA and forensic DNA analysis operate in synergy. Analysts not only determine stain origin and movement but also use genetic tools to confirm the identities of those involved. Integration with digital forensics, virtual reality scene reconstruction, and probabilistic genotyping further enhances the accuracy of courtroom presentations. This paper explores BPA's history, scientific principles, and ten diverse case studies that showcase how this evidence type continues to evolve as a powerful investigative tool.

2 Literature Review

Bloodstain pattern analysis (BPA) and forensic DNA profiling have been pillars of forensic science for decades, but their evolution has been shaped by a combination of scientific breakthroughs, courtroom scrutiny, and growing awareness of the potential for error. This review traces BPA's development from an observational technique to a validated scientific discipline, while also exploring DNA profiling's transformation into an indispensable investigative tool.

2.1 Early Foundations of BPA

The concept of reconstructing crime scenes using blood patterns dates back more than a century. In the late nineteenth century, Hans Gross, often called the "father of criminalistics," emphasized that crime scenes should be documented holistically, including any visible blood evidence (Gross 1904). Yet his observations lacked quantitative analysis. It was Victor Balthazard and his colleagues (1939) who first introduced a scientific framework for interpreting stain shapes and trajectories. Their mathematical calculations of impact angles represented a paradigm shift, positioning BPA as a quantifiable forensic science.

This work gained traction in Europe but only became widely practiced in North America through Herbert MacDonell's (1971) contributions. His publication *Flight Characteristics of Human Blood and Stain Patterns* formalized BPA terminology and introduced experimental validation. MacDonell also established one of the first structured BPA training programs for law enforcement, laying the foundation for modern expert certification.

2.2 Standardization and Growth

BPA grew significantly during the late twentieth century as crime scene investigators recognized its evidentiary potential. The International Association of Bloodstain Pattern Analysts (IABPA), founded in 1983, became a key organization for education and peer review, developing protocols that emphasized reproducibility and quality assurance. BPA also began to feature prominently in court cases, where analysts used stain interpretation to corroborate or challenge testimony.

2.3 DNA Profiling: A Forensic Revolution

While BPA was gaining structure, forensic DNA analysis revolutionized casework. In 1985, Alec Jeffreys discovered variable tandem repeat sequences, leading to the first DNA fingerprinting applications (Wambaugh 1995). The landmark Narborough murders in the UK demonstrated DNA's unprecedented discriminatory power, exonerating an innocent man and convicting the true perpetrator. This case marked a turning point, cementing DNA evidence as a gold standard for identification.

Early RFLP analysis, however, was time-intensive and required large DNA quantities. PCR technology, developed in the 1980s, transformed DNA testing by enabling exponential amplification of minute samples. STR profiling further refined this process, targeting specific loci for robust and statistically significant comparisons. Today, STR databases like Combined DNA Index System (CODIS) have millions of profiles, revolutionizing criminal investigations (Butler 2014).

2.4 Next-Generation Sequencing (NGS)

NGS represents a leap beyond STR profiling. It can simultaneously sequence STRs, SNPs, and mitochondrial DNA, providing rich genetic information from degraded or trace samples. Forensic scientists now use NGS to infer ancestry, phenotype predictions, and kinship relationships in cases where traditional STR profiles are insufficient. These advances are particularly valuable for cold cases where biological evidence has deteriorated over decades.

2.5 Probabilistic Genotyping

Another major development is probabilistic genotyping software, including STRmix and TrueAllele. Traditional interpretation of mixed DNA profiles relied on subjective analyst judgment, whereas probabilistic genotyping applies Bayesian statistical models to quantify likelihood ratios (Bright et al. 2016). These methods increase transparency and reduce human error, improving admissibility in court.

2.6 Validation Studies and Criticism

Although BPA is rooted in physics, critics have historically argued that it lacks sufficient error-rate studies. The 2009 National Research Council report (National Research Council 2009) emphasized the need for scientifically validated forensic disciplines, pointing out that BPA often relied too heavily on expert interpretation rather than empirical evidence. This criticism led to inter-laboratory proficiency testing, blind case reviews, and formal accreditation processes.

Studies now demonstrate that while BPA is generally reliable, analyst variability remains a concern. Wolson (2024) reported differences in conclusions among analysts interpreting complex patterns, highlighting the need for peer review. These findings have pushed BPA practitioners to adopt rigorous documentation protocols, including digital photography, laser scanning, and photogrammetry.

2.7 Integration of BPA and DNA Analysis

The combination of BPA and DNA profiling has transformed forensic investigations. BPA identifies bloodstain origins, directionality, and sequence of events, while DNA analysis identifies whose blood is present. This dual approach has been instrumental in solving cases involving multiple victims or suspects, especially when witness testimony is unreliable. Today, BPA often guides sample collection strategies, ensuring that investigators prioritize stains most likely to yield informative DNA profiles.

2.8 Ethical Considerations in Genealogy

The advent of forensic genealogy has sparked debates about privacy and consent. Cases like the Golden State Killer demonstrated genealogy's power but also raised ethical concerns: many individuals who submitted their DNA to recreational services never anticipated law enforcement use (Greytak et al. 2019). Policies around database opt-ins, informed consent, and data protection are evolving, reflecting society's struggle to balance privacy rights with justice.

2.9 Courtroom Influence

BPA's courtroom history has been shaped by both successes and controversies. Cases like *State v. Camm* exposed how overconfidence in expert testimony can lead to wrongful convictions. Conversely, high-profile cases have shown BPA's ability to reconstruct events more accurately than eyewitnesses, whose memories are notoriously unreliable. Courts now demand that BPA experts demonstrate training, certification, and adherence to standardized methodology, similar to the criteria applied to DNA experts.

3 Scientific Foundations

3.1 Bloodstain Pattern Analysis Principles

BPA classifies stains into three main categories:

1. *Passive stains* (formed by gravity),
2. *Transfer patterns* (contact or swipe marks), and
3. *Projected patterns* (resulting from external force, including gunshots or blunt trauma) (Bevel et.al, 2025).

Table 18.1 Classification of bloodstain patterns

Pattern type	Mechanism	Common features	Forensic relevance
Passive stains	Formed under gravity	Circular drops, pools, drip trails	Indicates position and movement of bleeding person
Transfer patterns	Contact between blood source and surface	Smears, wipes, swipes, handprints	Shows movement, struggle, or staging
Projected stains	External force applied to blood source	Cast-off arcs, impact spatter, arterial spurts	Determines weapon type, force, and sequence
Altered patterns	Stains modified post-deposition	Clotted, diluted, insect activity, voids	Indicates scene cleanup, environmental changes

Analysts calculate angles of impact by measuring stain width-to-length ratios and using trigonometry. Grouping these trajectories helps determine points of origin and reconstruct movement during the event.

Bloodstain patterns are broadly categorized into passive, transfer, projected, and altered stains, each with unique characteristics that provide critical context for crime scene reconstruction (Table 18.1).

3.2 Fluid Dynamics

Blood's higher viscosity compared to water influences droplet shape and size. Surface tension creates the distinct circular or elliptical stains seen at crime scenes. Variables like velocity, height, and angle determine stain morphology, while textured surfaces introduce distortion. Researchers replicate these factors in controlled environments to refine interpretation guidelines (James et al. 2005). Figure 18.1 illustrates the calculation of impact angles from stain morphology, a foundational method in BPA used to reconstruct trajectories.

3.3 Scene Documentation and 3D Modeling

Modern BPA integrates advanced documentation tools: laser scanners, photogrammetry, and computer-aided modeling. Digital reconstruction offers jurors visual clarity, reducing reliance on verbal descriptions alone.

3.4 DNA Extraction and Profiling

Blood evidence collection emphasizes contamination control. FTA cards, silica columns, and magnetic bead extractions preserve nucleic acids even in degraded conditions (Butler 2014). NGS platforms now enable comprehensive sequencing, including STRs, SNPs, and mitochondrial DNA, providing rich information for identity inference and ancestry.

Fig. 18.1 Blood Droplet Angle of impact calculation

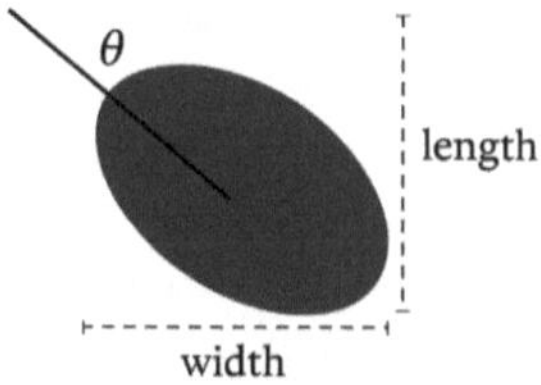

$$\theta = \arcsin \frac{width}{length}$$

Table 18.2 Summary of ten case studies in blood evidence interpretation

Case no.	Case type	Key techniques	Outcome	Lessons learned
1	Domestic homicide	3D laser scanning, DNA profiling	Conviction for intentional murder	Advanced visualization disproved self-defense claims
2	Staged crime scene	BPA, trace analysis	Suspect confession	Void patterns revealed manipulation
3	Gunshot trajectory	Laser rods, BPA	Conviction	Correlated spatter with autopsy trajectory
4	Cold case (NGS)	NGS, probabilistic genotyping	Solved after 30 years	Highlighted preservation value
5	Multivictim assault	Digital reconstruction	Multiple convictions	Differentiated attacker roles
6	Mass casualty	VR modeling, BPA	Training aid, conviction	Helped timeline reconstruction
7	Genealogy breakthrough	SNP profiling, BPA	Arrest & confession	Combined bioinformatics with BPA
8	Vehicle homicide	BPA, gunshot residue	Conviction	BPA effective in confined spaces
9	Outdoor scene	BPA, meteorological data	Timeline verified	Outdoor stains remain evidential
10	Wrongful conviction	Expert review	Exoneration	Showed risks of flawed BPA testimony

4 Case Studies

An overview of ten case studies, their investigative techniques, and outcomes is summarized in Table 18.2, illustrating BPA's versatility and impact on justice outcomes.

4.1 Domestic Homicide Reconstruction

This case demonstrates BPA's ability to dismantle self-defense claims. Responding officers found a woman dead in her living room, seated in a recliner with multiple blunt-force injuries. Her partner insisted that she had attacked him first and that her injuries were sustained in a chaotic struggle. Analysts documented several elliptical stains on the wall directly behind the victim's chair. Measurements showed consistent downward trajectories, indicating she was seated when attacked. Cast-off arcs on the ceiling and secondary spatter on furniture suggested repeated overhead swings of a blunt object. The scene was scanned with a FARO 3D laser scanner, and analysts plotted a precise point of origin for the blood patterns, reinforcing the conclusion that the victim never left her chair during the attack. DNA analysis confirmed all stains belonged solely to the victim, negating claims of mutual combat. These findings were critical at trial, and the visual 3D reconstruction proved compelling for jurors. The suspect was convicted of intentional homicide. This case highlights how combining BPA with DNA analysis and advanced visualization tools can transform a crime scene narrative, moving beyond anecdotal reconstructions to scientifically validated evidence (James et al. 2005).

4.2 Staged Crime Scene Exposure

In another high-profile case, a man reported a violent home invasion, claiming masked intruders attacked him and killed his spouse. Responding officers noticed inconsistencies: the husband's shirt bore several small stains that appeared to be low-velocity transfer patterns rather than high-energy spatter expected from a gunshot homicide. Analysts examined voids on the living room wall, indicating objects—later identified as family photos—had been removed post-incident. Swabs from those areas revealed diluted blood consistent with cleanup attempts. The stains on the suspect's clothing matched transfer patterns from dragging the victim's body. By recreating the scene, investigators demonstrated the implausibility of the husband's narrative. The defendant ultimately confessed, and BPA became a central feature in court testimony, showing how crime scene manipulation can be detected through meticulous analysis (Chancellor and Graham 2016).

4.3 Gunshot Crime Scene Trajectory Analysis

A homicide investigation involving a single gunshot wound illustrates how BPA complements ballistics. The defendant claimed he fired in self-defense from across the room. Analysts documented a fine mist of forward spatter on a nearby lamp and back spatter on the shooter's clothing, patterns typical of close-range gunfire. Using laser trajectory rods, they established a muzzle-to-target distance of approximately 30 centimeters, contradicting the defendant's statement. Additionally, the angle of impact on surrounding walls revealed the victim was seated at the time of the shooting. This detailed reconstruction, paired with medical examiner findings, proved critical to the prosecution's argument that the act was deliberate rather than defensive. The jury convicted the shooter of second-degree murder. This case underscores how precise trajectory mapping and integrated BPA-ballistic analysis strengthen evidence reliability (Bevel et.al, 2025).

4.4 Cold Case Resolution Through NGS

In the 1980s, a teenager's murder baffled investigators. The original case included multiple bloodstains on a bedsheet, but degradation rendered the profiles inconclusive. Decades later, NGS technology enabled analysts to generate complete STR and SNP profiles from previously uninterpretable evidence. Probabilistic genotyping separated mixed DNA from two individuals, one of whom matched a suspect initially excluded due to early testing limitations.

This breakthrough resulted in an arrest and conviction nearly 30 years after the crime. The case illustrates the enduring value of proper evidence preservation and demonstrates how integrating BPA findings with modern sequencing can change investigative outcomes. It also emphasizes the importance of reanalyzing historical evidence as forensic technology evolves (Butler 2014).

4.5 Multivictim Assault Reconstruction

A gang-related assault involving multiple perpetrators presented a complex challenge for investigators. Several victims were attacked simultaneously with blunt weapons, resulting in overlapping blood patterns. Analysts meticulously documented cast-off arcs on walls and ceilings, identifying three distinct points of origin corresponding to different weapon swings. Medical examiner reports on wound trajectories aligned with BPA findings, confirming the sequence of assaults. Investigators reconstructed the scene digitally to differentiate between strikes, allowing prosecutors to demonstrate premeditation and the roles of individual assailants.

This case illustrates BPA's utility in multiassailant investigations, showing how careful measurement, layering of stain trajectories, and collaboration between pathologists and BPA experts can clarify chaotic events.

4.6 Mass Casualty Shooting Event

Mass casualty scenes are among the most challenging to interpret due to overlapping stains, chaotic movement, and secondary transfer. In one public venue shooting, investigators used 3D laser scanning to map over 200 bloodstain points. Analysts identified multiple cast-off trails and overlapping impact spatters, which they combined with surveillance footage to reconstruct the shooter's movements. BPA confirmed the sequence of victims, helping prosecutors establish the timeline of the attack. The reconstruction was presented in virtual reality for training law enforcement on rapid-response strategies. This case exemplifies how BPA is expanding beyond courtrooms to inform tactical planning and policy, demonstrating its value in both investigation and prevention (Wolson 2024).

Advanced 3D scene modeling, as shown in Fig. 18.2, provides investigators and jurors with a clear, spatially accurate representation of blood evidence.

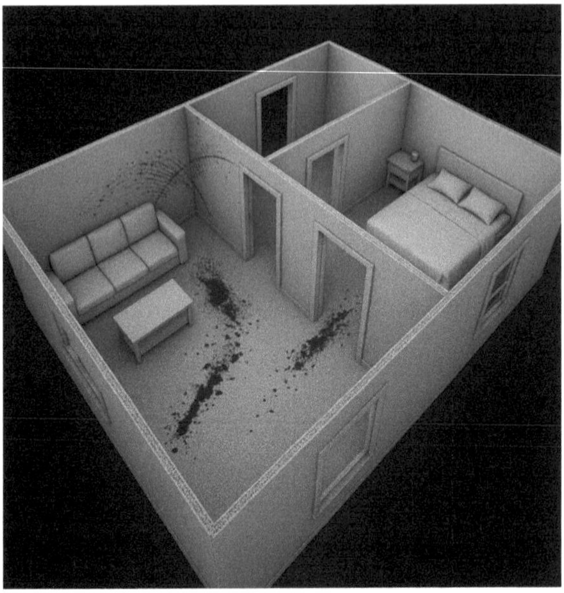

Fig. 18.2 3D Laser-scanned crime scene reconstruction

4.7 Genealogy-Driven Breakthrough

A cold sexual assault and homicide case from the early 1990s relied on a single degraded bloodstain. Analysts extracted a SNP profile and uploaded it to a public genealogy database, leading to a distant relative. Investigators constructed a family tree, eventually identifying the suspect, who confessed upon arrest. BPA findings from the scene, including transfer patterns and drag marks, corroborated the reconstructed sequence of events. The combination of genetic genealogy and BPA created a compelling narrative that solved a case once thought unsolvable. This case highlights the increasing convergence of molecular forensics, genealogical analysis, and traditional pattern interpretation, while also fueling ongoing privacy debates (Greytak et al. 2019).

4.8 Vehicle Homicide Reconstruction

In a homicide investigation occurring inside a moving car, BPA was essential. Analysts found bloodstains on the passenger-side window and door panel with radial projections, indicating a close-range gunshot while the victim was seated. Transfer stains on the gearshift confirmed the driver's involvement. By recreating trajectories, analysts disproved the driver's claim that the shooting occurred during a roadside altercation. This case demonstrates BPA's power in confined environments where limited evidence can still produce highly detailed reconstructions, especially when paired with gunshot residue analysis.

4.9 Outdoor Crime Scene Challenges

A homicide in a rural location presented unique challenges: the body was discovered days later after heavy rain, and much of the visible evidence was washed away. However, BPA experts examined residual stains on nearby rocks and vegetation, combining drying time analysis with meteorological data to approximate the time of deposition. Partial DNA profiles from weather-compromised stains matched the victim, helping investigators reconstruct the timeline. This case emphasizes that BPA, though often associated with indoor crime scenes, can provide valuable leads even in outdoor, environmentally compromised settings (Al-Soud and Rådström 2001).

4.10 Wrongful Conviction Reversal

The David Camm case is a cautionary tale about the misuse of BPA. In 2000, Camm, a former state trooper, was accused of murdering his family. Prosecutors claimed that tiny stains on his shirt were high-velocity back spatter, evidence they argued proved he fired the fatal shots. Later reviews revealed the stains were likely transfer or contact stains, and the original expert testimony lacked empirical support. After three trials and 13 years in prison, Camm was exonerated. This case has become a reference point in BPA training programs, emphasizing the need for validated methods, blind review, and transparency to prevent future miscarriages of justice (Wolson 2024).

Summary of Case Insights

These ten cases showcase the breadth of BPA applications—from proving intent in domestic homicides to solving decades-old cold cases and overturning wrongful convictions. They collectively emphasize:

- *Scientific rigor:* BPA has transitioned from a subjective practice to a validated forensic science.
- *Integration:* Combining BPA with DNA sequencing, probabilistic genotyping, and genealogy strengthens conclusions.
- *Courtroom visualization:* 3D scanning and VR presentations enhance juror comprehension.
- *Ethical responsibility:* Misinterpretation can have devastating consequences, necessitating strict standards.

5 Challenges

Despite its strong scientific foundation, the interpretation of blood evidence remains a discipline that requires extensive training, careful methodology, and constant validation to maintain its reliability. Misinterpretation risk remains significant, particularly when analysts lack standardized education or experience, and even well-trained experts can arrive at different conclusions when interpreting complex patterns. Studies have shown that inter-analyst variability persists, especially in scenes involving overlapping stains, altered patterns, or environmental interference (Wolson 2024). Outdoor crime scenes present particular difficulties, as weather, insects, and substrate absorption can dramatically alter stains before investigators arrive. Similarly, poorly lit or cluttered indoor environments can obscure evidence, making accurate reconstruction more challenging. Contamination is another persistent concern, whether from improper evidence handling, scene cleanup efforts, or secondary transfer, which can introduce misleading patterns. Even small mistakes during collection or documentation can compromise evidence integrity, emphasizing the need for strict adherence to protocols. These limitations underscore that while BPA is a powerful investigative tool, it is not infallible and must be interpreted with caution, ideally supported by peer review and other forensic disciplines.

6 Ethics and Law

The rapid expansion of forensic genealogy and advanced DNA sequencing has revolutionized cold case work, but it has also sparked ongoing debates about privacy, consent, and regulatory oversight. Genetic databases created for recreational use, such as those provided by consumer ancestry companies, were not initially intended for law enforcement investigations, raising questions about whether contributors have truly given informed consent for their data to be used in criminal cases (Greytak et al. 2019). The Golden State Killer investigation demonstrated the enormous value of genealogy in solving violent crimes, yet it also highlighted the urgent need for transparent policies that balance investigative benefits with civil liberties. Beyond genealogy, courts increasingly demand empirical validation of forensic methods, requiring experts to quantify error rates and adopt probabilistic reporting to avoid overstating certainty (National Research Council 2009). Misapplication of BPA has contributed to wrongful convictions in the past, emphasizing that ethical responsibility extends beyond technology and into testimony itself. Forensic practitioners must maintain objectivity, present findings with appropriate caution, and ensure that evolving methods are subjected to rigorous peer review and accreditation processes.

7 Future Directions

The future of blood evidence interpretation is being shaped by advancements in automation, artificial intelligence, and sequencing technologies that promise to enhance precision and reduce subjectivity. Machine learning algorithms trained on thousands of stain images are showing promise in categorizing patterns with high accuracy, offering valuable support to analysts while still requiring human oversight (Bevel et al. 2025). Portable sequencing platforms, such as Oxford Nanopore devices, are being explored for on-site DNA profiling, potentially allowing investigators to identify victims or suspects in real time, even in resource-limited or remote settings. Next-generation sequencing is also expanding the scope of forensic biology, with applications in ancestry inference, phenotypic prediction, and estimating time since deposition, giving investigators more investigative leads than ever before (Butler 2014). Virtual and augmented reality reconstructions are becoming increasingly common in courtrooms, offering immersive experiences that help jurors understand complex stain trajectories and event sequences. As these innovations emerge, the forensic community faces a responsibility to validate new methods thoroughly and integrate them into accredited workflows. If pursued carefully, these technologies will make BPA and DNA interpretation even more reliable, accessible, and central to solving crimes, continuing the field's evolution from an observational practice into a data-driven science.

8 Discussion

The interpretation of blood evidence continues to stand out as one of the most scientifically powerful tools in forensic science because of its ability to provide both identification and context in criminal investigations. Bloodstain pattern analysis (BPA), once considered a craft dependent on intuition and observation, has evolved into a mature scientific discipline grounded in physics, fluid dynamics, and mathematics, while advances in forensic DNA profiling have introduced a level of precision and reliability that was unimaginable only a few decades ago. Together, these fields now create a more complete narrative of crime scenes, allowing investigators to piece together events in a way that neither discipline could achieve alone. Blood tells a story not only through its genetic profile but also through the patterns and trajectories left behind, providing insight into movement, positioning, sequence of events, and even attempts to clean or stage a scene. The integration of DNA analysis with BPA means that every stain can potentially reveal both who was present and what occurred, leading to a multidimensional approach to crime scene reconstruction that has reshaped investigative strategies worldwide. The expansion of technologies such as probabilistic genotyping has strengthened this integration, allowing scientists to interpret mixed or degraded DNA samples with a high degree of statistical rigor, while advances in scene documentation—like three-dimensional laser scanning and virtual reality reconstructions—have made complex blood evidence more understandable to juries and investigators alike. This is particularly important given the weight jurors tend to place on forensic evidence, which makes scientific accuracy and transparency critical. The wrongful conviction of David Camm, often cited in BPA training programs, serves as a stark reminder of how overconfidence or misinterpretation can have devastating consequences, reinforcing the need for robust peer review, blind proficiency testing, and the careful communication of uncertainty. This has led to a noticeable cultural shift in forensic science, moving from absolute statements to evidence-based probabilities, reflecting a growing emphasis on transparency and the recognition that even scientifically grounded methods carry limitations. As a result, international accreditation standards such as

ISO/IEC 17025 now guide forensic laboratories in establishing stringent quality control measures, with BPA experts increasingly held to the same rigorous standards as DNA analysts. What has emerged is a forensic discipline that has embraced accountability, drawing on empirical studies, controlled experiments, and interlaboratory testing to validate interpretations. Yet challenges persist, especially in crime scenes that are chaotic, weather-affected, or deliberately manipulated. Outdoor scenes remain particularly complex due to environmental degradation, and overlapping stains in multivictim or mass casualty incidents demand a high level of expertise and teamwork to untangle. Advances in digital technologies and automation are helping to address these issues, with machine learning algorithms now being trained to classify stains and identify patterns from large datasets, reducing bias and increasing reproducibility, though human expertise remains essential to interpret subtle contextual cues that machines cannot yet replicate. Similarly, the growth of next-generation sequencing (NGS) has expanded the information that can be drawn from blood evidence, enabling ancestry inference, phenotypic predictions, and even time-since-deposition estimates, adding investigative value to stains that might previously have been overlooked. These innovations are changing the courtroom experience as well, allowing jurors and judges to visualize scenes more effectively and better understand the reasoning behind expert testimony. However, with these scientific advances comes a responsibility to address ethical and privacy concerns, especially regarding forensic genealogy, which has solved numerous cold cases but raised questions about consent and the use of consumer DNA databases by law enforcement. Striking a balance between investigative breakthroughs and individual rights is an ongoing conversation, and forensic practitioners are increasingly expected to participate in shaping policy as well as applying science. In parallel, training programs for BPA experts have become more structured, incorporating physics, biology, and statistical reasoning alongside field exercises, and emphasizing the importance of interdisciplinary collaboration with pathologists, DNA analysts, and crime scene technicians. Global research initiatives, including those led by the Forensic Technology Center of Excellence and similar bodies, continue to study error rates, improve analytical methods, and explore the integration of emerging tools like portable DNA sequencing and augmented reality reconstructions. The cumulative effect of these changes is a field that is not only more reliable but also more self-aware, constantly testing and refining its methods to maintain credibility in the eyes of courts, investigators, and the public. Ultimately, blood evidence is unique in that it embodies both hard science and narrative power, serving as a silent witness capable of revealing truths that even eyewitnesses cannot articulate. Its interpretation demands a balance of analytical precision, skepticism, and storytelling skill, as scientists must convey complex technical findings in a way that is clear and compelling without overstating certainty. As technology evolves, BPA and DNA analysis will continue to intersect more deeply, offering richer insights while also placing higher demands on practitioners to act as both scientists and communicators. The future of this discipline lies in its ability to maintain rigorous scientific standards, embrace new technologies responsibly, and remain grounded in ethical practice. By learning from past mistakes and leveraging innovation, blood evidence interpretation will continue to strengthen the foundation of forensic science, serving as a powerful bridge between scientific discovery and the pursuit of justice.

9 Conclusion

The interpretation of blood evidence has evolved into one of the most dynamic and scientifically grounded areas of forensic science, blending rigorous experimentation with practical investigation. From the early mathematical observations of blood droplet trajectories to today's integration of next-generation sequencing, probabilistic genotyping, and advanced 3D scene reconstructions, the discipline has undergone a remarkable transformation. Blood is unique in its ability to provide both a biological fingerprint and a narrative of events, serving as an unbiased witness that captures moments of violence or struggle more reliably than human recollection. The case studies discussed in this review highlight the versatility of BPA and DNA analysis, demonstrating their power in confirming or challenging testimony, uncovering deception, solving decades-old cold cases, and even correcting miscarriages of justice. Yet these advances also remind us that scientific credibility is not static; it relies on constant validation, transparency, and humility. The forensic community has embraced these responsibilities, implementing formal training programs, international accreditation standards, and technologies that reduce subjectivity and strengthen the interpretation process. At the same time, practitioners face new ethical questions, particularly around privacy in forensic genealogy and the responsible use of emerging tools. The future of blood evidence interpretation lies in a careful balance: embracing innovation while preserving rigorous methodology and ethical integrity. If pursued with diligence, this balance will ensure that blood evidence continues to provide reliable answers, strengthen public trust in forensic science, and serve its ultimate purpose—delivering justice by giving a voice to victims and clarity to complex investigations.

References

Al-Soud WA, Rådström P (2001) Purification and characterization of PCR-inhibitory components in blood cells. J Clin Microbiol 39(2):485–493. https://doi.org/10.1128/JCM.39.2.485-493.2001

Balthazard V, Piedelievre R, Desoille H, De Robert L (1939) Étude des gouttes de sang projetées. Annales de Médecine Légale 19:265–323

Bevel T, Gardner RM, Griffin TJ (2025). Bloodstain pattern analysis with an introduction to crime scene reconstruction, 4th edn. CRC Press.

Bright JA, Taylor D, McGovern C, Cooper S, Russell L, Buckleton J (2016) Developmental validation of STRmix, expert software for the interpretation of forensic DNA profiles. Forensic Sci Int Genet 23:226–239. https://doi.org/10.1016/j.fsigen.2016.05.007

Butler JM (2014) Advanced topics in forensic DNA typing: interpretation. Elsevier Academic Press

Chancellor AS, Graham GD (2016) Crime scene staging: investigating suspect misdirection of the crime scene. Charles C Thomas Publisher

Greytak EM, Moore C, Armentrout SL (2019) Genetic genealogy for cold case and active investigations. Forensic Sci Int 299:103–113. https://doi.org/10.1016/j.forsciint.2019.03.039

Gross H (1904) Handbuch für Untersuchungsrichter als System der Kriminalistik. 4. Aufl. J. Schweitzer

James SH, Kish PE, Sutton TP (2005) Principles of bloodstain pattern analysis: theory and practice. CRC Press

MacDonell HL (1971) Flight characteristics of human blood and stain patterns. U.S. Department of Justice, Law Enforcement Assistance Administration, National Institute of Law Enforcement and Criminal Justice

National Research Council (2009) Strengthening forensic science in the United States: a path forward. National Academies Press. https://doi.org/10.17226/12589

Wambaugh J (1995) The blooding: the dramatic true story of the first murder case solved by genetic "Fingerprinting". Bantam

Wolson TL (ed) (2024) Handbook of bloodstain pattern analysis. CRC Press

Legal Admissibility of Blood Evidence 19

Abstract

Blood evidence stands as one of the most potent and persuasive forms of forensic material presented in legal proceedings. It can trace the presence of individuals, determine timeframes, and establish or discredit alibis. However, the journey from the crime scene to the courtroom is fraught with scientific, legal, ethical, and procedural challenges. This paper critically examines the legal admissibility of blood evidence, focusing on the scientific standards required under U.S. law (e.g., Daubert), international legal comparisons, chain of custody principles, evolving forensic technologies like microbial DNA and touch DNA, and the ethical boundaries involving consent and privacy. The analysis highlights key court cases, emerging controversies like the CSI Effect, and recommendations for legal reform to ensure that blood evidence serves justice rather than undermines it. Ultimately, this paper argues that the legitimacy of blood evidence lies not only in its biological truth but in the legal system's ability to assess, regulate, and present it without bias or error.

Keywords

Blood evidence · Forensic science · Legal admissibility · Chain of custody · Ethical boundaries · Expert testimony

1 Introduction

Blood is a powerful symbol—both emotionally and legally. It evokes notions of life, violence, kinship, and guilt. Literature has long used blood to represent deep moral or existential truths, from Shakespeare's tormented Lady Macbeth to modern true-crime thrillers. Yet in the courtroom, symbolism has no standing. Blood is not poetry—it is *evidence*. And not just any evidence. It is *personal*, *biological*, and potentially *life-altering* for those involved in a legal case.

H. R. Dash et al., *Forensic Investigation of Blood Evidence*,
https://doi.org/10.1007/978-981-95-9183-1_19

In modern forensic science, blood can be used to identify suspects, confirm or dispute alibis, estimate the time of death, or reconstruct violent encounters. Thanks to DNA profiling, even a few nanograms of blood can uniquely link a person to a scene or victim. But for blood evidence to carry weight in court, it must do more than provide answers—it must withstand *legal scrutiny*. That means being collected, preserved, analyzed, and presented in a way that is both scientifically sound and legally admissible.

The *admissibility of blood evidence* is governed by strict procedural and legal standards, particularly in countries like the United States where constitutional rights to due process and privacy are robust. Blood evidence must be obtained lawfully—typically with a warrant or clear consent—and must be analyzed using *methods that are reliable, repeatable, and widely accepted in the scientific community*. If these conditions aren't met, the evidence can be challenged and even thrown out, no matter how seemingly damning it may be.

This tension—between the power of scientific evidence and the safeguards of the legal process—is at the heart of this paper. We will explore how courts assess blood evidence for admissibility, what scientific standards are required, and why *ethical and constitutional issues* matter just as much as laboratory results.

1.1 The Weight Blood Carries in Court

In criminal cases, blood is often the linchpin of forensic evidence. A drop of blood at the scene can suggest presence; blood under a victim's fingernails can imply a struggle; a high-velocity spatter on clothing might suggest proximity to an act of violence. In civil cases, too, blood evidence might support or undermine claims involving personal injury, paternity, or negligence.

Because of its persuasive nature, courts treat blood evidence with both interest and caution. Judges and juries alike tend to view DNA evidence—most often derived from blood—as extremely credible. This perception is not entirely misplaced. DNA has been called the "gold standard" of forensic science, with error rates as low as 1 in several billion when properly handled (Houck and Siegel 2009).

However, *forensic science is not infallible*, and blood evidence is no exception. Contamination, mislabeling, improper storage, lab technician errors, and even misinterpretation of data can all lead to *serious legal consequences*, including wrongful convictions (Inman and Rudin 2000). Indeed, the Innocence Project has exonerated over 300 individuals in the USA using DNA evidence, many of whom were originally convicted based on blood evidence that was later proven inaccurate or mishandled (Innocence Project 2023).

1.2 The Legal Gatekeepers of Science

In the USA, the admissibility of scientific evidence—including blood—has evolved significantly over the past few decades. Courts no longer simply assume that a scientific method is valid because it sounds technical or comes from a lab. Instead, courts use *legal standards* such as the *Daubert Standard* or the older *Frye Test* to evaluate whether the evidence should be admitted (Saunders 1997).

The *Daubert v. Merrell Dow Pharmaceuticals* (1993) ruling is the most influential of these legal tests. Under Daubert, judges serve as "gatekeepers" who must assess whether the scientific method used is valid and whether it has been properly applied in the case at hand (Saunders 1997). This includes considering whether the method has been tested, peer-reviewed, has known error rates, and is generally accepted in the scientific community (Giannelli 2017).

This legal framework makes it clear that courts are not laboratories—but they do require that *scientific rigor* accompany any evidence used to determine guilt or innocence. As such, forensic laboratories must meet high standards, and expert witnesses must be prepared to *defend both their findings and their methodology.*

1.3 A Crossroads of Science, Law, and Ethics

The legal admissibility of blood evidence is not just a technical issue—it's a deeply ethical one. Blood samples are *intimate biological data.* They contain not only identification markers but also potential indicators of health, heritage, and predisposition to disease. This raises concerns about *privacy, informed consent, and the scope of state power.*

When is it acceptable to take a person's blood? Who owns that data? Can it be stored in databases or used for unrelated investigations? These are questions increasingly being asked by privacy advocates, legal scholars, and bioethicists alike (Garrett 2016).

In *Missouri v. McNeely* (2013), the U.S. Supreme Court ruled that drawing blood without a warrant from a suspected drunk driver was a violation of the Fourth Amendment, unless exigent circumstances were present. This case affirmed the idea that even in the pursuit of justice, *bodily integrity cannot be ignored* (Whitaker 2019).

The debate extends beyond the USA In some countries, like the UK, genetic material from blood samples can be retained for years—even after suspects are acquitted—raising additional concerns about surveillance and civil liberties (Nuffield Council on Bioethics 2007).

In the twenty-first century, the courtroom is increasingly becoming a stage where science and law must interact with unprecedented complexity. As forensic technologies advance rapidly, legal systems are under growing pressure to determine not just what is scientifically possible, but what is *legally reliable, procedurally fair, and ethically justifiable*. Techniques such as *microbial forensics, touch DNA*, and *artificial intelligence-driven bloodstain analysis* offer new possibilities for truth-seeking—but they also introduce fresh challenges related to validation, transparency, and potential misuse.

This chapter addresses a critical juncture in both forensic science and legal practice. It seeks to bridge the widening gap between *scientific innovation* and *legal admissibility*, particularly as it relates to *blood evidence*, one of the most powerful and frequently contested forms of forensic material presented in court. As blood evidence continues to evolve in scope and sophistication, so too must the frameworks that govern its admissibility and interpretation.

Specifically, this chapter will examine:

- The *scientific criteria and legal standards* that underpin the admissibility of blood evidence in court;
- The importance of *chain of custody procedures* and the role of *expert testimony* in establishing evidentiary reliability;
- *Landmark judicial decisions* and *comparative international practices* that have shaped the current legal landscape;
- The implications of *emerging forensic technologies* and their potential to disrupt or reinforce existing evidentiary norms;
- The *ethical and constitutional considerations*, particularly around privacy, consent, and bodily autonomy, that define the boundaries of lawful evidence collection and use.

By engaging with these themes, the chapter aims to offer a nuanced and interdisciplinary perspective. The goal is not merely to catalog legal rules or scientific techniques, but to illuminate *why critical thinking, procedural rigor, and ethical vigilance are essential* when dealing with blood evidence in legal contexts. Ultimately, the chapter advocates for a model of forensic legal practice that is as *scientifically informed* as it is *constitutionally sound*, ensuring that the evidentiary use of blood remains a tool of justice—rather than a source of injustice.

Blood evidence is evolving. New technologies now allow for *microbial DNA* analysis from bloodstains—revealing not just who was present, but possibly *when* (Abdulhadi et al. 2025). While these methods are promising, their novelty raises admissibility issues. Courts are cautious when scientific methods lack reproducibility or peer-reviewed validation.

"Touch DNA" from blood—where DNA is recovered from surfaces only minimally contacted—also poses legal challenges. Its sensitivity makes it prone to contamination, and juries may be misled into overestimating its probative value (Holdren et al. 2016).

2 The Scientific Foundations Behind Legal Admissibility

In legal contexts, scientific evidence carries unique authority—often viewed by jurors and judges as objective, conclusive, and free from the biases that may affect human testimony. Among various forms of forensic science, *blood evidence* occupies a central role due to its biological specificity and potential to link individuals to crimes with extraordinary precision. However, the persuasive power of blood evidence can only be realized in court if it meets a threshold of *scientific reliability* and *legal relevance*. This is where the intersection of science and law becomes not only important but indispensable.

2.1 Defining Scientific Reliability in Legal Terms

In the United States and many other legal systems, courts do not automatically accept scientific evidence simply because it originates from a laboratory. Rather, the evidence must be shown to be reliable, properly applied, and relevant to the facts of the case. This requirement gained renewed emphasis after the *Daubert v. Merrell Dow Pharmaceuticals, Inc.*

decision in 1993. In this case, the U.S. Supreme Court established the *Daubert standard*, replacing the older *Frye test*, and set forth several criteria for assessing whether scientific testimony is admissible under Rule 702 of the Federal Rules of Evidence.

The *Daubert criteria* include:

1. *Testability*: Can the method be empirically tested and has it been tested?
2. *Peer Review and Publication*: Has the method been subjected to peer review and publication?
3. *Known or Potential Error Rate*: What is the rate of error, and is it within acceptable limits?
4. *Standards Controlling the Technique's Operation*: Are there established standards governing the methodology?
5. *General Acceptance*: Is the technique widely accepted in the relevant scientific community? (Giannelli 2017).

Blood analysis techniques such as *DNA profiling*, *blood typing*, and *serological testing* typically meet these standards. They are widely validated, tested, and accepted. However, more interpretive or subjective methods—such as *bloodstain pattern analysis (BPA)*—continue to face scrutiny.

To clarify how different jurisdictions approach scientific admissibility, see Table 19.1.

Table 19.1 Common legal standards for scientific evidence

Standard	Key criteria	Jurisdiction
Frye Standard	General acceptance in the scientific community	USA (some states)
Daubert Standard	Testability, peer review, error rates, standards, acceptance	U.S. Federal and many states
Mohan Test (Canada)	Relevance, necessity, reliability, qualified expert	Canada
EU Approach	Judgeâ€™s discretion + proportionality and rights-based review	European Union

2.2 The Case of DNA Profiling: A Scientific Gold Standard

Among all forms of blood analysis, *DNA profiling* is perhaps the most significant advancement in modern forensic science. Introduced into U.S. courts in the late 1980s, it has since become the foundation of countless criminal prosecutions and exonerations. DNA extracted from blood can uniquely identify individuals with a statistical probability of accuracy that is, under proper conditions, close to one in billions (Butler 2009).

DNA analysis meets all the Daubert criteria. It has been extensively tested, is supported by robust quality control procedures, and is subject to ongoing peer-reviewed research. Laboratories that perform forensic DNA testing are typically required to follow guidelines issued by the *Scientific Working Group on DNA Analysis Methods (SWGDAM)* and are often accredited by the *American Society of Crime Laboratory Directors/Laboratory Accreditation Board (ASCLD/LAB)*.

Still, problems can arise. If DNA samples are *contaminated*, *mislabeled*, or *interpreted by underqualified personnel*, the reliability of the evidence can be compromised. In such cases, even a gold-standard technology can produce misleading results (Inman and Rudin 2000).

2.3 Bloodstain Pattern Analysis (BPA): A Contested Science

In contrast to DNA profiling, *bloodstain pattern analysis* relies heavily on human interpretation. It involves examining the shape, size, and distribution of bloodstains to infer the events that produced them—such as the direction of an impact or the type of weapon used. While BPA has been used successfully in many investigations, its *subjectivity* and *lack of standardized methodology* have drawn increasing criticism from scientific and legal communities alike (Laber et al. 2014).

A 2009 report by the *National Research Council (NRC)* noted that there is limited empirical research supporting the reliability of BPA. It recommended the development of more rigorous training, standardized methods, and proficiency testing (NRC 2009). Until such improvements are widely adopted, courts may be justified in applying *heightened scrutiny* to bloodstain pattern evidence, especially when it is central to a case.

A comparison of key forensic blood techniques is provided in Table 19.2.

Table 19.2 Strengths and weaknesses of blood evidence techniques

Technique	Strengths	Weaknesses
DNA Profiling	Extremely high accuracy, strong validation	Contamination risk, lab errors, privacy issues
Blood Typing	Quick, inexpensive, useful for exclusion	Low discrimination power compared to DNA
Bloodstain Pattern Analysis	Can reconstruct crime events, visual evidence	Subjective interpretation, lack of standardization
Microbial DNA Analysis	May reveal timing and biological activity	Novel, limited validation, legal uncertainty

2.4 Laboratory Accreditation and Quality Assurance

Scientific methods do not operate in a vacuum; they are shaped by the environments in which they are practiced. That is why *laboratory accreditation* and *quality assurance protocols* are critical to the admissibility of blood evidence. Accreditation bodies like *ISO/IEC 17025* require forensic labs to implement standard operating procedures (SOPs), maintain detailed documentation, conduct regular audits, and ensure ongoing analyst competency.

Inconsistent lab practices have led to legal controversies and wrongful convictions. In the case of *State v. Schwartz (1990)*, the Minnesota Supreme Court excluded DNA evidence because the laboratory failed to adhere to established scientific protocols (Giannelli 2017). The ruling highlighted that even sound science, if poorly implemented, can be rendered inadmissible.

2.5 Scientific Standards Are Global—But Not Uniform

Scientific standards for admissibility vary across legal systems. In countries such as *Canada*, *Australia*, and *Germany*, judges generally exercise broad discretion in admitting scientific evidence. While many of these jurisdictions apply criteria similar to Daubert, they do so under different procedural frameworks. For instance, in Canada, courts weigh *"necessity, reliability, and absence of bias"*, as outlined in *R. v. Mohan (1994).*

This global diversity complicates cross-border investigations and extraditions where blood evidence may be challenged under different evidentiary rules. Forensic scientists and legal professionals must remain aware of these variations to ensure that their methods withstand not only scientific scrutiny but also *international legal review*.

2.6 Technological Advancement Outpacing Legal Infrastructure

One of the most significant challenges facing modern courts is the pace at which forensic technology evolves. New methods such as *microbial DNA analysis*, *touch DNA*, and *AI-based interpretation algorithms* are entering the forensic domain with limited legal precedent. These innovations are promising but often lack the validation and field-testing required for Daubert reliability assessments.

Courts, therefore, must walk a fine line between embracing scientific progress and upholding due process. Legal systems must be cautious not to become overly permissive simply because a technology is novel or appears scientifically advanced. At the same time, they must avoid dismissing useful evidence due to outdated procedural skepticism.

A comparative overview of reliability across techniques is illustrated in Fig. 19.1.

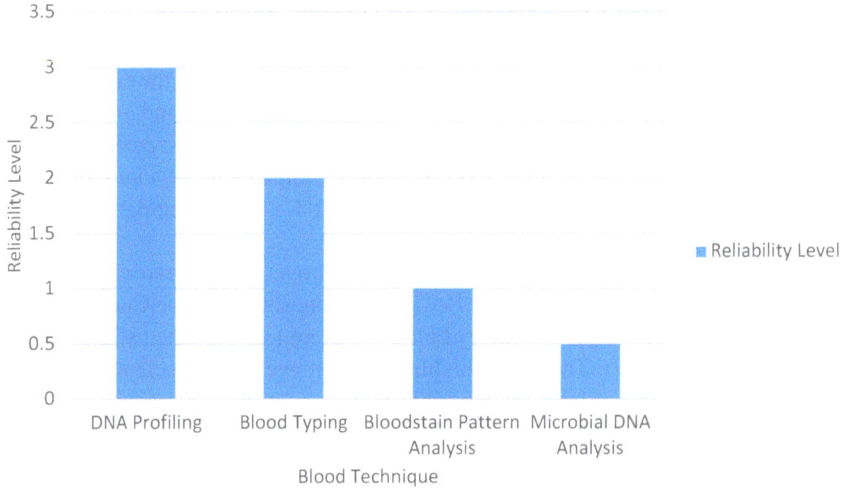

Fig. 19.1 Comparative reliability of blood evidence techniques

3 Chain of Custody: The Lifeline of Admissibility

No matter how advanced the science or powerful the technology, evidence is only as reliable as the process that brings it from the crime scene to the courtroom. For blood evidence in particular—given its susceptibility to contamination, degradation, or even deliberate tampering—the *chain of custody* serves as the backbone of admissibility. Without it, courts cannot be confident that the evidence presented at trial is the same as what was originally collected, or that it has remained free of interference.

3.1 Defining the Chain of Custody

The chain of custody refers to the *documented, unbroken record of the handling, transfer, and storage of evidence* from the moment it is collected until it is presented in court (Houck and Siegel 2009). Each step in the process must be accounted for, including:

- Who collected the blood sample, and under what conditions;
- How it was packaged, labeled, and sealed;
- Where and how it was transported;
- Who accessed it in the laboratory;
- The conditions of storage and analysis;
- Any transfers between labs, investigators, or agencies.

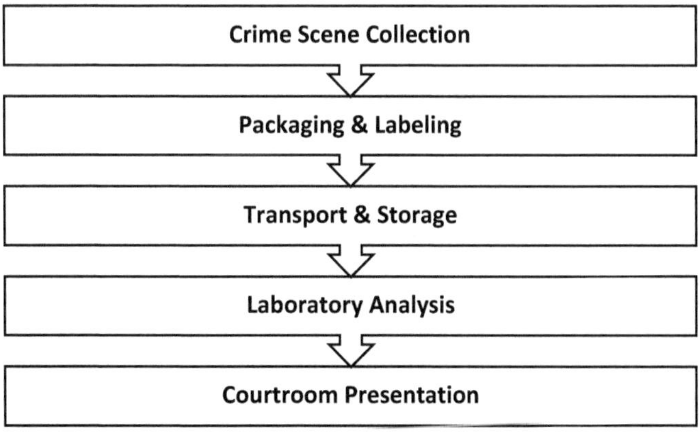

Fig. 19.2 Chain of custody process (flowchart showing step-by-step handling)

Every link in the chain must be clearly documented. Even a minor gap—such as a missing signature or an undocumented transfer—can give defense attorneys grounds to challenge the evidence.

Courts demand a clear chain of custody because evidence can be compromised in countless ways. Blood evidence may degrade if exposed to heat or moisture, may be mislabeled if proper documentation is not followed, or may be contaminated through improper handling. In some cases, allegations of deliberate tampering have led to public mistrust of forensic practices.

The *O.J. Simpson trial* is perhaps the most famous example of chain of custody becoming a focal point. Defense attorneys argued that Los Angeles police officers mishandled and possibly contaminated blood evidence, raising doubt about its integrity. While DNA evidence pointed strongly toward Simpson, the questions surrounding the chain of custody weakened the prosecution's case and helped the defense persuade the jury (Lynch et al. 2010). This high-profile example illustrates how failures in handling evidence can erode even the strongest forensic case. The process is summarized visually in Fig. 19.2.

3.2 Legal Standards for Chain of Custody

Courts generally apply a *reasonable assurance standard* to chain of custody. This means that the prosecution does not need to prove with absolute certainty that no tampering occurred, but they must demonstrate that the evidence is "reasonably likely to be what it purports to be" (Giannelli 2008). Minor clerical errors, such as a misspelled name or a slightly mis-timed entry, may not be enough to exclude evidence if the overall integrity is intact. However, substantial breaks in the record can render the evidence inadmissible.

3.3 The Human Element: Training and Errors

The chain of custody is only as strong as the people who maintain it. Errors often arise not from deliberate misconduct, but from inadequate training, lack of resources, or simple human oversight. Investigators under pressure at chaotic crime scenes may neglect proper sealing procedures, or lab technicians may fail to log every access to a freezer. Each mistake creates potential vulnerabilities that can be exploited in court.

This is why many jurisdictions have emphasized *training, auditing, and accreditation* for law enforcement officers and forensic laboratories. Accreditation programs like *ISO/IEC 17025* require rigorous documentation practices, regular audits, and accountability systems designed to minimize errors. Still, not all forensic facilities—especially in under-resourced regions—meet these standards, raising concerns about the reliability of evidence admitted in those jurisdictions (National Research Council 2009).

3.4 Chain of Custody in Digital and Cross-Border Contexts

In today's interconnected world, the chain of custody extends beyond physical evidence. Forensic laboratories increasingly use *digital tracking systems*, barcoding, and electronic evidence management platforms to reduce human error and enhance transparency. These systems provide time-stamped records of every transfer and access point, strengthening courtroom confidence in the integrity of evidence.

However, globalization introduces new challenges. In cross-border criminal investigations—such as terrorism, drug trafficking, or cyber-enabled crimes—blood evidence may pass through multiple countries, each with different evidentiary standards. This raises complex legal questions: What happens if a country's procedures do not meet the admissibility standards of the jurisdiction where the trial occurs? Can evidence be excluded simply because of international procedural differences? These questions highlight the growing need for *international harmonization* of forensic protocols (Jackson and Summers 2012).

3.5 Case Study: Innocence Through Chain of Custody

While lapses in chain of custody often lead to the exclusion of evidence, strict adherence can also safeguard justice. The *Innocence Project* has used preserved blood samples—carefully documented over decades—to exonerate wrongfully convicted individuals. In these cases, a robust chain of custody allowed for retesting using modern DNA techniques, demonstrating both the long-term value of careful evidence management and the dangers when procedures are neglected (Innocence Project 2023).

3.6 Chain of Custody as a Trust Mechanism

Ultimately, the chain of custody is more than a procedural requirement—it is a *trust mechanism*. It assures judges, juries, and the public that evidence has not been manipulated or corrupted. In a justice system that relies on both truth and perception, maintaining this trust is essential. Without confidence in the integrity of evidence, even the most sophisticated scientific analysis risks becoming meaningless in the eyes of the court.

4 Key Case Law Examples

Case law provides some of the clearest insights into how courts grapple with the admissibility of blood evidence. While statutes and procedural rules establish the framework, it is judicial decisions that bring those rules to life—testing them against real facts, flawed practices, and emerging science. This section highlights a series of influential cases, primarily from the United States but also from other jurisdictions, that illustrate the shifting standards and persistent challenges surrounding blood evidence in court.

4.1 *People v. Wesley (1988)*: The First Step for DNA from Blood

The case of *People v. Wesley* (1988) in New York marked one of the first times that DNA evidence derived from blood was admitted in an American courtroom. The defendant, accused of sexual assault, was linked to the crime through DNA testing performed on blood evidence. The trial court admitted the evidence, setting an important precedent for the use of DNA profiling in criminal cases (Filocoma 1994).

The significance of *Wesley* lies not only in its outcome but in the caution courts expressed. The judge acknowledged the promise of DNA technology but also emphasized the need for rigorous laboratory standards and careful judicial oversight. This dual message—embrace innovation, but verify reliability—became a hallmark of subsequent DNA admissibility cases (Lynch et al. 2010).

4.2 *State v. Schwartz (1990)*: A Warning Against Lax Laboratory Standards

Just 2 years later, the Minnesota Supreme Court in *State v. Schwartz* (1990) demonstrated the other side of the coin. Here, DNA evidence derived from blood was excluded because the prosecution failed to establish that the testing laboratory adhered to reliable scientific procedures. The court ruled that while DNA testing itself was scientifically valid, the *particular lab's methods were inadequate*.

This case underscored an important legal principle: *a valid scientific technique does not automatically guarantee admissibility*. Courts must also evaluate the *quality of execution*. If a laboratory's chain of custody, documentation, or contamination controls are suspect, even otherwise powerful evidence may be rejected (Giannelli 2017).

4.3 The O.J. Simpson Trial (1995): Chain of Custody in the Spotlight

Perhaps no case better illustrates the fragility of blood evidence in the public imagination than the O.J. Simpson trial. Blood samples collected from Simpson's home, car, and clothing were central to the prosecution's case. However, the defense successfully cast doubt on the integrity of these samples by highlighting alleged mishandling, poor documentation, and potential contamination. While the DNA analysis itself was scientifically robust, the jury ultimately found that the *chain of custody was too compromised* to support conviction (Lynch et al. 2010).

The Simpson trial demonstrated that the *perception of reliability* is just as important as the scientific facts. Even valid blood evidence can fail to persuade if the defense can show procedural sloppiness or possible contamination.

4.4 *Missouri v. McNeely (2013)*: Blood Draws and Constitutional Boundaries

Blood evidence is not only challenged on scientific grounds but also on constitutional ones. In *Missouri v. McNeely* (2013), the U.S. Supreme Court addressed whether police could draw blood from a suspected drunk driver without a warrant. The Court ruled that the Fourth Amendment's protection against unreasonable searches and seizures requires law enforcement to obtain a warrant, absent exigent circumstances (Whitaker 2019).

This decision placed blood evidence at the intersection of forensic science and human rights, emphasizing that even the most scientifically reliable evidence can be inadmissible if obtained unlawfully. It reinforced the principle that how evidence is collected is as important as what it proves (Garrett 2016).

4.5 *United States v. Chischilly (1994)*: Establishing DNA Reliability

In *United States v. Chischilly* (1994), the Ninth Circuit upheld the admissibility of DNA evidence derived from blood, affirming that the method was reliable under both the Frye and Daubert standards. The court stressed that DNA profiling had gained broad acceptance within the scientific community and that its methodologies were sufficiently standardized to meet legal thresholds.

This case further entrenched DNA testing as a routine and reliable tool in U.S. courts. Importantly, it also showed how appellate courts could provide guidance to trial courts uncertain about handling novel forensic science (Wiseman Jr 1994).

4.6 *Daubert v. Merrell Dow Pharmaceuticals (1993)*: Broader Implications for Forensic Evidence

Although not about blood evidence specifically, the Supreme Court's decision in *Daubert* fundamentally reshaped how scientific testimony—including blood evidence—is evaluated. By requiring judges to act as "gatekeepers" who assess the reliability of methods, the Court ensured that future cases involving blood analysis would face heightened scrutiny. In practice, this has meant that prosecutors and forensic experts must not only present results but also explain the scientific basis and error rates of their methods (Giannelli 2017).

4.7 International Perspectives

The treatment of blood evidence varies internationally, but similar themes emerge.

- United Kingdom: In *R v. Doheny and Adams* (1997), the Court of Appeal emphasized that DNA evidence from blood must be accompanied by clear statistical explanations. The court warned against "trial by science," highlighting the importance of communicating probabilities in a way that juries can understand (Muhamad 2010).
- India: Courts have increasingly admitted DNA from blood samples, but a 2025 study noted ongoing inconsistencies in forensic standards and infrastructure. Cases such as *State of Bombay v. Kathi Kalu Oghad* (1961) raised early questions about self-incrimination and bodily autonomy that remain relevant today (Khan and Katare 2025).
- France: French courts have historically been cautious about emerging forensic techniques. For instance, in cases involving toxicology from blood samples, judges often demand multiple confirmatory tests and expert consensus before admitting results (Kintz and Gheddar 2025).

Across these cases, several themes emerge:

1. *Scientific Reliability Is Not Enough*: Even when the underlying method (DNA analysis) is valid, poor execution can render evidence inadmissible. (*State v. Schwartz*).
2. *Procedural Integrity Matters*: Weaknesses in chain of custody or lab practices can undermine even strong DNA evidence (*O.J. Simpson* trial).
3. *Constitutional Safeguards Apply*: Evidence collected in violation of fundamental rights will not be admitted, no matter how scientifically probative (*Missouri v. McNeely*).

4. *Judges Are Gatekeepers*: Following *Daubert*, courts must carefully evaluate not only whether science works but whether it has been responsibly applied in a given case.
5. *International Variation Exists*: While DNA is globally recognized, differences in legal traditions mean that admissibility rules vary, especially regarding privacy and consent.

These lessons highlight that admissibility is a *multidimensional question.* It is not simply about whether blood evidence is scientifically reliable but also whether it is legally obtained, properly managed, ethically defensible, and communicated effectively to a lay audience.

5 The International Perspective: A Mixed Bag

While blood evidence is universally recognized as a powerful forensic tool, the rules governing its admissibility are far from uniform. Different legal systems—shaped by their histories, constitutions, and cultural attitudes toward science and privacy—approach blood evidence in markedly different ways. Comparing these approaches provides insight into how societies balance the demands of justice with the rights of individuals, and highlights both shared challenges and unique national concerns.

5.1 United States: Daubert, Frye, and Federalism

In the United States, admissibility is strongly shaped by the *federal structure* of the legal system. At the federal level, courts follow the *Daubert standard*, requiring that scientific evidence be relevant and reliable (Giannelli 2017). However, not all states have adopted Daubert. Some continue to use the older *Frye standard*, which requires only that the method be "generally accepted" in the scientific community. Others employ hybrid approaches that mix elements of both.

This diversity creates a patchwork of evidentiary standards. In one state, bloodstain pattern analysis may be admitted with minimal challenge; in another, it may be excluded for lack of scientific rigor. Critics argue that such inconsistency undermines fairness, particularly when defendants face different admissibility rules depending on jurisdiction (Mnookin et al. 2010). On the other hand, supporters contend that federalism allows courts to experiment and evolve in response to scientific advances.

5.2 United Kingdom: Emphasis on Jury Communication

In the United Kingdom, courts have generally been receptive to DNA evidence derived from blood, but they also emphasize the *communication of scientific findings to lay juries*. In *R v. Doheny and Adams* (1997), the Court of Appeal stressed that DNA evidence must be accompanied by clear statistical explanations. Expert witnesses must avoid overstating certainty and must contextualize results in terms that jurors can understand (Muhamad 2010).

This reflects a broader trend in British law toward *transparency and clarity*. Judges act as safeguards against "trial by science," ensuring that evidence is not admitted or relied upon without careful explanation. The UK's cautious stance has influenced other common law jurisdictions such as Australia and New Zealand, which similarly stress the need for juror comprehension (Edmond and Roberts 2017).

5.3 Canada: Reliability and Necessity

Canadian courts approach scientific evidence through the framework established in *R v. Mohan* (1994). To be admissible, expert testimony—including that derived from blood—must meet four criteria: relevance, necessity, absence of exclusionary rules, and proper qualification of the expert. Later cases added an explicit *reliability requirement*, aligning Canada's approach more closely with Daubert while retaining its own emphasis on judicial discretion (Paciocco and Stuesser 2005).

Canada also places strong emphasis on *Charter of Rights and Freedoms protections*, particularly Section 8, which safeguards against unreasonable search and seizure. As in the USA, this means that blood drawn without proper consent or judicial authorization may be excluded, even if scientifically valid.

5.4 India: Infrastructure Challenges

In India, blood evidence is admissible under the Indian Evidence Act, but practical challenges often undermine its effectiveness. A 2025 study by Khan and Katare highlighted the lack of standardized protocols across forensic laboratories, insufficient training of police officers collecting evidence, and delays that lead to degradation of samples (Khan and Katare 2025). Courts often accept blood evidence, but the variability in quality raises risks of wrongful conviction.

Another complicating factor is the absence of a comprehensive DNA database law, which has led to debates about privacy, consent, and potential misuse. While Indian courts recognize the probative value of blood evidence, the system's limitations highlight the need for *infrastructure reform and legislative clarity*.

5.5 France: Judicial Conservatism

French courts, operating under a civil law system, take a more conservative approach to forensic innovations. While DNA analysis from blood has long been accepted, newer techniques such as microbial forensics or probabilistic genotyping face *strict judicial skepticism*. French judges are more proactive in evaluating scientific validity than their common law counterparts, often requiring multiple expert opinions before admitting novel methods (Kintz and Gheddar 2025).

This conservatism can slow the adoption of cutting-edge science, but it also reflects a cultural and legal preference for *caution over enthusiasm*. In doing so, French courts attempt to avoid the risk of wrongful convictions based on unproven methods.

5.6 Germany: Balancing Utility and Rights

Germany's legal system, also civil law-based, integrates blood evidence within a strong framework of *constitutional rights*. The German Basic Law places high value on dignity and privacy, meaning that invasive procedures like blood draws must be justified by necessity and proportionality (Peters 2018). At the same time, German forensic laboratories are well-resourced and highly standardized, which enhances reliability once samples are lawfully obtained.

Notably, German courts have been early adopters of probabilistic reasoning in DNA interpretation, but they require rigorous documentation and transparency in expert reports. This approach balances the utility of advanced science with strict protection of defendants' rights.

5.7 International Harmonization: A Growing Need

Globalization has increased the need for *harmonized forensic standards*. Blood evidence is often relevant in transnational crimes such as terrorism, human trafficking, or war crimes. International tribunals, such as the International Criminal Court (ICC), must rely on evidence collected under diverse national systems. Without harmonized protocols, discrepancies in admissibility standards risk undermining prosecutions (Redmayne 2001).

Organizations like *Interpol* and the *European Network of Forensic Science Institutes (ENFSI)* have called for greater international standardization in evidence collection and handling. However, national sovereignty, differing constitutional protections, and resource disparities remain obstacles.

From this global survey, three lessons stand out:

1. *Common Standards, Different Priorities*: While most systems require reliability, they prioritize different aspects—whether it is juror comprehension in the UK, constitutional rights in the USA and Germany, or infrastructure in India.
2. *Caution vs. Innovation*: Some countries (e.g., France) prefer judicial conservatism, while others (e.g., the USA) are quicker to embrace new methods. Both approaches have strengths and weaknesses.
3. *Global Justice Requires Cooperation*: As cross-border cases expand, the lack of harmonized standards may increasingly complicate prosecutions, highlighting the urgency of international collaboration.

6 Ethical and Constitutional Boundaries

Behind the science of blood evidence lies a question that is just as important: how far can the law go in compelling someone to give up a piece of their body? Unlike fingerprints or photographs, which can be collected with minimal intrusion, drawing blood is invasive. It involves piercing the skin, collecting biological material, and potentially revealing more information than investigators originally sought. This raises issues of bodily autonomy, privacy, and constitutional protections.

6.1 Consent and Warrants

In many jurisdictions, the guiding principle is straightforward: a person's blood cannot be taken without either consent or judicial authorization. This requirement stems from broader constitutional safeguards against unreasonable searches and seizures. For example, in the United States, the Fourth Amendment requires that police obtain a warrant unless there are "exigent circumstances"—that is, emergencies where waiting for judicial approval would risk the loss of crucial evidence.

This principle was firmly reinforced in *Missouri v. McNeely (2013)*. In that case, police officers conducted a warrantless blood draw on a man suspected of drunk driving. The prosecution argued that because alcohol naturally dissipates in the bloodstream, waiting to secure a warrant risked losing evidence. The Supreme Court disagreed. Writing for the majority, Justice Sotomayor held that the natural metabolization of alcohol does not, by itself, create an exigency justifying bypassing a warrant. The decision clarified that routine DUI cases require a warrant for blood draws, emphasizing that bodily integrity cannot be sacrificed to convenience (Garrett 2016; Whitaker 2019).

6.2 Bodily Autonomy and Human Dignity

The ethical weight of blood evidence lies not only in legality but in the concept of bodily autonomy. Drawing blood without permission can feel like a violation of personal dignity. Scholars have argued that the state should tread carefully in demanding such invasive samples, especially since blood contains sensitive genetic information far beyond its immediate forensic use (Peters 2018). For instance, a sample taken for alcohol testing could theoretically reveal predispositions to disease, ancestry, or other private health details. Without clear boundaries, there is a risk that forensic science could slip into unauthorized surveillance.

6.3 International Perspectives

Different countries strike this balance in different ways. In Germany, constitutional jurisprudence places strong emphasis on human dignity, requiring that bodily intrusions like blood draws be proportionate and justified by serious state interests (Peters 2018). In India, the courts have admitted blood evidence but remain cautious about forced extractions, often requiring judicial oversight to protect against abuse (Khan and Katare 2025). Meanwhile, in the United Kingdom, consent plays a central role, though police may sometimes compel samples under statutory authority for serious crimes, provided safeguards are in place (Edmond and Roberts 2017).

6.4 Science, Procedure, and Fairness

Ultimately, the ethical and constitutional boundaries surrounding blood evidence remind us that admissibility is not purely a scientific question. It is equally about procedure and fairness. The law recognizes that evidence, no matter how powerful, cannot be collected at the expense of fundamental rights. Courts, therefore, serve as guardians not only of scientific validity but also of individual dignity.

As forensic technologies grow more sophisticated—revealing more information from ever smaller samples—the need for clear, ethical boundaries becomes even more urgent. The promise of justice that blood evidence brings must always be weighed against the principle that individuals are not mere data points for state collection. In this way, the admissibility of blood evidence is as much about protecting people as it is about prosecuting crimes.

7 The Problem of Overconfidence in "Scientific" Evidence

In contemporary courtrooms, science carries enormous persuasive weight. Jurors often see forensic science as objective, neutral, and nearly infallible. This confidence can be an asset when the evidence is truly strong and responsibly handled. However, it can also be a serious liability when scientific evidence is flawed, incomplete, or overstated. This phenomenon has come to be known as the "CSI Effect," named after the popular television franchise *CSI: Crime Scene Investigation*. The concern is that jurors who have been exposed to such shows may either expect forensic evidence in every case or place undue weight on scientific-sounding testimony, regardless of its actual reliability (Cole and Dioso-Villa 2008).

7.1 The CSI Effect in Practice

The CSI Effect manifests in two ways. First, jurors may hold unrealistic expectations about forensic science, believing that techniques like DNA analysis can provide absolute certainty in every case. When such evidence is absent, jurors may be reluctant to convict, even when other forms of evidence—such as eyewitness testimony or circumstantial facts—are strong. Second, when forensic evidence is presented, jurors may overvalue it, assuming that the presence of a DNA profile or bloodstain analysis automatically means guilt, without scrutinizing the underlying methodology or the potential for error (Shelton 2008).

In high-profile cases, defense attorneys often highlight this problem by attacking the credibility of forensic methods. In the O.J. Simpson trial, for instance, the defense did not need to disprove the DNA evidence outright. Instead, by raising doubts about handling errors and laboratory practices, they tapped into jurors' uncertainty about the science itself, demonstrating how confidence in scientific methods can quickly collapse when trust in procedure is shaken (Lynch et al. 2010).

7.2 Blood Evidence and Juror Perception

Blood evidence is especially vulnerable to these dynamics. DNA profiling from blood is widely seen as a gold standard of forensic science, but even it can be undermined by contamination, mislabeling, or interpretive error (Inman and Rudin 2000). Meanwhile, bloodstain pattern analysis, which is more subjective, risks being presented in court with the same aura of certainty as DNA. Jurors may not appreciate the difference between a highly validated technique and one that relies heavily on expert interpretation. Without proper explanation, the courtroom can blur the line between science and speculation.

7.3 Expert Witnesses as Translators

One solution lies in the role of expert witnesses. Judges increasingly require forensic experts to present their findings in terms lay jurors can understand, emphasizing error rates, limitations, and the probabilistic nature of results. This is especially important in cases involving blood evidence, where the difference between a strong DNA match and a contested bloodstain pattern analysis must be made clear. Transparency in testimony helps jurors weigh evidence appropriately, rather than assuming scientific evidence is always decisive (Edmond and Roberts 2017).

7.4 Maintaining Balance

The problem of overconfidence in science underscores a deeper truth: forensic evidence is not a magic bullet. It is a tool—powerful, but imperfect. Courts must balance the persuasive appeal of blood evidence with safeguards against misuse. That means admitting only scientifically sound methods, ensuring clear communication of results, and reminding jurors that science, like all evidence, must be evaluated critically. Only then can blood evidence fulfill its promise of advancing justice without distorting it.

8 Conclusion

The enduring value of blood evidence lies in its dual identity: it is at once a biological marker of human individuality and a legal instrument of truth. Few other forms of forensic material hold such sway in the courtroom. A drop of blood can link a suspect to a crime scene, exonerate the wrongly accused, or corroborate the testimony of victims. Yet, as powerful as it may be, blood evidence is not self-interpreting. Its journey from crime scene to courtroom is complex, and its admissibility rests on a delicate balance of *scientific reliability, legal procedure, and ethical responsibility*.

The history of blood evidence has shown us both its triumphs and its failures. Cases like *People v. Wesley* (1988) established the promise of DNA profiling, while the O.J. Simpson trial revealed how chain-of-custody lapses can erode public confidence. Decisions like *Missouri v. McNeely* (2013) further remind us that constitutional rights set firm boundaries on how far the state can go in obtaining bodily evidence, regardless of its scientific value (Filocoma 1994). Together, these examples demonstrate that *blood evidence is never evaluated in isolation*—it is always filtered through the lens of legal safeguards and public trust (Whitaker 2019).

As forensic technologies advance, these challenges only grow more urgent. Emerging methods such as microbial forensics and AI-assisted bloodstain analysis offer new opportunities for truth-seeking, but they also raise fresh concerns about validation, transparency, and potential misuse. The justice system must be proactive rather than reactive, adapting its evidentiary standards to keep pace with innovation while guarding against overconfidence in untested techniques (National Research Council 2009).

Moving forward, the credibility of blood evidence will depend on reform and vigilance. Several key measures can strengthen its role in court:

1. *Standardize protocols for collection and testing.* Consistency across jurisdictions ensures that evidence is handled with the same rigor everywhere, reducing disparities that can undermine fairness.
2. *Mandate laboratory accreditation and continuing education for forensic analysts.* Laboratories must not only meet international standards such as ISO/IEC 17025 but also commit to ongoing professional development. Science evolves, and so must the people who practice it.
3. *Adopt pretrial Daubert hearings more widely.* By vetting scientific evidence before trial, judges can resolve disputes about admissibility early, sparing jurors from confusion and ensuring only reliable methods are presented in court (Giannelli 2017).
4. *Educate jurors on the limitations and potential errors in blood analysis.* Transparency is essential. Jurors should understand that DNA profiling is probabilistic, that bloodstain pattern analysis can be subjective, and that no forensic method is immune from error (Shelton 2008).

Taken together, these measures highlight the central truth: *blood does not lie, but people and procedures can falter*. Ensuring the admissibility and credibility of blood evidence requires more than trusting in science. It requires building legal frameworks that respect rights, foster transparency, and continually test the boundaries of reliability.

Ultimately, justice depends not only on what blood reveals under a microscope but on how carefully the legal system listens, interprets, and safeguards that story. By uniting scientific rigor with constitutional protections, courts can ensure that blood evidence fulfills its greatest promise: to illuminate truth while preserving justice.

References

Abdulhadi BH, Latif AS, Khalah SA (2025) Microbiological forensics: new possibilities in criminal investigations. Central Asian J Med Nat Sci. https://cajmns.casjournal.org/index.php/CAJMNS/article/view/2877

Butler JM (2009) Fundamentals of forensic DNA typing. Elsevier Academic Press, Burlington, MA

Cole SA, Dioso-Villa R (2008) Investigating the "CSI effect": media and litigation crisis in criminal law. Stanford Law Rev 61(6):1335–1374

Edmond G, Roberts A (2017) The law commission's report on expert evidence in criminal proceedings. In: Expert evidence and scientific proof in criminal trials. Routledge, London. pp 533–551

Filocoma DA (1994) Unravelling the DNA controversy: *People v. Wesley*, a step in the right direction. JL Pol'y 3:537

Garrett BL (2016) Constitutional regulation of forensic evidence. Wash Lee L Rev 73:1147

Giannelli PC (2008) Forensic science and the ABA innocence report. Sw. UL Rev 37:993

Giannelli PC (2017) Forensic science: Daubert's failure. Case W Res L Rev 68:869

Holdren JP, Lander ES, Press W, Savitz M (2016) Forensic science in criminal courts: ensuring scientific validity of feature-comparison methods. President's Committee of Advisors on Science and Technology

Houck MM, Siegel JA (2009) Fundamentals of forensic science. Academic Press, Burlington, MA

Inman K, Rudin N (2000) Principles and practice of criminalistics: the profession of forensic science. CRC Press, Boca Raton, FL

Innocence Project (2023) DNA exonerations in the United States. https://innocenceproject.org/dna-exonerations-in-the-united-states/

Jackson JD, Summers SJ (2012) The internationalisation of criminal evidence: beyond the common law and civil law traditions. Cambridge University Press, Cambridge.

Khan R, Katare S (2025) Evaluating the multiple dimensions of DNA evidence in India. Int J Sci Technol 16(1):1

Kintz P, Gheddar L (2025) Hair testing for doping agents–not just a fight against sport authorities. Toxicologie Analytique et Clinique 37(1):S38–S39

Laber T, Kish P, Taylor M, Owens G, Osborne N, Curran J (2014) Reliability assessment of current methods in bloodstain pattern analysis. Final Report for the National Institute of Justice. US In: National Institute of Justice

Lynch M, Cole SA, McNally R, Jordan K (2010) Truth machine: the contentious history of DNA fingerprinting. University of Chicago Press, Chicago, IL

Mnookin JL, Cole SA, Dror IE, Fisher BA, Houck MM, Inman K et al (2010) The need for a research culture in the forensic sciences. UCLA Law Rev 58(3):725–779

Muhamad MMB (2010) Reliability and conclusiveness of DNA evidence in criminal trial. Malayan Law J 1:137

National Research Council (2009) Strengthening forensic science in the United States: a path forward. The National Academies Press
Nuffield Council on Bioethics (2007) The forensic use of bioinformation: ethical issues. https://www.nuffieldbioethics.org/publications/2007/forensic-use-of-bioinformation
Paciocco DM, Stuesser L (2005) The law of evidence. Irwin Law, Toronto, p 89
Peters A (2018) Between military deployment and democracy: use of force under the German constitution. J Use Force Int Law 5(2):246–294
Redmayne M (2001) Expert evidence and criminal justice. Oxford University Press, Oxford
Saunders R (1997) The circuit courts' application of Daubert v. Merrell Dow Pharmaceuticals, Inc. Drake L Rev 46:407
Shelton DE (2008) The ‘CSI effect’: does it really exist? Natl Inst Justice J 259:1–7
Whitaker QV (2019) An analysis of the relationship between Dui Arrests, Fatality Crashes, and Missouri V. Mcneely (Doctoral dissertation, Texas Southern University)
Wiseman TA Jr (1994) Judging the expert. Ohio State Law J 55:1105

MIX
Papier aus verantwortungsvollen Quellen
Paper from responsible sources
FSC® C105338

If you have any concerns about our products,
you can contact us on
ProductSafety@springernature.com

In case Publisher is established outside the EU,
the EU authorized representative is:
Springer Nature Customer Service Center GmbH
Europaplatz 3, 69115 Heidelberg, Germany

Printed by Libri Plureos GmbH
in Hamburg, Germany